SPRINGER
LAB MANUAL

Springer

Berlin
Heidelberg
New York
Barcelona
Budapest
Hong Kong
London
Milan
Paris
Santa Clara
Singapore
Tokyo

Bert Van Duijn   Anneke Wiltink (Eds.)

Signal Transduction –
Single Cell Techniques

With 113 Figures

Springer

BERT VAN DUIJN

Center for Phytotechnology RUL/TNO
Wassenaarseweg 64
2333 AL Leiden
The Netherlands

ANNEKE WILTINK

University of Amsterdam
Department of Physiology
Academic Medical Centre
Meibergdreef 15
1105 AZ Amsterdam
The Netherlands

ISBN 3-540-62563-1 Springer-Verlag Berlin Heidelberg New York

Library of Congress Cataloging-in-Publication Data

Signal transduction – single cell techniques / Bert Van Duijn, Anneke Wiltink, (eds.). p. cm. – (Springer lab manual) Includes bibliographical references and index.
   ISBN 3-540-62563-1 (wire-o-binding : alk. paper)
1. Cellular signal transduction – Laboratory manuals. 2. Cytology – Laboratory manuals. I. Duijn, Bert van, 1961– . II. Wiltink, Anneke, 1961– . III. Series. QP517.C45S556 1997 571.6 – dc21

© Springer-Verlag Berlin Heidelberg 1998
Printed in Germany

The use of general descriptive names, registered names, trademarks, etc. in this publication does not imply, even in the absence of a specific statement, that such names are exempt from the relevant protective laws and regulations and therefore free for general use.
Product liability: The publisher cannot guarantee the accuracy of any information about dosage and application thereof contained in this book. In every individual case the user must check such information by consulting the relevant literature.

Cover Design: design & production GmbH, Heidelberg

Typesetting: Mitterweger Werksatz GmbH, Plankstadt

SPIN 10523199   31/3137   5 4 3 2 1 0 – Printed on acid free paper –

Preface

Signal Transduction-Single Cell Techniques was compiled in response to a rapidly expanding field with an increasing number of possibilities and technical aspects. This laboratory manual provides not only insight into a variety of powerful single cell techniques, but also gives background information and step-by-step protocols for practical applications. In doing so, *Signal Transduction-Single Cell Techniques* is suitable for both single cell specialists and for researchers from outside the single cell field. In addition, the manual contains useful material to be used in teaching related courses.

To study signal transduction in single cells, one has to master many different techniques. Most of these techniques are complicated, contain many pitfalls and require a detailed knowledge of their physical basis. Here, the emphasis is placed on techniques used for handling of cells in experiments and on electrophysiological and fluorescence techniques. The various chapters deal with the theoretical background, the actual recording methods, and analysis of the recorded signals – every chapter being a complete guide. Pitfalls are indicated and useful tips provided.

In Part I, "Handling of Cells in Single Cell Experiments", there are guidelines for the construction of single cell measurement perfusion chambers, ideas for temperature control and microapplication of drugs, and the application of laser microsurgery. Part II, "Ion Channel and Membrane Potential Measurements Using the Patch-Clamp Technique", provides multiple examples of the application of different patch-clamp measurement configurations in various cell types from both animal and plant systems. A variety of methods is described in each chapter, enabling the reader to compare and choose the most suitable method for the desired application. In addition, a complete course guiding the reader (experimenter) through the theoretical and practical background of the patch-clamp technique using electrical simulation circuits (without the necessity of having a patch-clamp setup available) is included. Part III, "Fluorescence to Measure Intracellular Ions", introduces the use of fluorescent ion sensitive probes, and examples using flow cytometry, microfluorescence, ion imaging and confocal microscopy are given. Emphasis is put on calibration and validation of the different measurement techniques.

With this laboratory manual as a guide, one should be able to perform the experiments described in a well-equipped laboratory without the continuous support of an expert in the field.

Leiden, December 1996   BERT VAN DUIJN
                                                        ANNEKE WILTINK

Contents

Contributors

Fabienne Andris
Laboratoire de Physiologie Animale
Université Libre de Bruxelles
67 rue des Chevaux
1640 Rhode-St-Genèse
Belgium

Erika Baus
Laboratoire de Physiologie Animale
Université Libre de Bruxelles
67 rue des Chevaux
1640 Rhode-St-Genèse
Belgium

Laszlo Bene
Department of Biophysics
University Medical School Debrecen
Nagyerdei krt 98
4012 Debrecen
Hungary

Federica Bertaso
Dipartimento di Fisiologia e Biochimica Generali
Laboratorio di Elettrofisiologia
Università Statale di Milano
Via Celoria 26
I-20133 Milano
Italy

MARGREET BLOM-ZANDSTRA
AB-DLO
P.O. Box 14
6700 AA Wageningen
The Netherlands

PETTIE P. BOOIJ
Institute for Molecular and Biological Sciences
Faculty of Biology
Vrije Universiteit Amsterdam
De Boelelaan 1087
NL-1081 HV Amsterdam
The Netherlands

REMKO R. BOSCH
Department of Biochemistry
University of Nijmegen
P.O. Box 9101
6500 HB Nijmegen
The Netherlands

SANDOR DAMJANOVICH
Department of Biophysics
University Medical School Debrecen
Nagyerdei krt 98
4012 Debrecen
Hungary

ALBERTUS H. DE BOER
Institute for Molecular and Biological Sciences
Faculty of Biology
Vrije Universiteit Amsterdam
De Boelelaan 1087
NL-1081 HV Amsterdam
The Netherlands

ARIE DE VOS
Department of Physiology and Physiological Physics
Leiden University
P.O. Box 9604
2300RC Leiden
The Netherlands

RANDALL L. DUNCAN
Department of Orthopaedic Surgery
Physiology and Biophysics
Indiana University Medical Center
Clinical Building Suite 600
541 Clinical Drive
Indianapolis
Indiana 46202–5111
USA

MISA DZOLJIC
Department of Anesthesiology
AMC, University of Amsterdam
Meibergdreef 9
1105 AZ Amsterdam
The Netherlands

RACHEL ERRINGTON
Physiology Department
Parks Road
Oxford
OX1 3PT
UK

MARCEL T. FLIKWEERT
Kluyver Laboratory of Biotechnology
Department of Microbiology and Enzymology
Industrial Microbiology Section
Delft University of Technology
Julianalaan 67
2628 BC Delft
The Netherlands

MARK FRICKER
Dept. Plant Sciences
University of Oxford
South Parks Road
Oxford
OX1 3RB
UK

Reszo Gáspár jr.
Department of Biophysics
University Medical School Debrecen
Nagyerdei krt 98
4012 Debrecen
Hungary

Jochem Herrmann
ADIMEC Advanced Image Systems B.V.
Meerenakkerweg 1
Postbus 7909
5652 AR Eindhoven
The Netherlands

Can Ince
Department of Physiology
University of Amsterdam
Academic Medical Centre
Meibergdreef 15
1105 AZ Amsterdam
The Netherlands

Kei Inouye
Department of Botany
Division of Biological Science
Graduate School of Science
Kyoto University
Sakyo-ku
Kyoto 606–01
Japan

Attila Jenei
Department of Biophysics
University Medical School Debrecen
Nagyerdei krt 98
4012 Debrecen
Hungary

ZOLTAN KRASZNAI
Department of Biophysics
University Medical School Debrecen
Nagyerdei krt 98
4012 Debrecen
Hungary

JOLANDA LEMMERS
Department of Physiology
University of Nijmegen
P.O. Box 9101
6500 HB Nijmegen
The Netherlands

OBERDAN LEO
Laboratoire de Physiologie Animale
Université Libre de Bruxelles
67 rue des Chevaux
1640 Rhode-St-Genèse
Belgium

MIKE MAY
Laboratorium voor Genetica
Universiteit Gent
KL Ledeganckstraat 35
B-9000 Gent
Belgium

MICHELE MAZZANTI
Dipartimento di Fisiologia e Biochimica Generali
Laboratorio di Elettrofisiologia
Università Statale di Milano
Via Celoria 26
I-20133 Milano
Italy

HENK MIEDEMA
Biology Department
The Pennsylvania State University
208 Mueller Laboratory
University Park
PA 16802
USA

GYÖRGY PANYI
Department of Biophysics
University Medical School Debrecen
Nagyerdei krt 98
4012 Debrecen
Hungary

CARLO PIERI
Cytology Center
Research Department of Gerontology
Ancona
Italy

JAN H. RAVESLOOT
Department of Physiology
AMC, University of Amsterdam
Meibergdreef 15
1105 AZ Amsterdam
The Netherlands

WIM J.J.M. SCHEENEN
Department of Biomedical Sciences
University of Padova
Via Trieste 75
35121 Padova
Italy

ROLF L.L. SMEETS
Department of Biochemistry
University of Nijmegen
P.O. Box 9101
6500 HB Nijmegen
The Netherlands

MONIKA TLALKA
Dept. Plant Sciences
University of Oxford
South Parks Road
Oxford
OX1 3RB
UK

RAFFAELLA TONINI
Dipartimento di Fisiologia e Biochimica Generali
Laboratorio di Elettrofisiologia
Università Statale di Milano
Via Celoria 26
I-20133 Milano
Italy

JACQUES URBAIN
Laboratoire de Physiologie Animale
Université Libre de Bruxelles
67 rue des Chevaux
1640 Rhode-St-Genèse
Belgium

RUTGERIS J. VAN DEN BERG
Department of Physiology and Physiological Physics
Leiden University
P.O. Box 9604
2300RC Leiden
The Netherlands

BERT VAN DUIJN
Center for Phytotechnology RUL/TNO
Wassenaarseweg 64
2333 AL Leiden
The Netherlands

ANTONI VAN GINNEKEN
Department of Physiology
AMC, University of Amsterdam
Meibergdreef 15
1105 AZ Amsterdam
The Netherlands

ZOLTAN VARGA
Department of Biophysics
University Medical School Debrecen
Nagyerdei krt 98
4012 Debrecen
Hungary

MARIEKE W. VELDKAMP
Department of Physiology
AMC, University of Amsterdam
Meibergdreef 15
1105 AZ Amsterdam
The Netherlands

E. ETIENNE VERHEIJCK
Department of Physiology
AMC, University of Amsterdam
Meibergdreef 15
1105 AZ Amsterdam
The Netherlands

JOS A.H. VERHEUGEN
Neurobiologie Cellulaire
INSERM U261
Institut Pasteur
25 rue de Dr. Roux
75724 Paris Cedex 15
France

ARIE O. VERKERK
Department of Physiology
AMC, University of Amsterdam
Meibergdreef 15
1105 AZ Amsterdam
The Netherlands

PIET VIS
Department of Physiology
University of Nijmegen
P.O. Box 9101
6500 HB Nijmegen
The Netherlands

SAKE A. VOGELZANG
Vrije Universiteit
Fac. Biologie
Vakgroep Molecular and Cellular Biology
De Boelelaan 1087
1081 HV Amsterdam
The Netherlands

WYTSE J. WADMAN
Department of Experimental Zoology
University of Amsterdam
Kruislaan 320
1098 SM Amsterdam
The Netherlands

MEI WANG
Center for Phytotechnology RUL/TNO
Department of Plant Biotechnology
Wassenaarseweg 64
2333 AL Leiden
The Netherlands

ZHENG WANG
Department of Physiology and Physiological Physics
Leiden University
P.O. Box 9604
2300 RC Leiden
The Netherlands

ADAM F. WEIDEMA
Department of Physiology and Physiological Physics
Leiden University
P.O. Box 9604
2300 RC Leiden
The Netherlands

NICK WHITE
Dept. Plant Sciences
University of Oxford
South Parks Road
Oxford
OX1 3RB
UK

PETER H.G.M. WILLEMS
Department of Biochemistry
University of Nijmegen
P.O. Box 9101
6500 HB Nijmegen
The Netherlands

ANNEKE WILTINK
Department of Physiology
University of Amsterdam
Academic Medical Centre
Meibergdreef 15
1105 AZ Amsterdam
The Netherlands

JULIAN WOOD
Dept. Plant Sciences
University of Oxford
South Parks Road
Oxford
OX1 3RB
UK

DIRK L. YPEY
Department of Physiology and Physiological Physics
Leiden University
P.O. Box 9604
2300 RC Leiden
The Netherlands

List of Suppliers

Below the addresses of various companies referred to in various chapters are given.

Altai Nederland B.V.
Bedrijvenpark Twente 290
7602 KK Almelo
The Netherlands
phone: +31 546 574911
fax: +31 546 576006

Applied Imaging International Limited
Hylton Park
Wessington Way
Sunderland
Tyne & Wear
SR5 3HD
UK
phone: +44 (0)191 5160505
fax: +44 (0)191 5160512
e-mail: JS @ aii.co.uk

Axon Instruments, Inc.
1101 Chess Drive
Foster City, California 94404
USA
phone: +1-415-571-9400
fax: +1-415-571-9500
e-mail: sales @ axonet.com
Web: www.axonet.com

Bellco Glass, Inc.
340 Edrudo Road
Vineland, New Jersey 08360
USA
phone: +1–609 691 1075
fax: +1–609 691 3247
e-mail: sales @ bellcoglass.com

BIO-LOGIC Science Instruments
1 Rue de l'Europe
ZA de Font Ratel
F-38640 CLAIX
France
phone: +33–76 98 68 31
fax: +33–76 98 69 09
e-mail: Bio-Logic @ msn.com
Web: www.bio-logic.com

BIOSOFT
P.O. Box 10938
Ferguson, Missouri 63135
USA
phone: +1–314 524 8029
fax: +1–314 524 8129
e-mail: info @ biosoft.com
Web: www.biosoft.com

BIOSOFT
37 Cambridge Place
Cambridge CB2 INS
UK
phone: +44–1223 368622
fax: +44–1223 312873

Boehringer Mannheim B.V.
Postbus 1007
1300 BA Almere
The Netherlands
phone: +31–36 5394911
fax: +31–36 5394231
e-mail: biocheminfo.nl @ bmg.boehringer.com
Web: biochem.boehringer-mannheim.com

Carl Zeiss Jena GmbH
Zeiss Gruppe
Geschäftsbereich Mikroskopie
D-07740 Jena
Germany
phone: +49–364164 29 36
fax: +49–364164 31 44
Web: www.zeiss.de

Dagan Corporation
2855 Park Avenue
Minneapolis, Minnesota 55407
USA
phone: +1–612 827 5959
fax: +1–612 827 6535
e-mail: info @ dagan.com
Web: www.dagan.com

Difco Laboratories
P.O. Box 331058
Detroit, Michigan 48232–7058
USA

Ealing Electro optics
89 Doug Brown Way
Holliston, Massachusetts 01746
USA
phone: +1–508 429 8370
fax: +1–508 429 7893
Web: www.ealing.com

Fluka Chemie AG
Industriestrasse 25
D-9470 Buchs
Germany

Glaxo Wellcome B.V.
P.O. Box 780
3700 AT Zeist
The Netherlands
phone: +31–3069 31800
fax: +31–3069 38471

Glaxo Wellcome Inc.
Five Moore Drive
P.O. Box 13398
Research Triangle Park
North Carolina 27709
USA
phone: +1–919 2482100
fax: +1–919 2482381
phoneex: 802813

World Precision Instruments
Liegnitzer Strasse 15
D-10999 Berlin
Germany
phone: +49–30 618 8845
fax: +49–30 618 8670
e-mail: 100604.2573 @ compuserve.com

Imperial Laboratories (Europe) Ltd.
West Portway
Andover
Hampshire
SP10 3LF
UK
phone: +44–1264 33 33 11
fax: +44–1264 33 24 12

List-electronic
Pfungstaedter Strasse 20
D-64297 Darmstadt
Germany
phone: +49–6151 56000
fax: +49–6151 56060
e-mail: List-electronic @ t-online.de

Life Technologies, Inc.
8717 Grovemont Circle
P.O. Box 6009
Galthersburg, Maryland 20884–9980
USA
phone: +1–301 840 4114
fax: +1–301 670 8539
Web: www.lifetech.com

LORENZ Messgerätebau
Bundesstrasse 116
D-37191 Katlenburg-Lindau
Germany
phone: +49–5556 1093
fax: +49–5556 5002

LSI Laser Science In.
26 Landsdowne St.
Cambridge, Massachusetts 02139
USA
phone: +1–617 225–2929
fax: +1–617 225–5002

Merck KGaA
Frankfurter Strasse 250
D-64293 Darmstadt
Germany
e-mail: service @ merck.de
Web: www.merck.de

Molecular Probes Europe B.V.
Poortgebouw, Rijnsburgerweg 10
2333 AA Leiden
The Netherlands
phone: +31–71 5233378
fax: +31–71 5233419
e-mail: 76225.2203 @ compuserve.com
Web: www.probes.com

Molecular Probes Inc.
PO Box 22010
Eugene, Oregon 97402–0414
USA
phone: +1–541 465 8300
fax: +1–541 344 6504
e-mail: order @ probes.mhs.compuserve.com
Web: www.probes.com

National Instruments
6504 Bridge Point Parkway
Austin, Texas 78730–5039
USA
phone: +1–512 794–0100
fax: +1–512 794–8411
e-mail: info @ natinst.com
Web: www.natinst.com

Olympus Optical Co. Ltd.
1–3, 2-Chome
Shinjuku
Shinjuku-ku
Tokyo
Japan
phone: +81–3–3340–2136

Olympus Optical Co. (Europa) GmbH
Postfach 104908 20034
Wendenstrasse 14–16
D-20097 Hamburg
Germany
phone: +49–40–237730

Olympus America Inc.
2 Corporate Center Drive
Melville, New York 11747–3157
USA
phone: +1–516–844–5000

Omega Optical
P.O. Box 573
3 Grove Street
Brattleboro, Vermont 05302
USA
phone: +1–802 254 2690
fax: +1–802 254 3937
e-mail: omega @ sover.net
Web: www.sover.net

Sigma-Aldrich Chemie BV
Stationsplein 4
P.O. Box 27
3330 AA Zwijndrecht
The Netherlands
phone: +31–06–0229088
fax: +31–06–0229089

Sutter Instrument Co.
40 Leveroni Court
Novato, California 94949
USA
phone: +1–415 833 0128
fax: +1–415 883 0572
e-mail: info @ sutter.com
Web: www.sutter.com

Part I

Handling of Cells
in Single Cell Experiments

Chapter 1

Maintaining Cells Under the Microscope

CAN INCE

Background

Cell physiological investigations in single living cells have provided much insight into the underlying mechanisms of the functional activities of cells and their components in health and disease. The advantage of investigating living cells is that it provides the investigator with the opportunity of following the sequence of physiological events in time. This is in contrast to a histological approach in which a physiological process is frozen in time and analyzed in detail. In addition, physiological studies on populations of cells fail to detect the kinetics of these processes in the individual cell and the intercellular differences which are no doubt present.

Techniques which are used for studying individual living cells can be generally summarized as including membrane electrophysiological, optical spectroscopic and mechanical measurements. To ultimately gain insight into how the process under investigation reflects the situation in vivo, however, care has to be taken to create environment conditions for the cell that are comparable to those in vivo. Factors which need to be taken into account include: physical conditions (temperature, gas pressures, humidity and osmolality), chemical environment (composition of the medium, presence of unwanted toxins), mechanical aspects (flow, pressure and tension) and cellular paramenters (effect of isolation procedure, use of transformed cells, influence of neighboring cells, metabolism).

Environmental Conditions

Temperature is one of the most important physical parameters to control since physiological processes are highly temperature-dependent. The temperature of in vitro animal systems is usually the body temperature

Temperature

of the animal under study, although room temperature measurements are often carried out in patch clamp experiments. In studies done at room temperature, however, there is often insufficient information provided about the subsequent effects on cellular function. A temperature control system that can be applied in microscopic studies is described in Chap. 3.

Oxygen

The level of gas partial pressures in bathing solutions is also a major determinant of cell function. Oxygen, for example, is sometimes kept at atmospheric level or is sometimes applied at maximal levels to avoid dysoxia. There is thus a need to define oxygen pressures since too little or too much oxygen can cause cell dysfunction. To mimic ischemic conditions in vitro exotic gas mixtures may be applied but care should be taken when extrapolating the results from such studies to the in vivo condition since ischemia is a complex interaction involving many factors.

Medium composition

The chemical composition of perfusate solution is always a source of much discussion. Media used in tissue culture studies include serum-supplemented tissue culture medium, often supplemented with antibiotics as well. These complex cocktails are meant to mimic the composition of blood plasma with antibiotics substituting for the absence of the immunological control of the bathing medium. Deviation from such media can lead to cell cultures that are no longer viable. Therefore, tissue culture recipes are often strictly adhered to. By contrast, single cell physiological investigations mainly use buffered salt solutions, the composition of which is largely dictated by the experimental goals and techniques used.

Mechanical conditions

Mechanical conditions experienced by the cell in vivo differ markedly from those imposed on the cell in a Petri dish on a microscope. Mechanical factors such as stretch and pressure as well as cell-glass or cell-plastic interaction can influence cell function markedly. In addition there is the glass micropipette-cell interaction to take into account in patch clamp studies.

Cell isolation

Cells can undergo changes when they are taken out of their natural in vivo environment for single cell investigations. Cell isolation procedures using enzymatic treatments (especially in plant cells and animal cells that are tightly connected in a tissue) can disrupt cellular processes such that a period of culture may be required to enable recovery before use.

Nonetheless, reports of changes in ion channel composition as a function of culture time argue that it may be wise to use freshly isolated cells. Also, cells in vivo are under continuous and strict social control. This control can be mechanical (coupled muscle cells), electrical (nerve cell network) or paracrine (vascular cell systems) and needs to be taken into account when studying isolated cells.

Transformed cell lines are easy to obtain and thus are often used in single cell investigations. In extrapolating results from such cell lines to in vivo conditions, however, consideration should be taken of the possibility of alterations in, or even loss of, specific cellular functions in transformed cells. **Cell lines**

Measurement Chambers

To keep cells alive and viable for single cell investigations a physiological environment has to be created on the stage of a microscope. This is conventionally achieved by use of special chambers designed for this purpose. Two primary physical parameters which these chambers control are temperature and gas pressures. Gas pressures which need to be regulated include pO_2 for supporting cellular respiration and pCO_2 for pH control. Chambers in which such conditions need to be met can be divided into those that are open and those that are closed.

Open chambers are used in electrophysiological experiments in which direct access to cells with electrodes is required; closed chambers are used for optical measurements such as fluorescence and time lapse studies. Open chambers can be divided into those requiring continuous perfusion and those that mimic tissue culture conditions present in incubators (i.e., no-flow). The essential difference is the manner in which temperature control is maintained and the presence or absence of mechanical stimulation of cells by the flow of the perfusate. In perfusion chambers the temperature and gas pressures of the perfusate are controlled externally whereas no-flow chambers have to ensure temperature by heating of the chamber itself. Here temperature gradients, osmolality and sterility need to be adequately maintained. Figure 1.1 shows an example of a micro-CO_2-incubator which can be used to control both temperature and gas exchange in a Petri dish. A detailed description of this incubator can be found elsewhere (Ince et al. 1983). **Open chambers**

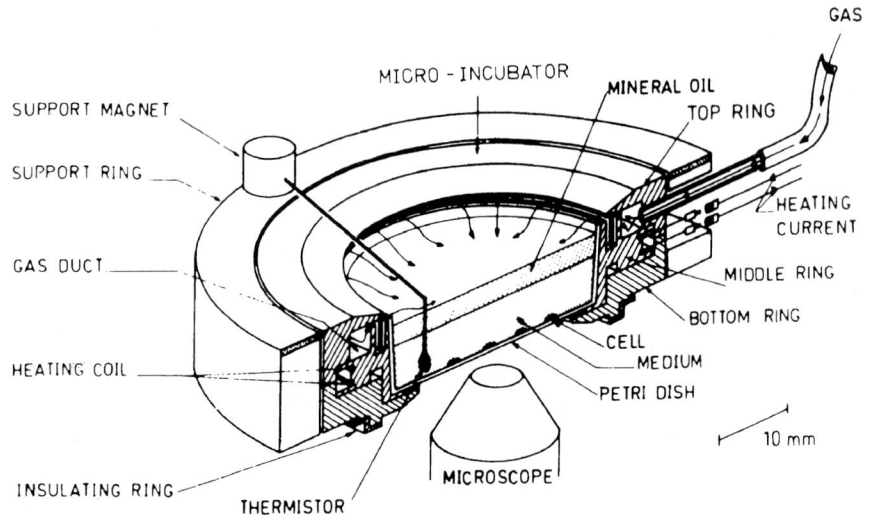

Fig. 1.1. Example of a micro-CO_2-incubater which can be used to control both temperature and gas exchange in a Petri dish. Cross-sectional view. (Reprinted with permission from Ince et al. 1983)

Visual control

Having created an optimal environment on the stage of the microscope, good control over cell function needs to be ensured throughout the course of the experiments. In particular, good visual control is essential. Petri dishes, often used in electrophysiological experiments, give poor visual control and coverslips suited for microscopy are recommended instead. These allow the use of optimal microscopic control (100x) objectives and a variety of microscopic techniques. The superior visual control provided by glass coverslips allows immediate detection of any disruption in cell morphology such as bleb formation or Brownian motion of intracellular particles, the first morphological signs of cell degeneration.

A specially designed open-bottom Teflon culture dish is shown in Fig. 1.2. This dish allows the use of thin (0.18 mm) glass coverslips – and thus visual control at high magnification – for setups in which Petri dishes would normally be used.

Fig. 1.2 A, B. A specially designed open-bottom Teflon culture dish. **A** Telescopic representation of the unit. **B** Cross-section of one side of the assembled unit. The glass coverslip (*1*) is placed between the silicon rubber ring (*2*) and the aluminium securing ring (*3*) and secured to the teflon dish (*4*) by a bayonet catch. (Reprinted with permission from Ince et al. 1985)

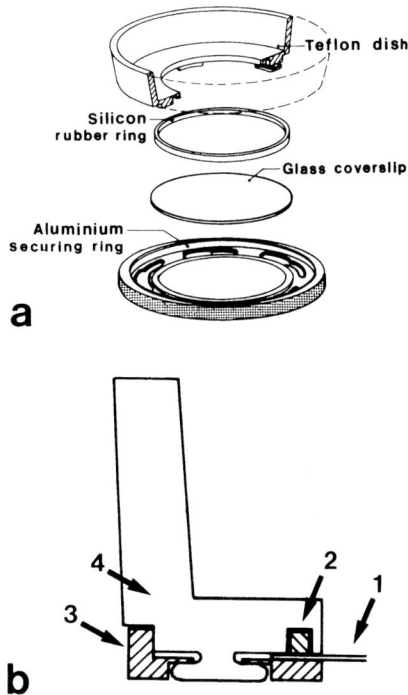

Teflon dish

Silicon rubber ring

Glass coverslip

Aluminium securing ring

a

b

Closed perfusion chambers are generally used for optical investigations not requiring physical contact to be made with the cells. In such chambers accessibility is a problem, but the environmental control offered by closed chambers is superior to that of open chambers. Besides ensuring the above mentioned physical parameters, care needs to be given when choosing the chemical composition of the perfusion medium since different cell types can have quite specific nutritional needs. Figure 1.3 shows the closed version of the culture dish shown in Fig. 1.2.

Closed chambers

Fig. 1.3 A, B. The components of the closed version of the micro-perfusion chamber (**A**) with a close-up for securing glass cover-slips (**B**). **A** The mainframe of the chamber (*1*) is made of Eriflon, a hard durable type of Teflon. A glass coverslip with adherent cells (*2*) can be attached to the top or the bottom of the chamber by a metal securing ring (*3*). Perfusion plugs (*4*) or other types types of plug (*5*) with or without probes can be inserted into conically drilled holes on the side of the chamber. Once assembled, the chamber can be placed into a brace (*6*) to prevent leakage when high internal pressure occurs under high perfusion flow rates. **B** Detail showing how the glass coverslips are secured on the mainframe. A glass coverslip (*1*) is placed on the inner silicon rubber O ring (*2*) and the stainless steel securing ring (*3*) is snapped over the coverslip. Sealing is achieved by the pressure exerted by the larger outer O ring (*4*) on the conically drilled stainless steel ring. (Reprinted with permission from Ince et al. 1990)

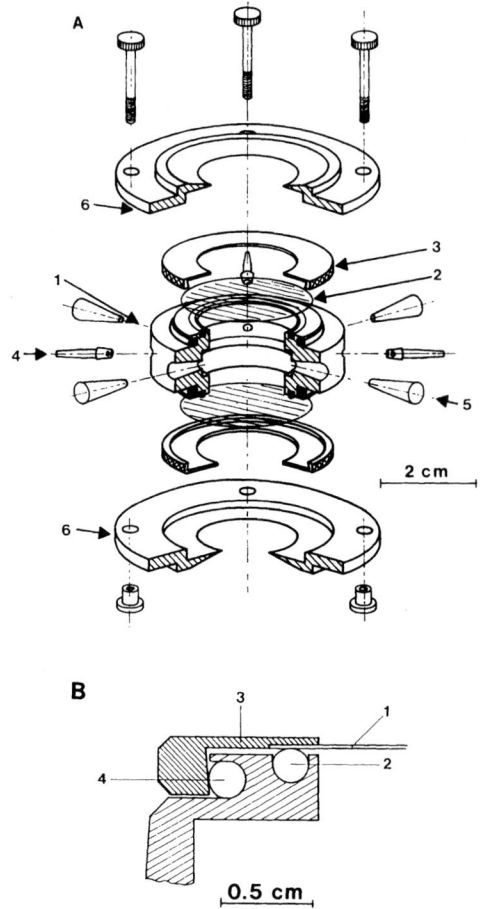

References

Ince C, Beekman RE, Verschragen G (1990) A micro-perfusion chamber for single-cell fluorescence measurements. J Immunol Meth 128:227–234

Ince C, van Dissel JT, Diesselhoff-den Dulk MMC (1985) A Teflon culture dish for high magnification observations and measurements from single cells. Pflügers Arch 203:240–244

Ince C, Ypey DL, Diesselhof-den Dulk MMC, Visser JAM, de Vos A, van Furth R (1983) Micro-CO_2-incubator for use on a microscope. J Immunol Meth 60:269–275

Chapter 2

Simple Perfusion Chambers for Single Cell Measurements

Kei Inouye

Background

Perfusion chambers are widely used for studying responses of cells to external stimuli. Because of the simplicity of its principle, perfusion chambers are often designed and constructed in laboratories. However, making a good perfusion chamber is not so simple a matter as it might appear. For instance, use of a perfusion chamber of an inappropriate design could result in a failure of sufficient stimulation of the cells.

A very simple yet efficient perfusion chamber is described by Swanson (1989). Briefly, the coverslip with attached cells is mounted cell-side down on fragments of coverslips and partially sealed with a paraffin-like material which can be melt by heating. With the same sealing material, two wells are made on both sides of the mounted coverslip for addition and removal of solutions. When many samples are to be examined, however, making a chamber each time will be tedious, and more permanent chambers may be desired. This chapter describes some examples of permanent perfusion chambers that can be made fairly easily and have been tested for their operation in detail. Figures 2.1 and 2.2 show examples of such chambers. Although they have been originally designed and later improved for the use in cytosolic pH measurements in *Dictyostelium* (Inouye 1988), they will be useful for diverse purposes in a wide spectrum of organisms.

The main features of these chambers are described below.

A. Simple set-up and operation

Main features

B. Complete exchange of the solution possible within a few seconds

C. Only a very small amount of materials needed for perfusion

D. Can be easily modified and adapted for specific requirements

E. Economy

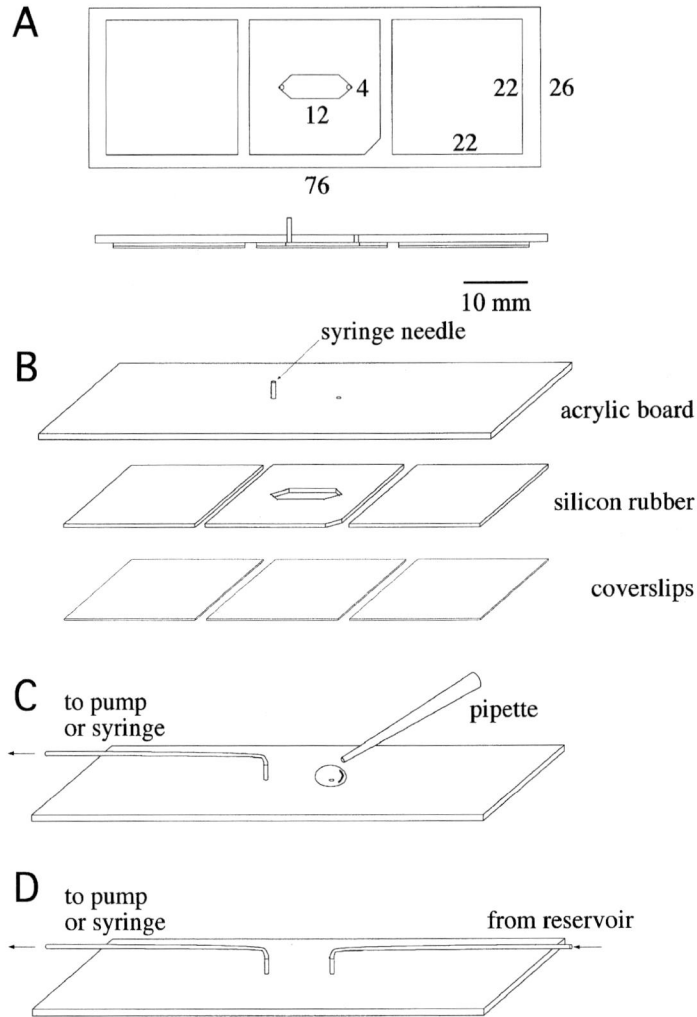

Fig. 2.1 A–D. A perfusion chamber for use with an inverted microscope. **A** Top view and cross-section. **B** A schematic representation showing all components of the chamber and the material of each component. **C** The configuration for direct application of the solution to the chamber orifice. **D** The configuration for perfusion from reservoirs: another piece of syringe needle needs to be glued to the hole on the right-hand side of the top unit

Point B is essential when one wants to make a quantitative measurement of a rapid response of the cells to external stimuli, while point C becomes important when a response of the cells to a precious substance needs to

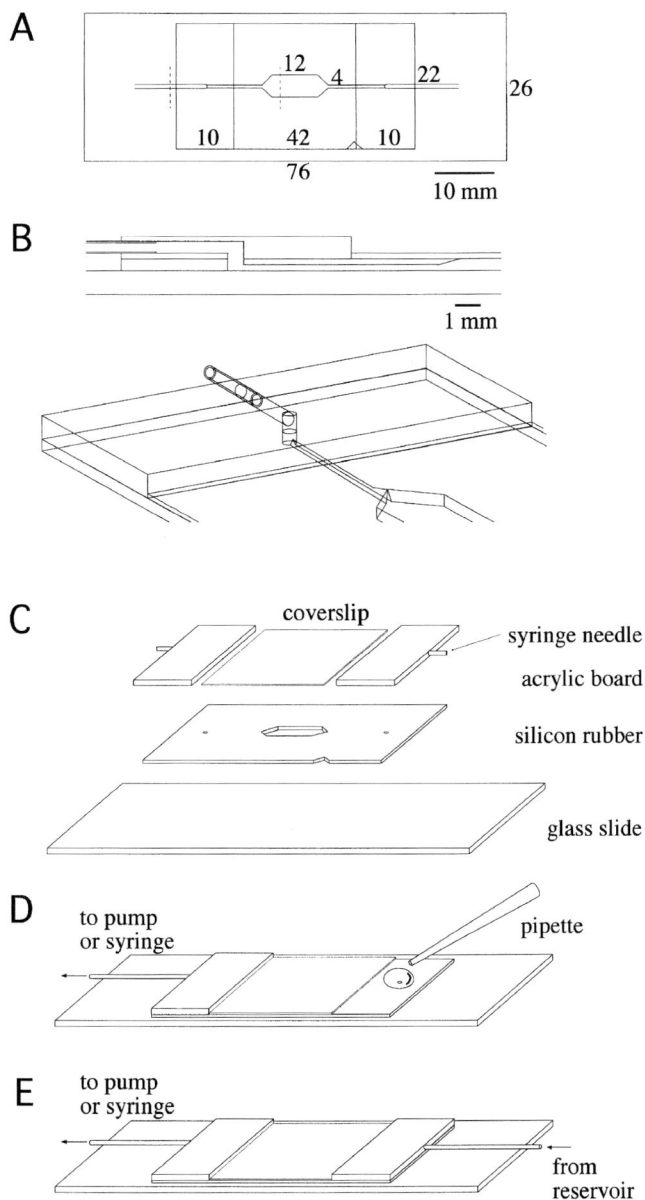

Fig. 2.2A–E. A perfusion chamber for use with an upright microscope. A Top view. B Cross-section and transparent view of the part of the chamber indicated between the two *dashed lines* in A. C A schematic representation showing all components of the chamber and the material of each component. D The configuration for direct application of the solution to the chamber orifice. E The configuration for perfusion from reservoirs

be investigated. Point E may be particularly important when the budget is tight.

In the following, I will first show some data demonstrating the performance of the chambers, and then give instructions for making and using the chambers.

Materials and Procedure

2.1
Characteristics of the Chambers

Factors affecting the rate of solution exchange

Ideally, the bathing solution should be instantly and completely replaced by a new solution when required. In practice, however, there is always a delay, and more importantly the change of the solution can never be stepwise but more-or-less gradual. Figure 2.3A shows example recordings to illustrate the time-course of solution exchange as determined by perfusing buffer solutions with or without a fluorescence dye. The recordings shown represent different flow rates and different lengths of the tubing that connects the chamber and the reservoir of the solutions.

Note that not only the lag (T_1) but the speed of solution replacement (T_2) is proportional to the length (Fig 2.3B) and the diameter (not shown) of the connecting tubing. The large T_2 values for the long and thick tubings are the result of the inhomogeneous flow inside the tubing due to the friction on its inner surface.

Naturally, the speed of solution exchange is proportional to the rate of perfusion (Fig. 2.3C) and inversely proportional to the width and depth of the chamber (data not shown).

The time when the new solution actualy reaches the cells on the coverslip was estimated by monitoring the pH-dependent change of fluorescence from a fluorescein-coated coverslip upon exchange of buffer (without dye) from low pH to high pH. The results indicate that the time required for solution exchange (both T_1 and T_2) determined at the surface of the coverslip could be considerably longer than the values estimated from bulk flows (Fig. 2.3B, C triangles).

Upper limits to the flow rate

There is a limit for the rate of perfusion that can be used without affecting the cells inside. A fast flow of the solution could cause deformation of firmly adhered cells (Liu et al. 1994), which in turn may have various effects on the cells. In fact, shear stress of about 10–20 dynes/cm^2 (1–2 Pa) has been shown to cause a variety of effects in vascular endo-

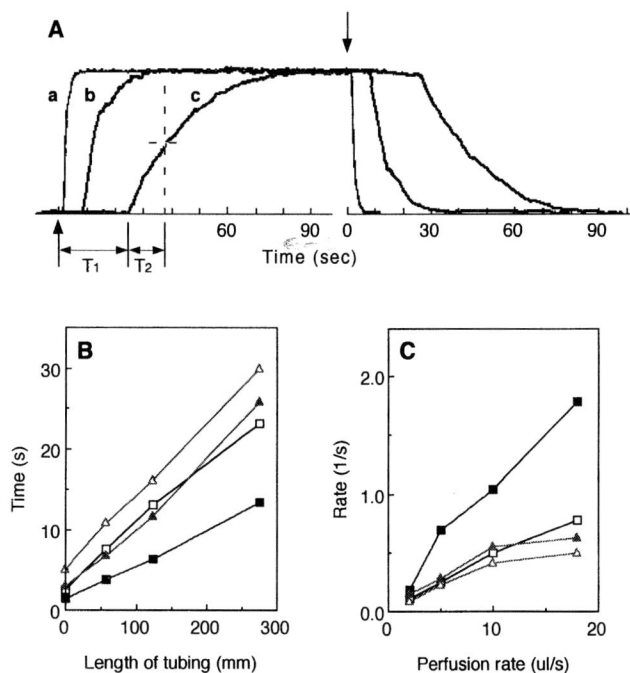

Fig. 2.3A–C. A Time course of solution exchange in a perfusion chamber of the type shown in Fig. 2.2. The chamber was first perfused with buffer (pH 7.5), then switched to 5 µM carboxyfluorescein in the same buffer, and then switched back to buffer alone. The fluorescence intensity of the solution (ex 492 nm, em 530 nm) was continuously monitored at the centre of the chamber using a Reichert Zetopan fluorescence microscope equipped with a photomultiplier, and recorded on a chart recorder. The solutions were perfused continuously using a peristaltic pump, except for short pauses to switch solutions. Typical recordings of fluorescence are shown. The rate of perfusion was 18 µl/s (*a*) and 2 µl/s (*b, c*). The length of the tubing for the incoming solution was 58 mm (*a, b*) and 275 mm (*c*). *Arrows* indicate the times when the pump was restarted after the tubes containing the perfusing solutions were switched. T_1 represents the lag before fluorescence intensity starts to rise, T_2 the time required for the fluorescence intensity to reach the half-maximum after the lag. **B** Dependence of T_1 (*open symbols*) and T_2 (*closed symbols*) on the length of the tubing for the incoming solution as determined by fluorescence of the solution (*squares*) and by pH-dependent changes of the fluorescence from the coverslip coated with FITC-γ-aminopropyltriethoxysilane (*triangles*). The inner diameter of the tubing is 0.57 mm and the rate of perfusion is 2 µl/s. **C** Dependence of $1/T_1$ (*open symbols*) and $1/T_2$ (*closed symbols*) on the rate of perfusion. The same symbols are used as in **B**. The length of the tubing for the incoming solution is 58 mm

thelial cells, including changes of the cytosolic Ca^{2+} and pH (Schwarz et al. 1992; Ziegelstein et al. 1992). Weakly adhered cells may be ripped off the substratum. Vegetative amoebae of *Dictyostelium* completely come off the substratum within 30 s at shear stresses smaller than 30 dynes/cm^2 while starved cells are resistant to much higher shear stress (based on the data of Yabuno 1971).

Using the Poiseuille's law, shear stress (τ in dynes/cm^2) at the glass surface of the chamber (ignoring the effect of the side walls of silicon rubber) is calculated to be $6\eta Q/d^2 w$, where η is the viscosity of the solution in Poise (0.01 for water), Q the flow rate in ml/s, d and w the depth and width of the chamber in cm. For the chambers shown above, τ (dynes/cm^2) $= 0.06 \times Q$ (μl/s) or just Q (ml/min). It is advisable to use a flow rate well below 10 ml/min or 167 μl/s to avoid shear stress close to 10 dynes/cm^2.

Use of an open-top chamber

With the highest perfusion rate tested (18 μl/s) without upstream tubing, it still took 1.4 s for a half-maximal replacement as measured on the surface of the substratum. If even faster exchange of the solution is required, direct application of the solution from a micropipette tip can be attempted. The pipette must be quickly brought into the solution using a manipulator so that its tip is accurately positioned near the cells, and immediately a pressure is applied to the pipette for delivery of the perfusing solution. An open-top chamber required for this type of experiment can easily be made by replacing the top unit of the chamber shown in Fig. 2.1 by one with a single large hole of the shape identical to the one in the silicon rubber sheet. Overflow of the solution is drawn from the edge of the hole. With this configuration, however, shear stress exerted on the cells is difficult to control. The same type of chamber can also be used for localized stimulation of cells using a micropipette of a small tip diameter.

2.2
Construction of the Chambers

Materials

A thin acrylic (Perspex, Plexiglas) board (1 mm thickness), a sheet of silicon rubber (0.5 mm thickness), glass coverslips (no. 1, thickness 0.12–0.17 mm, 22×22 mm), a syringe needle (22-G to 27-G), glue (e.g., Araldite), flexible rubber tubing that fits the syringe needle, and a small syringe (1 ml).

Tools

Only common tools likely to be found in or around every laboratory are required for making the chambers. The only exception may be a small hand-held drill (which looks like a dissection needle) with a "bit" matching the diameter of the syringe needle for making holes in the acrylic board. For the chamber for an upright microscope, a razor blade broken to ca. 1 cm long along the blade is used for making the grooves, and a tiny "cork borer" is used for making small holes in the silicon rubber sheet. The latter is made by cutting and sharpening the syringe needle attached to the syringe.

- Construction   **Notes**
  For the most part, Figs. 2.1 and 2.2 are self-explanatory. The measures shown are only for a guide and need not be exactly as indicated, and any modification can be made to suit specific purposes. Note that the shape of the hole in the centre of rubber sheet is important, as shapes like a rectangle cause stagnation of the solution at its corners. Also, the thickness of the rubber sheet should be as small as possible to reduce the capacity of the chamber. Some additional explanations are given in the following.

- Silicon Rubber
  - The central hole of the silicon rubber sheet measures 4 mm (width) × 12 mm (span). The length of the widest part is 8 mm, so the area is 40 mm^2 and the capacity 20 µl. Small indentations at the edge of the rubber sheet are to make removal of the coverslip easier.
  - To make the silicon rubber piece for an upright microscope, make tiny holes at 10 mm from the edges of the central hole and cut out grooves of 5 mm long and 0.3 mm wide connecting the small holes and the edges of the central hole (Fig. 2.2B). This can be done by using a piece of razor blade (10 mm long) held horizontal along its blade and slanted by ca. 30° sideways. The cross-section of the grooves will be triangular. If possible, deepen the grooves towards the central hole as shown in Fig. 2.2B.
  - Wash all pieces of silicon rubber in boiling water with several changes of water to remove any harmful ingredients of the material.

- Syringe Needle
  - When cutting the syringe needle, first put it on the syringe tightly, then cut it from near its tip. If a pair of pliers is used for cutting, the cut ends get flattened and need to be opened by grinding the cut ends using a grinder or sandpaper. Smoothe all the cut ends of the syringe needles using sandpaper. The small pieces of needle get lost easily, so handle with care.

- Drilling and Gluing
  - It is easier to drill holes manually, as an electric drill is difficult to handle when making such small holes. If a drill is not available, an intact syringe needle can also be used, but lot of patience is required.
  - When gluing the small pieces of the syringe needle, make sure that the glue does not obstruct the hole.

- Accurate Positioning (Upright Chamber)
  - The small holes on both sides of the silicon rubber sheet and the holes on the acrylic pieces must be accurately positioned so that they form a continuous hole when assembled. This can be done by first making the holes on the acrylic pieces, then assembling the parts, with the coverslip, on a clean sheet of paper (instead of the glass slide) so that the exact positions for the tiny holes can be marked on the rubber sheet from underside. It is important to allow a fraction of mm between the acrylic pieces and the coverslip when determining the positions of the holes, because the size of coverslips is not exactly the same and a slightly larger coverslip would not fit in between the acrylic pieces after the chamber is completed.

- Assembly
  - After the glue has solidified, wash the slide and the rubber sheets with detergent and rinse them with distilled water. Any remaining dust will weaken the adhesion of the silicon rubber to the glass and plastic surfaces.
  - Assemble all the parts as shown in Fig. 2.1 or 2.2. The coverslips must stick tightly to the rubber sheets without leaving air connecting the inside and outside of the chamber. The two coverslips on the both sides of the chamber shown in Fig. 2.1 are to hold it horizontal and need not be removed after use.

About the material

Perspex (acrylic resin) is used as the top unit of the chamber for an inverted microscope since it allows cutting and drilling without special tools. However, it emits some fluorescence even by excitation light of long wavelengths. This of course causes a serious problem when a low-level fluorescence needs to be measured. In many applications, however, the fluorescence is weak enough so the background fluorescence is negligible compared to the signals from the cells, because the distance of 0.5 mm between the cells and the plastic surface allows dispersion of both excitation light and fluorescence. For other plastic materials of good optical properties, polycarbonate is even more fluorescent while polystyrene (which is used in plastic Petri dishes) emits much less fluorescence and can be used instead of Perspex. Note, however, that it cracks easily. Glass slide can be used if facilities are available to drill tiny holes in hard glass. In that case, siliconize the glass slide after making the holes (dip the glass slide in 5 % methyltriethoxysilane in ethanol, air dry, and bake at 100 °C for 1 h).

Some may prefer a simplified version of the above chambers consisting of a silicon rubber sheet with a central hole with two 27-G needles stuck on its both sides. Although this is much easier to make, it is prone to leakage and will be useful only for quick examinations. The silicon rubber (because of the needles inside) needs to be rather thick (more than 2 mm) to obtain sufficient adhesion to glass. A syringe directly attached to either needle is very difficult to use, and more stable operation will be achieved if the syringe is connected to the needle via a rubber tubing.

Simplified chamber

2.3
Operation

A coverslip with the cells that have been incubated on the bottom of a Petri dish can be directly placed on the chamber unit for perfusion. For cells in suspension, put a small volume of cell suspension (up to 5 µl) in the centre of a coverslip and allow the cells to adhere. Then place the coverslip on the chamber unit. Alternatively, put 20 µl of cell suspension on the orifice of the chamber and draw it into the chamber using a syringe or a pump. For continuous perfusion, place exactly in position the upstream unit that has been filled with the solution to be perfused.

For exchange of solutions without using a tubing for introducing the solution, apply 100 µl of solution and draw at a rate not higher than 20 µl/s (i.e., allowing 5 s or more). This would achieve 99.9 % replacement of the solution at the centre of the chamber within 5 s, 90 % replacement at the surface of the coverslip within 5 s, and 50 % replacement at the surface of the coverslip within 1.4 s. If a chamber of a capacity of x µl (including the upstream capillary part and tubing) is used, washing and perfusion solutions should at least be $5x$ µl (see p. 12).

As discussed above, better results would be obtained by using a chamber of a small capacity (both depth and area) without a tubing for introducing the solution. If use of a tubing is necessary, use a thin tubing of a minimum length. If a short period of perfusion (e.g. 3 min) at a moderate rate (e.g. 5 µl/s) is enough, an Eppendorf tube containing 1 ml of the perfusion solution can be laid close to the chamber on the microscope stage (it won't spill), in which case a tubing of only 5 cm should be enough.

Maintenance

After measurement, wash the inner part of the chamber by perfusing distilled water, then the middle piece of the coverslips is removed. Wash the remaining part of the chamber (with the coverslips on the sides attached) with detergent and water, dry it, and keep it in a dry place. When it is not used for a long period, remove all the coverslips. Note that acrylic resin does not tolerate temperatures higher than 70 °C.

References

Inouye K (1988) Differences in cytoplasmic pH and the sensitivity to acid load between prespore cells and prestalk cells of *Dictyostelium*. J Cell Sci 91:109–115

Liu SQ, Yen M, Fung YC (1994) On measuring the third dimension of cultured endothelial cells in shear flow. Proc Natl Acad Sci USA 91:8782–8786

Schwarz G, Callewaert G, Droogmans G, Nilius B (1992) Shear stress-induced calcium transients in endothelial cells from human umbilical cord veins. J Physiol Lond 1458:527–538

Swanson J (1989) Fluorescent labeling of endocytic compartments. Meth Cell Biol 29:137–151

Yabuno K (1971) Changes in cellular adhesiveness during the development of the slime mold *Dictyostelium discoideum*. Dev Growth and Differ 13:181–190

Ziegelstein RC, Cheng L, Capogrossi MC (1992) Flow-dependent cytosolic acidification of vascular endothelial cells. Science 258:656–659

Chapter 3

Temperature Control on the Stage of the Microscope

Arie de Vos, Bert Van Duijn, Anneke Wiltink

Background

Temperature control is an important issue for applications in single cell research. The need to control temperature during experiments arises because cell responses are temperature-dependent. Especially in applications using fluorescent probes, adequate control of the temperature is critical, since dissociation constants of fluorescent probes such as Fura-2 are also highly dependent on temperature. (Lattanzio 1990, for review see Moore 1990). Video microscopy setups are highly susceptible to heat loss due to the extensive interface of the incubation chamber with the surroundings. One approach to controlling temperature is to heat the entire setup; however, for practical reasons, such as the need to have access to the incubation chamber during the experiment, this may not be feasible. Another approach is to heat only certain essential components of the setup; however, in this case it is vital to examine carefully whether the desired temperature is actually reached and if temperature control is constant in the vicinity of the cells in the incubation chamber. Oscillations in the temperature due to inadequate control may cause apparent oscillations in, e.g., intracellular calcium concentrations measured using Fura-2.

To control temperature in our video microscopy setup, we devised a system in which temperature is controlled at several points in the setup during perfusion of the incubation fluid. In addition to controlling the temperature of the incoming fluid, we use an oil-immersion objective that is heated. Temperature in the vicinity of the cells is measured by very small microsensors. The system can be used to control the temperature in the vicinity of cells during single cell research. Here, an overview of our temperature control system is presented. The pitfalls in temperature control are also discussed.

Temperature control system

Materials and Procedure

Description of the Temperature Control System

The temperature control system is used in combination with a setup for microfluorescence or electrophysiology (Wiltink et al. 1995) and a perfusion system. For single cell research, cells attached to glass cover-slips (diameter 24 mm) can be used. During the experiments, these are mounted in a Teflon culture dish especially devised for single cell research (Ince et al. 1985). The incubation medium is perfused with a constant flow velocity. An overview of the system is given in Fig. 3.1.

Flow profile   If a round incubation chamber is used together with a perfusion system, the flow profile is unpredictable and may cause temperature as well as concentration differences in the incubation medium. Therefore, we designed an insert for the culture dish that ameliorates flow properties

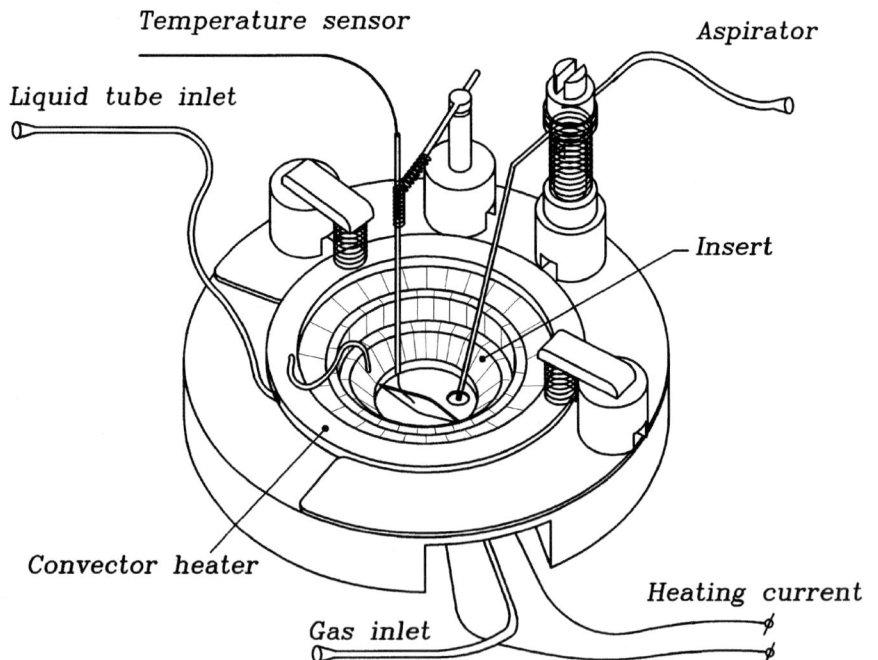

Fig. 3.1. Overview of the temperature control system in the perfused culture dish

Fig. 3.2. Stainless steel insert to improve flow profile of the culture dish

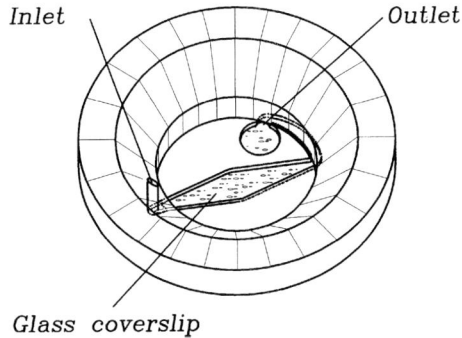

Inlet   *Outlet*

Glass coverslip

and temperature control near the cells (Fig. 3.2). The insert is made of stainless steel surrounding a flow channel. The dimensions of this channel are such that the flow profile near the cells is laminar, as was determined using ink. The advantage of stainless steel is that the heat conductance is good, and it poses no toxicity hazard to living cells.

Convector Heater

The culture dish is placed in a micro-incubator (Ince et al. 1983) (Fig. 3.1). Using this setup with unstirred incubation medium creates a considerable temperature gradient from the outside of the dish towards the center. Therefore, we modified the original micro-incubator by placing a convector with a tube on top of it. Through this tube, the incoming perfusion fluid passes. The convector heater is then used to elevate both the temperature of the culture dish and that of the incoming incubation fluid (Fig. 3.3). During experiments, perfusion with a fixed fluid flow of, e.g., 1 ml/min should be realized using a pumping system that results in a continuous flow. Fluctuating flow profiles, such as would arise using a roller pump, should be avoided as they cause temperature fluctuations in the vicinity of the cells.

Temperature control unit

The temperature control unit consists of a 4-channel PID controller (Multitemp, Multitronics, Amersfoort, The Netherlands) that is connected to the oil immersion objective and the convector heater via a power supply. Thermocouples (type K, Thermo Electric, Warmond, The Netherlands) were used to monitor the temperature of the oil immersion objective, the convector heater and the vicinity of the cells during the experiments. It is advantageous to choose a PID controller that has an autotuning function to avoid the tedious task of manually determining the proper PID values. The chosen power supply should be compatible with the PID controller and preferably generate a DC current, since AC

Fig. 3.3. Convector heater

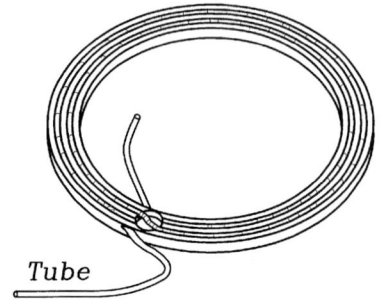

Convector heater

Tube

currents may interfere with, e.g., patchclamp measurements. Thermocouples of very small diameter (0.25 mm) were chosen to avoid a temperature leak through the sensor wires, which may occur if the sensors are much bigger than the incubation volume. If the room temperature is lower than the temperature of the incubation medium, an erroneously low temperature would be measured.

Oil immersion objectives

Temperature leaks may also occur via the oil immersion objective (achrostigmat 40x/1.3 PH2 objective from Zeiss), making it necessary to control the temperature of this part of the setup as well. To accomplish this, we devised a metal heating ring for the objective (Fig. 3.4).

Note. Check with your supplier if the objective you use can be heated.

Fig. 3.4. Heated objective

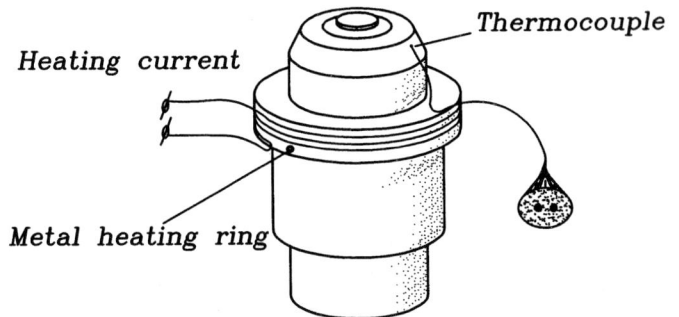

Thermocouple

Heating current

Metal heating ring

Comments

Both commercially available and "home-made" temperature control systems can be used in single cell research. In either case, it is very important that the temperature at different parts of the system be monitored in order to tightly control the desired temperature in the vicinity of the cells.

References

Ince C, Ypey DL, Diesselhoff-den Dulk MCM, Visser JAM, De Vos A, Van Furth R (1983). Micro-CO_2-incubator for use on a microscope. J Immun Methods 60: 269–275

Ince C, Dissel JT, Diesselhof-den Dulk MMC (1985) A Teflon culture dish for high-magnification microscopy and measurements in single cells. Pflügers Arch 403: 240–244

Lattanzio FA (1990) The effects of pH and temperature on fluorescent calcium indicators as determined with chelex-100 and EDTA buffer systems. Biochem Biophys Res Comm 171: 102–108

Moore EDW, Becker PL, Fogarty KE, Williams DA, Fay FS (1990) Ca^{2+} imaging in single living cells: Theoretical and practical issues. Cell Calcium 11: 157–179

Wiltink A, Nijweide PJ, Scheenen WJJM, Ypey DL, Van Duijn B (1995) Cell membrane stretch in osteoclasts triggers a self-reinforcing calcium entry pathway. Pflügers Arch (Eur J Physiol) 429: 663–671

Chapter 4

A Modified U-Tube Drug Application System

Misa Dzoljic

Background

In electrophysiological or microfluorescence experiments, application of drugs is a necessity when studying receptors gated by these substances. This requires a system allowing local and rapid application of the desired drug. Moreover, when the effects of more than one drug are studied on the same cell, the demands on the system become even more complicated. Most drug application systems are based on a U-tube drug application device. The essential idea of this U-shaped loop is depicted in Fig. 4.1.

Fig. 4.1. The U-tube application system

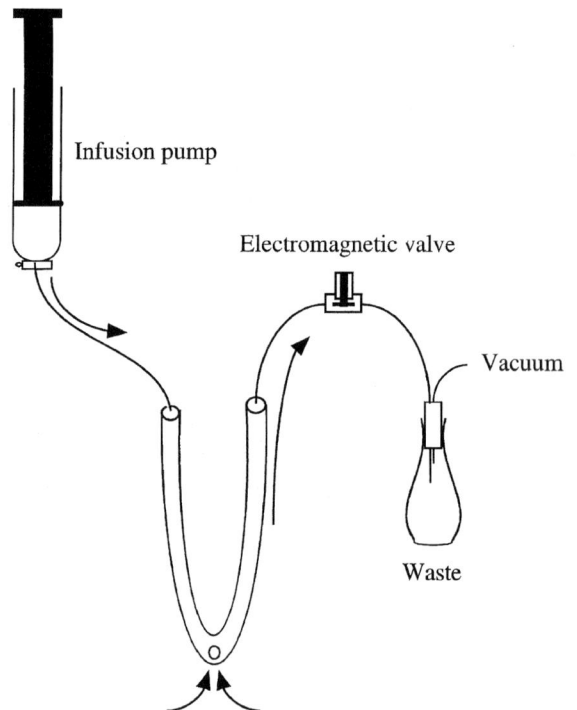

Infusion pump

Electromagnetic valve

Vacuum

Waste

A reservoir (usually a syringe with a pump) is connected to one side of the U-tube which has an opening of approximately 100 µM in diameter at its end. The other side is connected to a vacuum. In the open state, a valve (usually an electromagnetic valve that can be controlled by a computer) allows flow of the fluid, supplied by the reservoir, through the suction line. Closing the valve causes the flow to escape through the opening in the U-tube thereby changing the medium near the opening. Thus, if the cell is placed in the proximity of the U-tube opening, the medium around the cell will rapidly be exchanged for medium containing the agonist. Opening of the valve again will remove the agonist-containing solution around the cell. While many variations exist on this basic idea, the system, in its more sophisticated form, has been described by Bormann (Bormann 1992). This system has proven its usefulness in many experiments but has a disadvantage in that application of different drugs within one experiment is complicated. Obviously when different applications are necessary some kind of distribution system (or selector dial) should be placed between the different reservoirs and the U-tube. Depending on the speed of perfusion, the diameter of the tubing and the distance between the selector and the U-tube, the dead space (space between the selector and the opening in the U-tube) may cause a problem if rapid exchange between different drugs is necessary. After changing the drug, one should wait until this space has been filled with the new fluid before allowing the next application to take place. Changing the application pipette fluid in 10–100 ms is impossible with this system.

A modification of the U-tube system was based on an adaptation described by Vreugdenhil (Vreugdenhil 1994), in which the U-tube is replaced by a slightly modified pipette holder. Figure 4.2 shows this pipette holder and a glass pipette with an opening of 100 µm. The pipette holder allows a quartz tube with a tip of 50 µm to be inserted within the outer glass pipette. The outer glass pipette is held under negative pressure, which can be disrupted by an electromagnetic valve. The application technique remains similar to the U-tube system, with the advantage that a more compact setup is achieved; however, the disadvantages remain the same. There is still a dead space to account for if more reservoirs are connected to one inner tube. More importantly, due to the tube-in-pipette construction, small particles (like cellular debris) sucked up from the extracellular fluid can obstruct the suction in the outer pipette giving rise to uncontrollable application of the agonist. The need to have greater control over the application, to avoid accidental applications, and to have a small dead space led us to construct an application system based on a controllable pressure and vacuum generator.

Principle of operation

Fig. 4.2. The pipette holder with the glass pipette containing an inner quartz tube containing the drug to be applied to the cell. The tubing at the top of the pipette holder ensures that a continuous negative pressure can be applied to the outer pipette. Positive pressure is applied on the inner tube, while the negative pressure in the outer glass pipette ensures that no spillage from the drug reaches the cell. Closing of the electromagnetic valve shuts off the suction and the drug is applied to the cell. Reopening of the valve actively removes the drug again

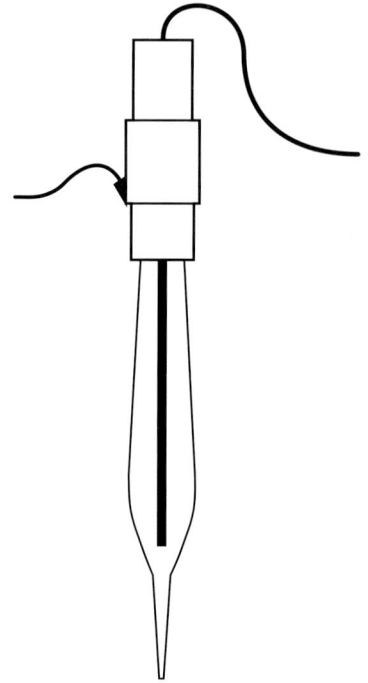

Materials and Procedure

The Application Pipette

The application pipette is in essence the same as the one used by Vreugdenhil (1994). The outer pipette is a glass pipette with an outer diameter of 1 mm that has been pulled into a tip of approximately 100 μm diameter. The pipette is placed in a pipette holder as depicted in Fig. 4.3. Within the outer pipette, three inner tubes (164 μm OD, MicroFil CMF 34 G, World Precision Instruments, Sarasota, Florida, USA) can be positioned. Using a silicone band the compartment of the outer pipette is sealed air-tight, allowing for negative pressure to be applied through a special connection. The inner tubes are connected to polythene tubing (1.5 meters in length, 1 mm O.D., 0.6 mm I.D.) and filled with extracellular fluid containing the dissolved drugs. The polythene tubing forms the reservoir for the application fluid.

Note. Visual control over the application is, certainly in the beginning, advisable. Adding phenol red (phenolsulfonphthalein; Sigma, St. Louis,

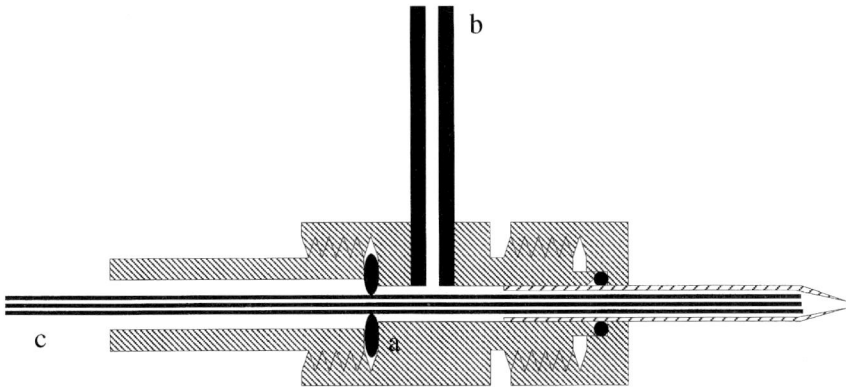

Fig. 4.3. The pipette holder with three inner tubes and a glass outer tube. A silicone band (*a*) ensures that the compartment of the outer pipette is sealed air-tight, allowing for negative pressure applied through a special connection (*b*). The inner tubes (*c*) are filled with extracellular fluids containing the dissolved drugs and connected to the pico-injector

USA) to the application fluid makes this possible. After addition of this indicator (approx. 0.1 mM) the pH should be re-adjusted. Since most drugs are dissolved in small quantities, we found it useful to prepare a stock solution of extracellular fluid already containing phenol red and with the pH adjusted (do not forget to filter the solution before storage or use). Since phenol red (or impurities) might cause interference with the measurements, additional experiments without phenol red are necessary. We found no interference in experiments with the $GABA_A$ or NMDA receptors.

Note. Before the application fluid is injected into the tubing it should be at room temperature. If cold application fluid is warmed to room temperature while already in the tubing, bubbles will form which are difficult to evacuate.

Once the tubing is filled and the application pipette is mounted above the measurement chamber it will be moved only with micromanipulators. The tubing is connected to a 4-way selector which is operated manually. This selector is connected to the pico-injector.

The Pico-injector

The pressure to the inner tubes (with the application fluid) as well as the vacuum applied on the glass outer pipette is regulated with a PLI-100 pico-injector (Medical Systems Corp., Greenvale, NY, USA). This device, developed for intracellular microinjection, allows application of regulated pressure for a selected period of time. Additionally, it is possible to apply continuous pressure and there is an internal vacuum pump.

The pico-injector, connected to the selector with tubing, keeps the tubing under continuous pressure. This pressure can be regulated between 0.7 and 70 kP. The pressure should be adjusted to overcome the resistance of the tubing, yielding a slight flow out of the inner tube. This should be done for all the three tubes. This so-called balance pressure, applied to the delivery pipette between the injections, prevents clogging as well as dilution of the injected material by capillary action.

Note. It is obviously convenient if the resistance in all three (inner) tubes is the same. The same tubing, having the same length and made out of the same material, should be used for all three tubes.

Suction is continuously applied to the outer pipette (Fig. 4.3b). It should be adjusted in such a way as to remove all of the agonist accumulated within the outer pipette but also some of the fluid in the measurement chamber.

Finally, the injection pressure should be adjusted. This pressure forces the application fluid (usually the extracellular fluid with a drug), despite the continuous suction, through the outer pipette into the fluid near the outer pipette (where hopefully a cell is present). Triggering of the injection pressure is accomplished with an externally applied electrical signal controlled by a computer (TTL pulse). Obviously, this pressure should be high enough to overcome the relative vacuum in the outer pipette but not so high as to damage the cell. Additionally, one should try to exchange the fluid near the cell only, since the drug should also be removed after the application. Very often a continuous flow of extracellular fluid is present in the measurement chamber during the experiments. This flow can be directed in such a way as to help remove the application fluid. The entire application cycle is shown in Fig. 4.4.

Controllable application pressure and controllable suction of the outer pipette, in combination with continuous extracellular flow, ensure a rapid and controllable system that allows application of three different drugs within very short time intervals.

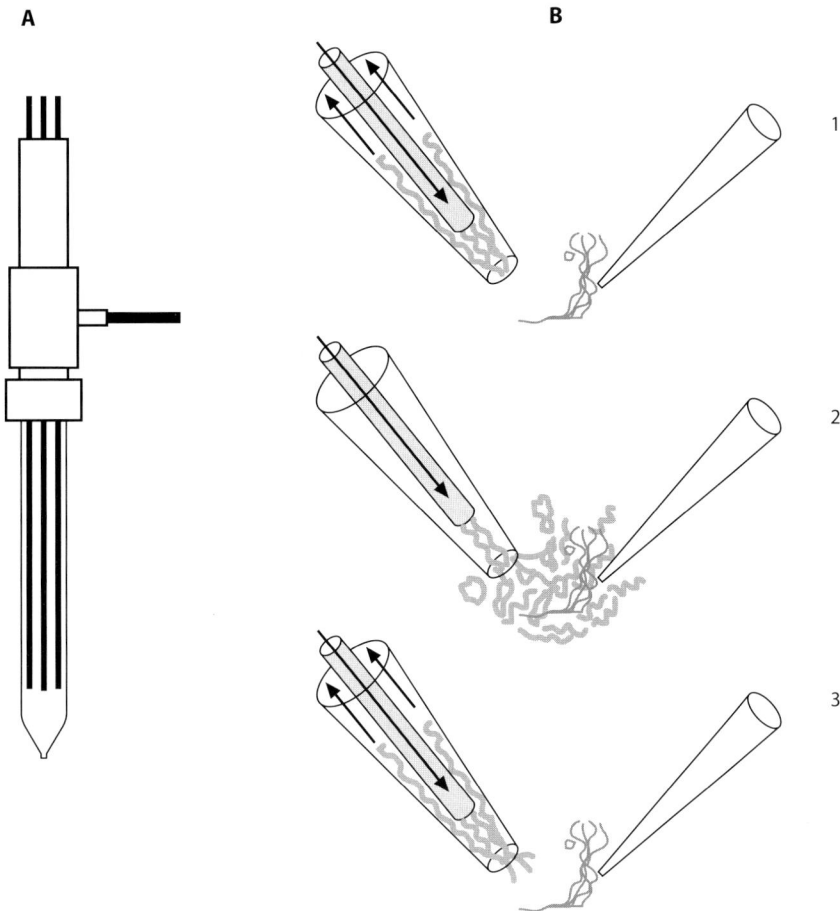

Fig. 4.4A, B. A The pipette holder with the glass pippete containing three inner tubes. **B** The steady state of the system. There is positive pressure on the inner tube, while the negative pressure in the outer glass pipette ensures that no spillage from the drug reaches the cell (*1*). Additional positive pressure causes application of the drug to the cell (*2*). Continuous suction applied to the outer pipette ensures that the applied drug is removed once the additional pressure is stopped (*3*). Up to three inner pipettes can be accommodated within one outer pipette. For the sake of clarity only one inner pipette is depicted

Note. Rinse the entire system with distilled water at the end of the experiment. Dried solutions will not only cause changes in resistance but may also cause irreversible clogging of the small tubing.

References

Bormann J (1992). U-tube drug application. In: Practical electrophysiological methods. Kettenmann H, Grantyn R (eds) Wiley-Liss, New York, pp 136–140

Vreugdenhil M (1994). Intrinsic membrane properties of hippocampal pyramidal neurons after kindling epileptogenesis. Ph. D. thesis, University of Amsterdam, The Netherlands

Chapter 5

Laser Microsurgery as a Tool in Single Cell Research

Pettie P. Booij and Albertus H. De Boer

Background

The patch-clamp technique is a powerful tool enabling the study of function and regulation of single channel currents and kinetics associated with the gating of ion channels. Successful application of the patch-clamp technique relies on the formation of a high resistance seal between the plasma membrane and the patch-clamp pipette. To gain access to the plasma membrane of plant cells, the cell wall must first be removed.

Conventionally, enzymes are used to digest the cell wall of plants to produce spherical protoplasts (Cocking 1960). However, enzymatic digestion of the cell wall can have serious disadvantages. Firstly, there is increasing evidence that cell wall-degrading enzymes produce elicitors, in the form of cell wall fragments, during breakdown of the cellulose matrix which may induce expression of defence genes in these cells (Rincon and Boss 1987). Secondly, proteases present in the cocktail of cell wall-degrading enzymes may damage the extracelluar moiety of the membrane proteins (e.g., ion channels, membrane receptors). Thirdly, problems with identification of the protoplast can arise because information on the original localization in the intact tissue is lost. Fourthly, the original orientation is lost when protoplasts are isolated enzymatically. This problem is of particular importance in the study of ion channel activity, distribution and regulation in polarizing and polarized cells.

To overcome these problems, a new technique has recently been developed, laser microsurgery (Greulich and Weber 1992). This technique uses an UV laser microbeam to induce precisely localized submicrometer perforations in the cell wall of plants (Taylor and Brownlee 1992). Laser microsurgery can be used for the removal of the cell wall in a variety of cell types (De Boer et al. 1994). Here, we present a method to release protoplasts from *Vicia sativa* root hair cells using laser microsurgery. Root hairs are highly polarized and specialized cells and the original orientation and localization are features which must not be lost during the release of root hair protoplasts.

Root hairs of leguminous plants are the target of nodulation factors (Nod factors), signal molecules produced by *Rhizobium* bacteria, which bind to receptors at the root hair plasma membrane. Since Nod factors are lipo-oligosaccharides it is difficult to assign their specific effects to enzymatically obtained protoplasts. Therefore, we used the laser released protoplasts for patch-clamp experiments in which we look at the effect of Nod factors.

Principles and application

Protoplasts of *Vicia sativa* root hair cells can be released using laser microsurgery. First, the root hair has to be plasmolysed in order to withdraw the plasma membrane from the cell wall. This is done to prevent damage to the plasma membrane by the laser pulse. After plasmolysis the cell wall is removed locally with a few laser pulses, each lasting a few nanoseconds. After the cell wall is perforated the protoplast swells, fills the plasmolytic space and partially protrudes through the laser hole. These expelled root hair protoplasts can be used for patch-clamp experiments.

Figure 5.1 represents a short outline of the release of root hair protoplasts using laser microsurgery.

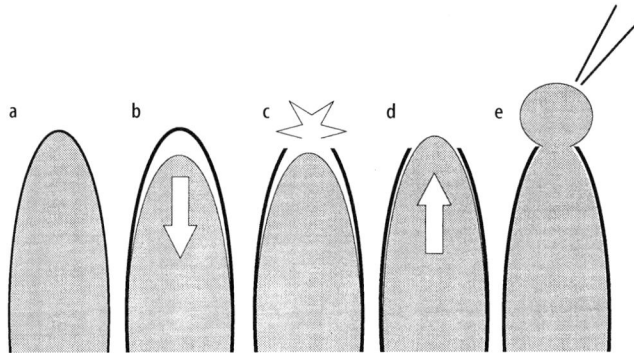

Fig. 5.1 A–E. Release of the root hair protoplasts using the technique of laser microsurgery. The *thick black line* represents the cell wall and the *shaded area* represents the cytoplasm surrounded by the plasma membrane (*thin black line*). **A** Root hair. Only the distal part of the root hair is shown. **B** The root hair is plasmolysed in order to withdraw the plasma membrane from the cell wall. **C** With a few pulses of a UV laser, a hole is cut in the cell wall of the root hair tip. **D** The protoplast swells immediately and fills the apical plasmolytic space. **E** The protoplast partially protrudes through the laser hole. This results in the formation of a small bleb of plasma membrane, which is now accessible for the glass patch-clamp pipette

Materials

- Inverted microscope (e.g., Axiovert 35, Carl Zeiss Inc.) with quartz optics.
- A UV nitrogen laser (VSL 337 ND, LSI, Cambridge, MA, USA). The laser characteristics are: wavelength 337.1 nm, frequency 1–20 Hz, pulse energy >250 µJ, beam divergence 0.3 mRad, pulse duration a few nanoseconds. The laser is coupled into the microscope via the epifluorescence path (Fig. 5.2).
  The cross-section of the laser pulse in the focal plane is a circular disk with a diameter near that of the UV laser light; 0.3–0.5 µm.
- A beam expander (Fig. 5.2) (Carl Zeiss Inc.), necessary to focus the laser light through the objective (ultrafluar x100, numerical aperture 1.25, glycerine) of the microscope to its diffraction limit. Using the beam expander it is possible to place the laser and microscope on the same vibration-free table.

  Note. Removal of lens 3 and 4 (Fig. 5.2) results in a higher energy output in the focal plane of the objective, because each time the light beam is focused energy is lost.

- A color video camera (e.g., Sony CCD/RGB DCX-151p) and a color monitor (e.g., Sony Triniton PVM-1442QM).
- Bath perfusion device (e.g., Lorenz, Lindau, Federal Republic of Germany). This device monitors the level of the medium in the bath and is coupled to a controller. In this way the level of the bath solution can be set and controlled during addition of new bath medium. New bath medium is added with a peristaltic pump with an inlet below the water level in the bath, ensuring minimal disturbance of the root hairs.
- Linear gradient mixer. This mixer is used to create a linear gradient from one solution to another. Solutions from a linear gradient mixer, which consists of two connected compartments (total volume 20 ml), flow into the bath using a peristaltic pump.

Buffer A: **Buffers**
- 5 mM 2-[N-morpholino]ethanesulfonic acid (MES)/KOH, pH 6
- 10 mM $CaCl_2$
- 10 mM sucrose
- Total osmolarity: 70 mOsm

Fig. 5.2. Path of the laser beam into the microscope via the beam expander and the epifluorescence path

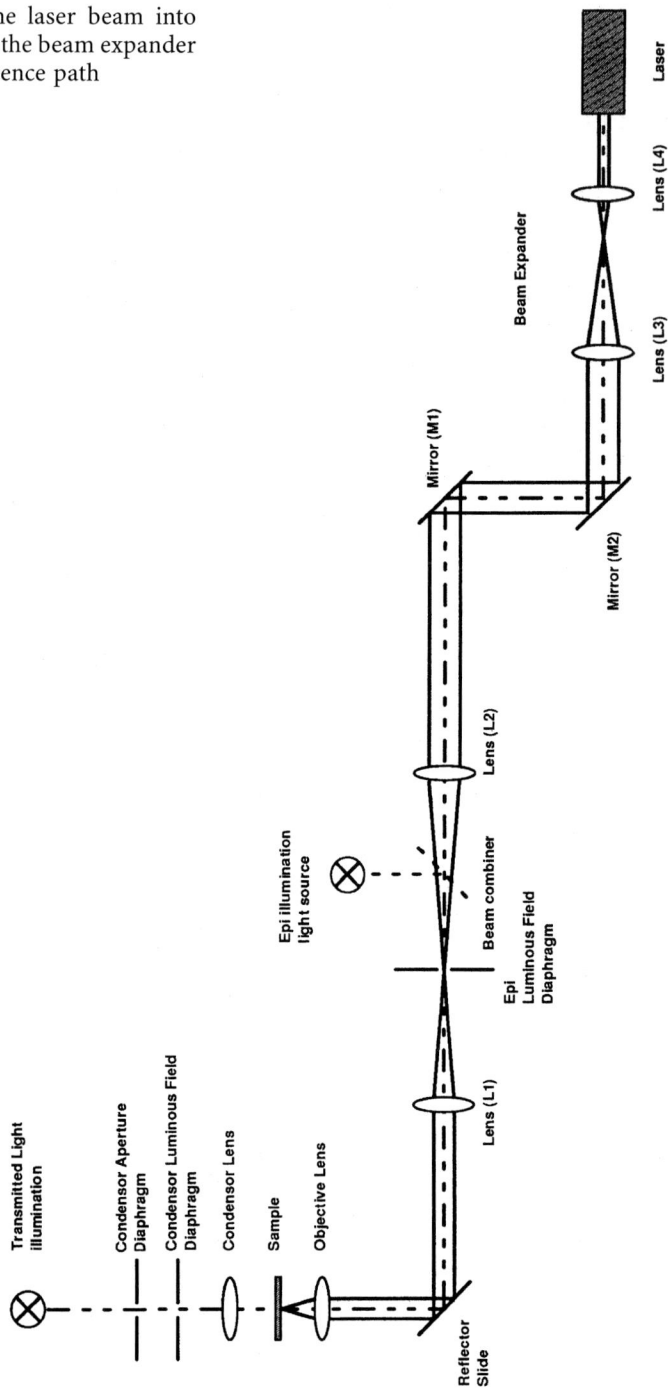

Buffer B:
- 5 mM MES/KOH, pH 6
- 10 mM CaCl$_2$
- 10 mM sucrose
- 495 mM sorbitol
- Total osmolarity: 500 mOsm

Surface sterilize the *Vicia sativa* subspecies *nigra* (Vetch) seeds by **Plant material** immersion in concentrated sulfuric acid (20 min) followed by rinsing with tap water for 5 h (Van Brussel et al. 1982). Germinate the seeds on 1 % agar plates. Incubate the plates at 4 °C for 3–7 days, followed by 2–3 days incubation at 20 °C (75 % humidity) in the dark. Transfer the seedlings after germination to modified Fåhraeus slides (Bhuvaneswari and Solheim 1985) (Fig. 5.3A). The Fåhraeus slides are prepared by affixing a 24 × 40 mm coverslip centered to a microscope slide. Drops of Xantopren L (BayerDental, Leverkussen, Germany) applied at the corners glue the coverslip to the slide; leave a space of 1 mm between the slide and coverslip using pieces of heat resistant plastic of 1 mm thickness. Place the Fåhraeus slides, with the roots of the seedlings between the microscope slide and the coverslip, upright in small trays containing Fåhraeus medium (in mM: CaCl$_2$ 0.90, MgSO$_4 \cdot$7H$_2$O 0.49, KH$_2$PO$_4$ 0.74, Na$_2$HPO$_4$.2H$_2$O 0.84, C$_6$H$_5$FeO$_7 \cdot$5H$_2$O 0.015, Traces: Mn, Cu, Zn, B, Mo; pH 6.5, Fåhraeus 1957). Grow the plants for 2 or 3 days at 20 °C, 75 % humidity with a light period of 16 h. Transfer the slides to Petri dish after1 day, the roots should have grown 1 cm, and put them in a slanted position (angle 15°); replace the medium by fresh medium (Fig. 5.3B). After 1 day in a slanted position, the roots have developed new root hairs on the elongating part of the root.

Note. Each Fåhraeus slide can contain one to three seedlings.

Note. The root hairs can also be used for experiments on the third day but then the medium has to be replaced with fresh Fåhraeus medium on the second day.

Procedure

1. Determine the location of the focused laser microbeam by shooting a hole in the top surface of a glass cover slide. The energy density in the focus of the laser beam is sufficiently high to etch the glass. Mark this location with a marker pen on the monitor screen.

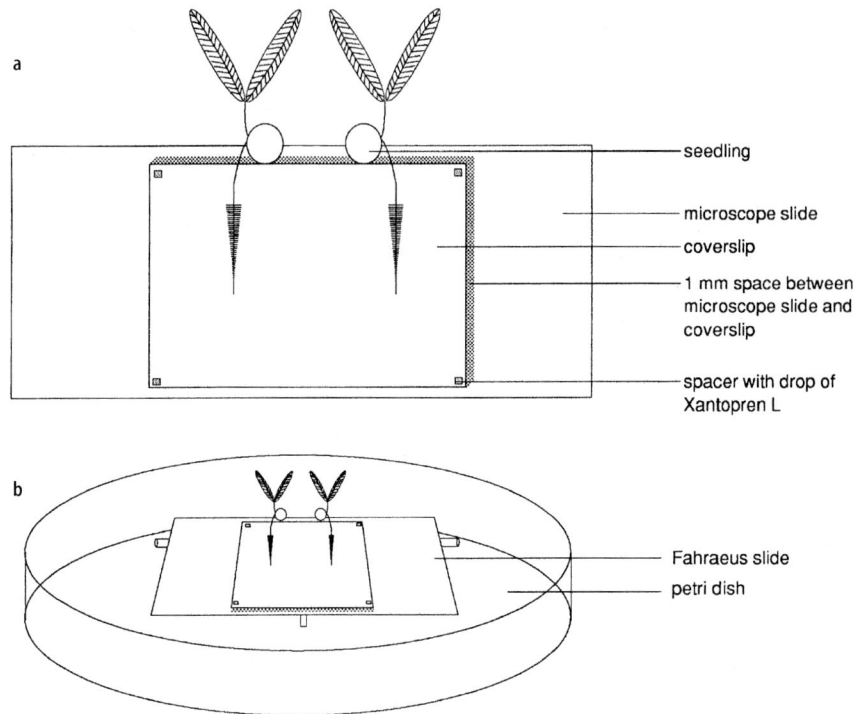

Fig. 5.3A, B. A A Fåhraeus slide. **B** Fåhraeus slide in Petri dish in a slanted position

2. Cut the root from the seedling and fix the root into the bath chamber with clips.

Plasmolysis   **3.** Plasmolyse the root hairs:

    a. Gradient method: Perfuse the bath with a linear gradient from buffer A to B. For the linear gradient, take 8 ml of both buffers and pump the whole gradient at a rate of $0.4\,\text{ml}\,\text{min}^{-1}$ into the bath.

    b. Step method:

      – 5 min Buffer A

      – 5 min 50 % Buffer A + 50 % Buffer B (285 mOsm)

      – 5 min Buffer B

Note. Using the step method there is no need for the peristaltic pump; the buffers can be added directly to the bath chamber.

Note. The gradient method is the best method to start with because the change in medium osmolarity is slow, giving the root time to adapt to the changing osmolarity. However, sometimes the step method works as well as the gradient method and then becomes the method of choice since it is faster.

The root hairs are sufficiently plasmolysed when the plasma membrane is withdrawn at least 6 μm from the cell wall. When the plasma membrane is too close to the cell wall it will be damaged when the cell wall is cut with the laser pulse.

Note. The extent of the plasmolysis depends on the osmolarity of Buffer B.

Note. When the cell withdraws from the cell wall during plasmolysis, many small patches of cytoplasm surrounded by plasma membrane (subprotoplasts) remain attached to the cell wall in the tip of the root hair. In other cell types, e.g., tissue culture cells and cells from *Elodea densa*, no subprotoplasts are formed during plasmolysis (De Boer et al. 1994).

4. Focus on the young root hairs, between 3 and 6 μm long, which have developed at the elongating part of the root. These are the youngest and most suitable for laser microsurgery. Position the tip of a root hair with the manipulators of the microscope over the laser point, marked on the monitor screen, and cut the cell wall while moving the microscope table over the fixed microbeam. Root hairs absorb the energy of the laser readily, so a few pulses are usually sufficient to cut a hole in the cell wall. However, cutting the cell wall of many other cell types can be more difficult because of the low absorption efficiency. The absorption efficiency can be improved by attaching molecules that absorb light at around 340 nm to the cell wall matrix. Calcofluor White (CFW, Sigma Chemicals Co., St. Louis, MO, U.S.A.) is an example of such a molecule and has proved to be useful (De Boer et al. 1994). CFW has an absorption maximum around 345 nm and a high affinity for cell wall components, i.e., cellulose, chitin, and other polysaccharides containing β1–4 linkages.

Cut Cell Wall

Note. CFW is prepared as a stock solution (1 % w/v) in water and is dissolved by heating to 90 °C. Before use, CFW is diluted to a final concentration of 0.001 %–0.01 % in the bath solution. CFW tends to precipitate at lower pH in combination with divalent cations.

5. After rupturing the cell wall, the protoplast swells immediately and fills the apical plasmolytic space. This is the space which has formed between the cell wall and the plasma membrane during plasmolysis. The protoplast protrudes through the laser hole and forms a bleb of

plasma membrane. Sometimes extrusion of the cell from the root hair is so fast that the plasma membrane ruptures. If the plasma membrane stays intact, cytoplasm and vacuoles stream from the protoplast inside the cell wall envelope into the extracelluar protoplast and thus an expelled protoplast is formed. Most of the time the nucleus remains in the root hair protoplast. The expelled protoplasts remain attached to the protoplast inside the cell wall envelope. Sometimes a thin cytoplasmic bridge between the protoplast inside the cell wall envelope and the expelled protoplast can be seen. The diameter of the expelled protoplast ranges from 5 to 10 µm.

Note. The speed of extrusion of the cytoplasm depends upon the calcium concentration in the bath medium. When the concentration is between 1 and 10 mM most protoplasts remain intact. With 0.1 mM Ca^{2+} or less in the bath medium most protoplasts rupture.

Note. Sometimes the protoplasts do not swell after a hole is cut in the cell wall. When the osmolarity is lowered to 260–300 mOsm, most protoplasts will be expelled in 5–30 min.

Wash   6. Replace the bath medium. If this new medium is very different from the plasmolysis medium with respect to osmolarity or ion concentration/composition, use the gradient mixer to create a linear gradient from the original to the final solution. At the end of the linear gradient wash with a few milliliters of the final bath medium.

Note. Wash carefully, because the expelled protoplasts are fragile.

7. The expelled protoplasts remain intact for several hours and cytoplasmic streaming is present in the protoplasts throughout the experiment. Cytoplasmic streaming can be used as a visible sign of viability.

8. The expelled protoplasts can be used for patch-clamp measurements. Figure 5.4 shows an example of a patch-clamp experiment, in the cell-attached patch configuration, on a laser released *Vicia sativa* root hair protoplast.

Comments

- A comprehensive description of UV laser beam release of protoplasts from *Medicago sativa* (alfalfa) root hairs has been given by Kurkdjian et al. (1993).

- Other applications of the laser microsurgery technique have been described by Weber and Greulich (1992).

Fig. 5.4A, B. A Patch-clamp recordings from a laser released *Vicia sativa* root hair protoplast in the cell attached patch config-uration. Bath solution (in mM): 10 MES, 9.58 Tris, 5 sucrose, 2 MgCl$_2$, 10 CaCl$_2$, 10 KCl, 335 sorbitol, pH 7.2, 400 mOsm. Pipette solution (in mM): 10 MES, 2.92 Tris, 2 MgCl$_2$, 1 CaCl$_2$, 100 KCl, 179 sorbitol, pH 5.6, 360 mOsm. **A** Re-cordings from unitary currents at negative pi-pette potentials (V$_{pip}$). The holding potential was 0 mV. **B** The correspond-ing current-voltage (I–V) relationship of the single channel currents indicates a single channel conduct-ance of about 18 pS. Pulse potentials were applied ranging from 0 to −180 mV

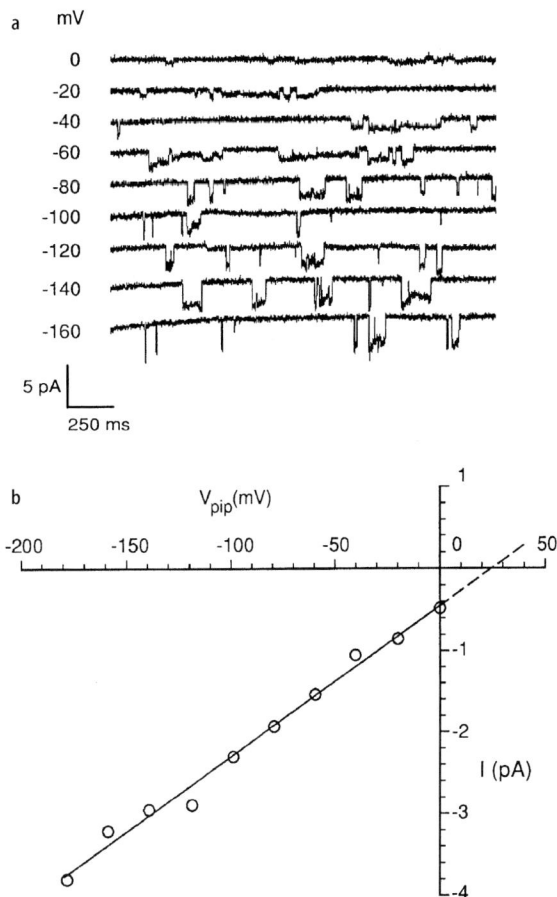

References

Bhuvaneswari TV, Solheim B (1985) Root hair deformation in the white clover/*Rhizo-bium trifolii* symbiosis. Physiol Plant 63:25–34

Cocking EC (1960) A method for the isolation of plant protoplasts and vacuoles. Nature 187:962–963

De Boer AH, Van Duijn B, Giesberg P, Wegner L, Obermeyer G, Köhler K, Linz KW (1994) Laser microsurgery: a versatile tool in plant (electro) physiology. Proto-plasma 178:1–10

Fåhraeus G (1957) The infection of clover root hairs by nodule bacteria is studied by a simple glass technique. J Gen Microbiol 16:374–381

Greulich KO, Weber G (1992) The light microscope on its way from an analytical to a preparative tool. J Microsc 167:127–151

Kurkdjian A, Leitz G, Manigault P, Harim A, Greulich KO (1993) Non-enzymatic access to the plasma membrane of *Medicago* root hairs by laser microsurgery. J Cell Science 105:263–268

Rincon M, Boss WF (1987) Myo-inositol triphosphate mobilizes calcium from fusogenic carrot (*Daucus carota* L.) protoplasts. Plant Physiol 83:395–398

Taylor AR, Brownlee C (1992) Localized patch clamping of plasma membrane of a polarized plant cell. Plant Physiol 99:1686–1688

Van Brussel AAN, Zaat SAJ, Canter Cremers HCJ, Wijffelman CA, Pees E, Tak T, Lugtenberg BJJ (1982) Small leguminosae as test plants for nodulation of *Rhizobium leguminosarum* and other rhizobia and agrobacteria harbouring a leguminosarum Sym-plasmid. Plant Sci Lett 27:317–325

Weber G, Greulich KO (1992) Manipulation of cells, organelles, and genomes by laser microbeam and optical trap. Int Rev Cytol 133:1–41

Ion Channel
and Membrane Potential Measurements Using
the Patch-Clamp Technique

Introduction to the Measurement Technique

Chapter 6

Significance of Ion Channels and Membrane Potential Changes in Cells

SANDOR DAMJANOVICH

Background

Ion channels are integral membrane proteins spanning membranes. These proteins have virtual holes inside (which actually serve as tunnels) allowing ions to pass through, thereby circumventing the hydrophobic barrier of the lipid bilayer that separates cell interior from extracellular space. The cations and anions pass through their specialized channels in a strictly regulated way, in which the diameter and fixed charges in the actual ion tunnel are important, but not exclusive, restricting structural characteristics. Ion channels of cells from the nervous system and muscles are well studied and display enormous diversity (Lewis and Cahalan 1988; Pieri et al. 1989; Jan and Jan 1990).

Based on the results of very efficient electrophysiological (voltage clamp, patch clamp) and spectroscopic studies, which revealed ion channels in a great variety of cells, it can be stated that ion channel activities and membrane potential changes are common characteristics of every cell.

Ion channels have three major types of regulatory mechanisms, which are found in most cells, namely; ligand, voltage and G protein (GTP-binding protein) regulation.

Ion specificity

The anion or cation specificity of ion channels, although far from being absolute, depends mainly on the size and charge of the individual cations and anions. The size of the ions highly depends on the hydrate shell surrounding the "naked" ion, due to dipole-dipole interactions with the dipolar solvent (especially water) molecules. The thickness of the hydrate shell does not necessarily depend on the size of the ion. A well known example is the comparison of the hydrated Na^+ and K^+ ions. The naked Na^+ ion is smaller than the K^+ ion, according to their respective positions occupied in the periodic system of atoms. However, due to the greater hydrate shell of Na^+ ions, they have a larger diameter and consequently a diffusion rate that is two orders of magnitude slower than that of hydrated K^+ ions. Thus, a relatively small difference in size causes a 100-fold difference in the permeability constants across membranes in favor of the K^+ ions. We also have to keep in mind that the average squared distance covered by a diffusing particle has a linear relationship to diffusion constants and time.

Diseases A number of diseases, among them genetic ones, are thought to be linked, perhaps causally, to impairments of ion channel activities. For example, cystic fibrosis may very well be due to impaired regulation of single chloride channels (Rich et al. 1990). Autoimmune diseases, such as systemic lupus erythematosus (SLE), diabetes mellitus, allergic arthritis and encephalomyelitis, are accompanied by altered potassium channel activities (Grissmer et al. 1990a, b). However, it remains to be determined whether these alterations in the occurrence, rate and functional properties of a particular channel type are indeed causative or whether their identification will only lead to good diagnostic markers (Damjanovich et al. 1994).

On the Origin of the Membrane Potential

A well known series of physicochemical interactions across a semipermeable membrane explain how the transmembrane electric potential is generated. The three major components of the transmembrane potential are:

1. Donnan potential, which can be calculated from the uneven ion distribution across the (semipermeable) plasma membrane.

2. Diffusion potential of particular ions (eminently among them K^+ ions), which is supposed to be a major contributor to the resting potential (and also the action potentials of excitable cells). Diffusion of K^+ ions is believed to be a decisive contributing factor since it alone generates a potential value across the membrane of over $-100\,\text{mV}$. In contrast to this, the contribution of the Donnan potential is only in the order of $10-20\,\text{mV}$, as can be calculated from the Nernst equation.

Note. Exceptions are only those cells in which there are no K^+ channels in the plasma membrane. For example a variation of Ehrlich ascites tumor cells, Lettre cells, have no K^+ ion channels and the major contributing component of the membrane potential is from the Na^+/K^+ ATPase enzyme (Bashford and Pasternak 1986).

3. The ion pumping activities can also modulate membrane potential values. Some of them can be electrogenic, e.g., the Na^+/K^+ ATPase, and can drive ions "uphill", i.e., against the electrochemical gradients, exploiting the readily available chemical potential from metabolite catabolism, in the cited case that of ATP dephosphorylation.

Note. It is of interest that by reverting the ion concentrations across such membranes the diffusion of K^+ ions can "drive" the same enzyme to synthezise ATP molecules from ADP.

The electrogenic proton pump alone can also generate membrane potential in the cellular slime mold *Dictyostelium discoideum* (Van Duijn and Vogelzang 1989).

The many apparent contradictions of the explanations of membrane potentials have their origin in the fact that different cell types maintain their membrane potentials by relying on the three major components to different extents. Energy-dependent pumping mechanisms can also provide extremely sensitive regulation of plasma membrane potential by maintaining **pseudo-Donnan systems.** In contrast to the real Donnan system, in which electrochemical potential is generated by nonpenetrating protein anions, pseudo-Donnan equilibrium can be generated and maintained by a continuously active ion pump that will create a lack of equilibrium of ions across the membrane. It is easy to accept that any perturbation of the pumping mechanisms will result in an immediate change of equilibrium potential due to the uncompensated ion diffusion. This rapid ion concentration and membrane potential regulation is driven by the passive diffusion of ions along the concentration gradient.

It is a remarkable fact that changes in membrane potential very sensitively affect viability and also the metabolic processes of cells in general. This underlines the theoretical and practical aspects and the importance of the membrane potential and ion transport processes that will be discussed below.

Ion Concentration Differences Across Membranes

It is generally accepted that biological membranes separate compartments with different ion concentrations.

In closed (thermodynamic) systems, sooner or later an equilibrium will be formed in ion concentrations and membrane potential along the two sides of a membrane. Of course, a state before equilibrium is reached is a nonequilibrium state which tends to shift towards the equilibrium.

These processes, which involve a free enthalpy decrease, will take place spontaneously. Any nonequilibrium process can be maintained in a stationary way as long as the effects of the spontaneous processes have been compensated. Biological membranes can be considered as parts of open thermodynamic systems in which a high number of events are made possible utilizing either energy-demanding metabolic steps or passive spontaneous processes which take place in such a way that the thermodynamic potential is decreased. Living cells are capable of maintaining such processes, which render unequilibrated ion concentrations and potentials stationary.

Of course, the measurable potential differences (the actual membrane potentials) between compartments under normal conditions are not necessarily stationary, but may sometimes display sudden changes, depending on environmental factors. These changes may rapidly follow alterations in membrane permeability parameters or those of ion pump activities. Changes in membrane potential may accompany physiological or pathophysiological processes exerting a perturbation on cell metabolism.

The Origin of Electrochemical Potentials: The Nernst Equation

Let us consider a semipermeable membrane separating two electrolyte solutions of unequal composition (compartments I and II) and start with the simple case of only one membrane-permeable ion species, e.g., K^+. The K^+ ions tend to diffuse along their concentration gradient from the solution of high K^+ concentration through the membrane into the solution of low K^+ concentration. However, they cannot reach the bulk of the low concentration solution, because they are held back by impermeable anions (proteins are the most likely candidates for this role) left behind at the opposite membrane side. Thus two thin layers of excess positive and negative charges will be created at either sides of the membrane surfaces.

In case of only one membrane-permeable ion species an equilibrium state is established in which ion diffusion across the membrane is abolished by the created diffusion potential. At equilibrium the change in free energy when one mole of K^+ ions pass from compartment I to compartment II is zero:

$$\Delta G = \mu_c^{II} - \mu_c^{I} = 0 \tag{1}$$

In this equation the μ_cs are the electrochemical potentials for compartments II and I, respectively, consisting of a chemical component μ and an electrical component $zF\psi$ (with z as the valence of the ionic species, F as the Faraday constant, and ψ as the electrical potential). Substituting the appropriate expressions for the electrochemical potentials we obtain:

$$RT \ln c_c^I + z_c\, F\, \psi^I = \mu_c^0 + RT \ln c_c^{II} + z_c \tag{2}$$

Solving this equation for the electrical potential difference gives:

$$\Delta\psi = \psi^{II} - \psi^I = \frac{RT}{z_c F}\, \ln\, \frac{c_c^I}{c_c^{II}} \tag{3}$$

Hence, electrical potential difference is proportional to the logarithm of the ratio of the two concentrations. In the case of cations z, it is positive and the electrical potential is higher on the more dilute side of the membrane. Equilibrium is attained because the buildup of electrical potential on the dilute side of the membrane raises the electrochemical potential of the solution in the more diluted compartment to that of the more concentrated solution in the other compartment.

What is the Origin of the Donnan Potential?

As is known, cytoplasm of cells contain a large number of proteins which cannot permeate the membrane in resting cells. Polyvalent anions, like proteins, have a net negative charge under physiological conditions. The electrical neutrality of the cytoplasm is maintained by cations. These are mostly small-sized metallic ions which can permeate the plasma membrane to varying extents, depending largely on the specific ion channel activities of the membrane. The cytoplasm also contains a significant number and variety of anions which due to their size can also permeate the membrane through channels.

In order to investigate such a system, let us analyze the model shown in Fig. 6.1.

The extracellular space is represented by compartment II. Compartments I and II are separated by a semipermeable membrane. The membrane is permeable to Na^+ and Cl^- ions and impermeable to A^{n-} anions, which have n unit negative charge. The system is in equilibrium when the electrochemical potential of the two parts is identical. Let us label the valency of the ith ion with z_i, and the concentration with C_i. The electrochemical potential is:

$$\mu_i = \mu_i^0 + RT \ln c_i + z_i\, F\, \psi \tag{4}$$

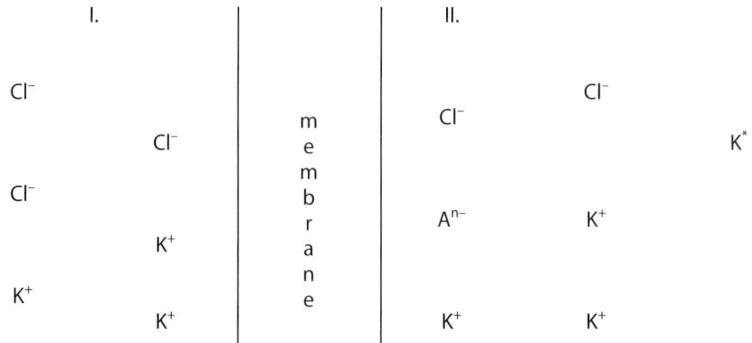

Fig. 6.1. Model system displaying the origin of the Donnan potential

where μ_i^0 is the standard chemical potential (which serves as a reference point), $F = 96500$ C, (i.e., the Faraday constant, which is the charge of a mole univalent ions), ψ is the electric potential generated between the two compartments. The conditions of the equilibrium are as follows:

$$\Delta \mu_{Na} = RT \ln \frac{[Na^+]^{II}}{[Na^+]^{I}} + F \Delta \psi = 0 \tag{5}$$

$$\Delta \mu_{Cl} = RT \ln \frac{[Cl^-]^{II}}{[Cl^-]^{I}} - F \Delta \psi = 0 \tag{6}$$

Thus the membrane potential at equilibrium ($\Delta \psi$) is:

$$\Delta \psi = - \frac{RT}{F} \ln \frac{[Na^+]^{II}}{[Na^+]^{I}} = \frac{RT}{F} \ln \frac{[Cl^-]^{II}}{[Cl^-]^{I}} \tag{7}$$

$$r = \frac{[Na^+]^{II}}{[Na^+]^{I}} = \frac{[Cl^-]^{I}}{[Cl^-]^{II}} \tag{8}$$

The parameter r gives the ratio of the equilibrium concentration of the permeating ions. The electrochemical and electric potential differences are normalized in these cases to compartment I, representing the extracellular space. For compartments I and II, electrical neutrality must be maintained:

$$[Na^+]^{I} = [Cl^-]^{I} = c \tag{9}$$

$$[Na^+]^{II} = [Cl^-]^{II} + na \tag{10}$$

where na is the nonpermeating anion concentration. From the above equation, after some simple algebraic calculations:

$$r = \frac{na + \sqrt{(4c^2 + na^2)}}{2c} \tag{11}$$

The correlations between concentrations of the two compartments are:

$$[Na^+]^{II} > [Na^{+]II} = [Cl^-]^I > [CL^-]^{II} \tag{12}$$

It is obvious from the definition of r that when r is not equal to 1, there is always a membrane potential which is different from 0. This is called the Donnan potential. The r is equal to 1 only if $na = 0$. If one has non-permeating anions in the system, r is larger than 1. It also means that a negative Donnan potential is generated for a plasma membrane as a consequence of the nonpermeating anions. It also follows from the basic equations that the value of r (and that of the Donnan potential) increases with the intracellular anion concentration and has an inverse relationship with the extracellular salt concentration.

The value of the Donnan potential does not (directly) depend on metabolic processes, thus metabolic inhibitors do not influence its size. This is a basic difference with any other kind of nonequilibrium potential. Additional permeating electrolytes given to a Donnan system will be distributed according to the Donnan equilibrium and the original ion distribution will only be slightly perturbed.

Diffusion Potential

When a given concentration of an electrolyte (e.g., a HCl solution, which can be considered a solution of H^+ and Cl^- ions) is overlaid on top of a higher concentration of the same electrolyte, the positive (H^+) and negative (Cl^-) ions start diffusing towards the lower concentrations due to the concentration differences. As we introduced above, our electrolyte can be taken as H^+Cl^-. Hence one has to reckon with the higher mobility, i.e., the significantly larger diffusion constant, of the protons than of the Cl^- ions. The protons will "run" ahead of the heavier chloride ions. Thus an imbalance in the different charges will be generated. In the transient domain at the lower concentration layer a relative excess of H^+ ions will be generated. This part of the layer will have more positive charges than the other, where the Cl^- ions will be lagging behind the H^+. This potential obviously deserves the name of diffusion potential. The size of the potential has a direct correlation with the relative mobility difference between the ions. The charge of the more mobil ion will dominate the lower concentration domains and vice versa. The definition of ion mobility is the diffusion speed of a particle in a field of unit force. Naturally, a diffusion potential is transient, because the concentrations will be sooner or later equilibrated in a steady state system. The significance of the measurable diffusion potentials is further complicated by the fact that they may frequently be perturbed by convection streams. Taking into account more than one type of permeating cation and anion, the mathematical analysis will become more complex. Omitting details after a more complicated , but basically similar, series of calculations as before, the final outcome is given by the Goldman-Hodgkin-Katz equation:

$$\Delta\psi = \frac{RT}{zF}\ln\frac{p_{Na}[Na^+]^{II} + p_K[K^+]^{II} + p_{Cl}[Cl^-]^I}{p_{Na}[Na^+]^I + p_K[K^+]^I + p_{Cl}[Cl^-]^{II}} \tag{13}$$

In which P_i represents the permeability of the membrane for the specified ion.

It is clear from this equation that those ions which have higher permeability and unequal distribution will contribute to a greater extent to the generation of a diffusion potential. This fact gives the K^+ ions a great advantage over the Na^+ ions in the generation and regulation of resting membrane potential due to their approximately 200-fold higher diffusion speed rendered possible by the thinner hydrate shell. In the presence of ionophores with specificity to particular ions, due to the high permeability such contribution can be very great (e.g., ionomycin and Ca^{2+} ions). Of course, nonpermeating ions will not contribute to the above equation.

The ion movements across biological membranes in general, and different pumping activities and transporting, cotransporting or antiporting of ions in particular, can be electrogenic or electronically silent. Electronically silent pumps may transport ions with different charges in the same direction, or antiport pumps ions with the same electric charge to opposing directions. As examples, the Na^+/Cl^- pumps or the Na^+/H^+ and K^+/H^+ antiports can be mentioned. The first type is known in the gall bladder, while the other two can be found in lymphocytes or in cells of the stomach wall. The Na^+/Cl^- pump might also influence the membrane potential indirectly by changing the ion concentrations in the different cellular compartments. Electrogenic pumps are characterized by net charge fluxes, in opposite direction to the existing potential difference.

At the end of this brief and far from complete analysis, we feel it necessary to emphasize again that biological variations on the origin of the plasma membrane potential do not belong to the class of rare biological phenomena. When we compare electric phenomena of different cells, we should not be disturbed by a number of apparent contradictions with data obtained by others studying other cell types.

References

Bashford CL, Pasternak CA (1986) Plasma membrane potential of some animal cells is generated by ion pumping, not by ion gradients. Trends Biochem Sci 11: 113–116

Damjanovich S, Edidin M, Szöllösi J, Trón L (1994) Mobility and proximity in biological membranes. Chapter 6. Ion channels and membrane potential changes in lymphocytes. CRC, Pearl River, NY, pp 225–326

Grissmer S, Hanson DC, Natoli EJ, Cahalan MD, Chandy KG (1990a) CD4$^-$CD8$^-$ T cells from mice with collagen arthritis display aberrant expression of type l K$^+$ channels. J Immunol 145: 2105–2109

Grissmer S, Dethlef B, Wasmoth JJ, Godlin AL, Gutman GA, Cahalan MD, Chandy KG (1990b) Expression and chromosomal localization of a lymphocyte K$^+$ channel gene. Proc Natl Acad Sci USA 87: 9411–9415

Jan LY, Jan YN (1990) How might the diversity of potassium channels be generated? Trends Neurosci 13: 415–419

Lewis RJ, Cahalan MD (1988) Plasticity of ion channels: parallels between the nervous and the immune systems. Trends Neurosci. 11: 214–218

Pieri C, Recchioni R, Moroni F, Balkay L, Marian T, Tron L, Damjanovich S. (1989) Ligand and voltage gated sodium channels may regulate electrogenic pump activity in human, mouse and rat lymphocytes. Biochem Biophys Res Comm 160: 999–1002

Rich DP, Anderson MP, Gregory RJ, Cheng SH, Paul S, Jefferson DM, McCann JD, Klinger KW, Smith AE, Welch MJ (1990) Expression of cystic fibrosis transmembrane conductance regulator corrects defective chloride channel regulation in cystic fibrosis airway epithelial cells. Nature 347: 358–363

Van DuijnB, Vogelzang SA (1989) The membrane potential of the cellular slime mold *Dictyostelium discoideum* is mainly generated by an electrogenic proton pump. Biochim Biophys Acta 983: 186–192

Practical Introduction to Patch Clamping by Simulation Experiments with Simple Electrical Circuits

Dirk L. Ypey

Background

This laboratory course has been developed for those who are interested in a research project in membrane electrophysiology, but who lack experience in doing electrical measurements.

The excercises are designed to simulate carrying out an electrophysiological experiment. Each exercise is first stated as a membrane electrophysiological problem. Then the relevant electrical circuit is given and the corresponding equations are derived. Subsequently, experiments are described and then illustrated with a few oscilloscope records. Finally, general comments are given on the outcome of the experiments and their implications. For comparison, all the results and answers are given in the Appendix. The electrophysiological description of the exercises acquaints you with the jargon of the field. You do not need to understand all the equations before you do the exercises! Thinking as well as experimenting will contribute to your understanding. Eventually, understanding will improve by doing real patch-clamp experiments.

The minimally required knowledge to follow this course is high school-level biology (in particular physiology of nerve and muscle cells, bioelectricity), physics (electricity, Ohm's law, etc.) and mathematics (algebra, differential equations, exponential functions). Many college-level physiology textbooks are available which are usually good enough to familiarize oneself with the important bioelectrical phenomena such as action potentials, synaptic potentials and single ion channel activity. However, didactic texts written by expert electrophysiologists, including the inventors of the patch-clamp technique Neher and Sakmann (1992), as well as the Nobel laureate lectures of these two researchers (Neher 1992; Sakmann 1992) are strongly recommended.

Materials

- A set of resistors (0.01, 0.1, 1, 10, 100 and 1000 KΩ), at least four of each. Carbon composition resistors of 1/4 W are cheap and will suffice.
- A set of capacitors (.01 μF – 10 mF). The higher capacitance values are probably polarized (electrolytic) capacitors.
- Two batteries of 1.5 V with attached wires (or in a holder with wires) to connect the batteries to the circuit.
- Two linear potentiometers of 10 KΩ (hand-controlled variable resistors). Solder one wire to one of the two fixed resistance terminals and another wire to the variable resistance terminal (the central one) for connection to the circuit. Include a 1 KΩ resistor in one of the two terminal wires to prevent short circuiting of the membrane circuits.
- A circuit board to build the circuits. A "bread board" or "prototype board," for design, educational or test purposes, is a plastic board with rows of holes for inserting components. The holes are internally connected: vertical rows for vertical connections and horizontal rows for horizontal connections. This board makes it possible to quickly build a circuit without soldering connections. It can be obtained from electrical component shops or electrical suppliers. We used a simple board (Project Board K-102 from Altai Nederland BV, Almelo, The Netherlands) and mounted it on a metal plate with sockets for 4 mm plug/ BNC connection to the pulse generator, amplifier and oscilloscope. If no bread board type circuit board is available, one could solder the circuit together or hold it together with clamps.
- Isolated wires with stripped endings to make the connections between the components on the circuit board.
- A forceps, to hold the tiny wires and the components and push and pull them into and from the holes of the circuit board.
- Ohm/Volt/Ampere meter with small alligator clamps for measurements in the circuits or on the components.
- Leads (BNC and single-end) and connectors to make the connections between the circuit and the instruments.
- Solder gun and some other electronics workshop tools and supplies (screw drivers, cutter, matches, wires, etc.) to establish circuit constructions and connections and to do repair work.
- A function generator and/or voltage pulse source to stimulate the circuits with voltage or current pulses.
- A two-channel oscilloscope to measure the responses of the circuits to voltage or current stimulation. A storage scope would be very useful, but is not required.

Components and equipment

– If the oscilloscope does not have a differential input, a differential amplifier is needed in order to be able to measure voltage drops across resistors in the circuit without grounding the circuit.
– An oscilloscope camera, printer or plotter to record the observations (optional). If these are not available, one could use computer facilities to make figures or – in the worst case – one could make sketches from the screen of the oscilloscope.

Furthermore, it is always useful to have the instruction manuals available for the equipment in use.

Getting started   Become familiar with all the equipment and components listed so that they can easily be used during this course:

1. Identify the color codes of the resistors and check the values with the use of the Ohm meter. Electronics shops can provide you with a listing of the color codes.

2. Check the battery values with the Volt meter and/or the oscilloscope.

3. Learn to operate the function generator and the oscilloscope by testing the function generator with the oscilloscope and vice versa.

4. Learn to operate the differential amplifier (if available) by displaying the output of the amplifier on the oscilloscope and testing the gain (and probably the filtering) for small amplitude input signals from the function generator.

5. Identify the connections inside the circuit board with the use of the Ohm meter.

Note. If you get stuck during the measurements, disconnect everything and rebuild the circuit. If you are still having problems, try to identify the whole circular structure, i.e., the circuit, through which the current is supposed to flow. Without a circuit no current flow! Then try to understand what is to be measured and how it is to be done. If still in trouble, call for assistance from an expert patch-clamper or an electrical/electrotechnical engineer! Or read a good book on patch-clamp techniques and the electronics involved (Standen et al. 1987; Sherman-Gold 1993).

Procedure

7.1
Current-Clamp/Voltage-Recording Exercises

Membrane Potential Changes Due to Channel Opening and Closing

The role of membrane conductance (or resistance) and capacity in membrane potential changes of a membrane can be elegantly demonstrated by a simple equivalent circuit experiment, simulating the opening and closing of only two ion channels in a membrane. Such conditions may occur during real patch-clamp experiments (cf. Ravesloot et al. 1994). We consider the cell under ideal voltage recording conditions (high input impedance voltage amplifier, so that no significant current is drawn from the circuit). Electrophysiologists call this condition current-clamp, even if the applied current is 0 Ampere.

Note. The microscopic condition studied in the present exercise (single channel opening and closing) may also serve as a model for sudden, multichannel, large conductance changes of a cell membrane, as occur, for example, during the generation of action potentials or during postsynaptic potentials.

The basic circuit considered (Fig. 7.1a, b) is a parallel resistance (or parallel conductance) circuit of three branches: (1) a K^+ conductance branch with the resistor Rk and Nernst potential Ek, (2) a leakage resistance branch with a leakage resistor $R1$ and a leakage diffusion potential $E1$ (often zero, as in this case) and (3) a membrane capacitance (Cm) branch. The channel opening and closing behavior (see gates in Fig. 7.1a) is realized by the switches Sk and Sl (Fig. 7.1b), which can be operated manually during the experiment to switch the channels on (open) or off (close). These switches need not to be hardware components of the circuit, but do symbolize the making of the connection of Rk or $R1$ to the circuit by manually inserting the top ends of these resistors into the circuit board. The membrane potential Vm is equal to the potential across the capacitor (Vcm) and is measured via the patch pipet in the whole-cell mode. The Ag/AgCl electrode (spiral symbol) in the pipet is the measuring electrode, which connects the inside of the cell membrane, via the intracellular and pipet solution, to the measuring (plus) input of the meter (amplifier, oscilloscope). The grounded Ag/AgCl electrode in the bath is the reference electrode, which connects the outside of the cell membrane to the minus input of the meter via the bath.

Circuit

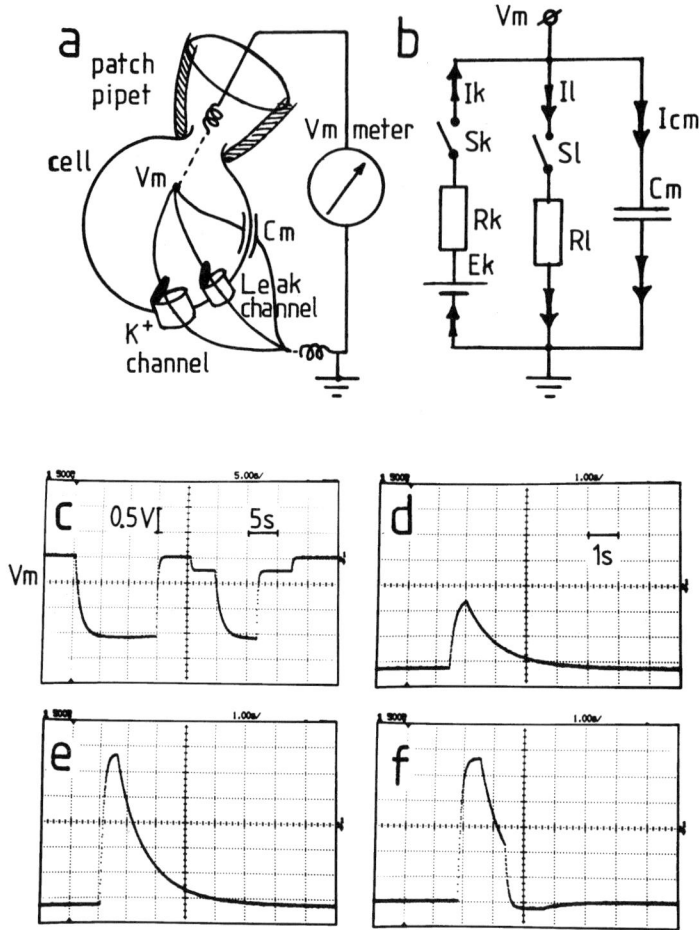

Fig. 7.1a–f. Membrane potential (*Vm*) changes of a cell under current-clamp conditions due to ion channel opening and closing. The measurement configuration is illustrated in **a**, the electric circuit derived from **a** is given in **b**, example oscilloscope records are displayed in **c–f**. Ion channel switching and component values in the records are as decribed in the description of Experiment 1. Record **c** displays *Vm* changes upon channel switching in various channel-open combinations; record **d** "simulates" an excitatory postsynaptic potential; record **e** (leak conductance substituted for a sodium conductance) "simulates" an action potential with only sodium conductance increase and decrease; record **f** "simulates" an action potential with an extra repolarization mechanism: activation of extra *Gk*. The afterhyperpolarization after the action potential occurred because the normal resting *Vm* is more postive than *Ek* because of an added leak branch of 10 K. Record **a** is on a time base of 5 s/div, records **d–f** are on 1 s/div. *Vertical scales* of all records are 0.5 V/div. Ground potentials of the records are indicated by an *arrow* at the right sides of the frames: 1 division above the middle line in **c**, the middle lines in **d–f**.

Fig. 7.1a illustrates this measurement mode (whole-cell configuration; cf. Hamill et al. 1981) and the way the circuit is derived from the measurement configuration (compare Fig. 7.1a with 7.1b). The currents in the branches are indicated in Fig. 7.1b as Ik, Il and Icm. The K$^+$ channel may represent the main channel responsible for the resting membrane potential of the cell, while the leakage channel may be an excitatory transmitter-operated channel.

Note. In this and all subsequent circuits of this course, a certain current flow direction (of postive charges) is chosen with consistent battery polarities, i.e., with a dominant positive battery (+ at intracellular side) in the leftmost branch. This means that the current in the other branches is cell-outward. Since the derivations of the equations (see below) are general, any battery size or polarity can be plugged into the equations to find the right voltages and currents in the circuits. Any local voltage at a certain point in the circuit is defined with respect to extracellular zero (ground) potential. Resulting positive currents are cell-outward and negative currents are inward, consistent with polarity conventions in electrophysiology.

Since no current can escape into the Vm amplifier circuit (ideal voltage recording) or does enter the circuit (zero current-clamp), current leaving any of the branches must enter the other branches. Thus, according to Kirchof's current law, we may write (see Fig. 7.1b)

Equations

$$Ik = Il + Icm \qquad (1)$$

Ohm's law for the resistive currents states that

$$Ik = (Ek - Vm)/Rk \qquad (2)$$

and

$$Il = Vm/Rl \qquad (3)$$

For the capacitive current Icm we have the equation

$$Icm = Cm.dVm/dt \qquad (4)$$

When Eqs. (2)–(4) are substituted into Eq. (1) we obtain the simple, first-order linear differential equation

$$(Ek - Vm)/Rk = Vm/Rl + Cm.dVm/dt \qquad (5)$$

For the stationary condition, $dVm/dt = 0$, we find the stationary value of

$$Vm, Vms, Vms = Rl.Ek/(Rk + Rl) \qquad (6)$$

By rearranging terms we can rewrite differential Eq. (5) into an expression which includes Vms and better illustrates how dVm/dt depends on Vm:

$$dVm/dt = -(1/tm).[Vm-Vms] \qquad (7)$$

with

$$tm = Rkl.Cm \qquad (8)$$

and

$$Rkl = Rk.Rl/(Rk + Rl) \qquad (9)$$

The solution for the time behavior of Vm upon channel on and off switching can be derived from Eq. (7) and is

$$Vm(t) = Vms + (Vm(0)-Vms).\exp(-t/tm) \qquad (10)$$

If one of the two channels closes or is closed, Eqs. (7)–(10) apply assuming that the closed channel has an infinitely large resistance.

If we would have derived these equations for the case that the leakage branch includes a diffusion potential El between Rl and ground, only the equation for Vms would have been different (cf. Eq. 6) and would be

$$Vms = (Rl.Ek + Rk.E1)/(Rk + Rl) \qquad (11)$$

For two arbitrary branches 1 and 2, Eq. (11) is

$$Vms = (R1.E2 + R2.E1)/(R1 + R2) \qquad (12)$$

We may need Eq. (12) for later experiments. Equation (10) remains the same for two branches 1 and 2, but Eqs. (8) and (9) should be adapted by writing 1 for k and 2 for l.

Equation (12) becomes Eq. (13), if the expression is rewritten in terms of the conductances $G1 = 1/R1$ and $G2 = 1/R2$

$$Vms = (G1.E1 + G2.E2)/(G1 + G1) \qquad (13)$$

Experiment 1: RC Behavior of a 2-Channel Membrane

Make the circuit of Fig. 7.1 on the circuit board and connect $Vcm = Vm$ to the input of the oscilloscope. Take $Rk = 1\,K$, $Rl = 200\,\Omega$ (two $100\,\Omega$ resistors in series), $Ek = -1.5\,V$ (a standard battery) and $Cm = 1\,mF$. If Cm is a polarized condensor, put the negative side up (at the intracellular side). Choose a continuous (trigger switch on "line") sweep speed of about 1 s/div and a recording sensitivity of 0.5 V/div. Determine the zero volt line and make sure that $Vcm = 0\,V$. If that is not the case, discharge Cm by switching Sl on for a few seconds. Do not forget to switch Sl off again.

Now follow the steps in the next protocol:

1. Switch Sk on and monitor $Vcm(t)$ until Vcm is constant (at Ek).

2. Switch Sk off and observe that Vcm now does not change and remains at its original stationary value (or declines very, very slowly).

3. Switch Sl on and watch the rapid return of Vcm to $Vcm = 0$.

4. Switch Sk on again (with Sl still on) and see Vm rapidly moving to a small negative stationary value.

5. Switch Sl off, while Sk remains on. Watch the slow further development of Vcm toward Ek.

6. Switch Sl on and see Vcm returning to the same stationary level as in step 4.

7. Switch Sk off to see Vm rapidly returning to zero. Compare the observations with Fig. 7.1c.

8. Repeat steps 1–7 without Cm or with smaller and larger Cm values to establish the role of Cm in Vm changes.

Comments

These introductory experiments have familiarized you with the conductance changes underlying membrane potential changes, both on a microscopic and a macroscopic level. In addition, the retarding role of the membrane capacitance in these potential changes was illustrated. For further experiments (cf. Fig. 7.1d–f), see Appendix.

The values of the resistors chosen were such that voltage recording could be done with the oscilloscope, without the problem of affecting the experiment by the measurement. If you would do the same experiment with $Rk = 1\,M$, $Rl = 200\,K$ and $Cm = 1\,\mu F$ (smaller Cm, otherwise the experiment lasts too long!), the recorded Vm values would no longer be predicted by the derived equation, because the input resistance of the oscilloscope (1 M) would draw significant current from the circuit. You may do this experiment to discover the importance of using high-input resistance voltage amplifiers in doing membrane potential recording experiments, since single cell membrane resistances may be very high (about $1\,G\Omega$; cf. Neher 1992)!

The value of Cm used (10 mF) was much higher than one will ever see for a single cell (10–100 pF). This allowed us to do a "slow movie" experiment manually. However, the slow time constants of 0.2–1 s observed in our equivalent circuit experiment may occasionally occur in real cells (e.g., up to 1 GΩ; 100 pF = 1 s, cf. Ravesloot et al. 1994). Thus, scaling down the resistances while scaling up capacitance maintains physiologically occurring membrane time constants, although most time constants are around 1 ms.

Single-channel opening and closing in the multichannel membrane of the living cell occurs randomly and results in only small (about 1 mV) membrane potential fluctuations, called membrane noise (Verveen and DeFelice 1974). Experiment 1 illustrates its origin!

In the present experiments the conductance changes of the cell were "hand-made". In reality these changes are caused by membrane potential changes as a result of stimulating current from external sources. The next experiment explains what happens in a membrane during current stimulation.

Current-Clamp Stimulation

A resting nerve or muscle cell may locally change its membrane potential because it receives current from an "active" neighbor patch of membrane, i.e., a patch where the membrane potential is changing from the resting condition. The difference in membrane potential between the membrane areas is the cause of a local circuit current, which makes the active membrane share its membrane potential change with the neighboring membrane area. What is here the stimulus for the stimulated membrane, a current change or a voltage change, or both? It is clear that, since the stimulated membrane is electrically coupled to the stimulating membrane, the coupled membrane patches both affect the stimulating current and the size of the changing voltages at the stimulating and stimulated membrane areas.

Definitions: current- and voltage-clamp   Membrane electrophysiologists are used to investigating the properties of excitable membranes under two extreme conditions of stimulation. They either stimulate with constant currents (**current-clamp**) or with constant voltages (**voltage-clamp**), independent of the properties of the stimulated cells. This makes it easier to interpret the recorded responses. During current-clamp experiments membrane potential changes as a result of the applied currents are recorded, for example to study action potential properties of nerve cells. During voltage-clamp experiments membrane currents, flowing under the influence of the applied potentials, are recorded, for example to study the ionic mechanism underlying the action potential. Although these extreme stimulating conditions may be approached in vivo, the usual situation seems to be an intermediate one, but more at the current-clamp side.

Here we do a few electrical experiments to make ourselves familiar with current-clamp experiments. The current-clamp condition seems closer to in vivo conditions and has always been easier to realize experimentally. Historically, it preceded voltage-clamp. We will see that a perfect current-clamp experiment requires an ideal current source, i.e., a high voltage source (with $Vcc \gg Vm$, see Fig. 7.2) in series with a high current-clamp resistance ($Rcc \gg Rm$), so that the injected current Icc becomes independent of Vm and Rm.

Circuit

Figure 7.2 illustrates both the experimental measurement condition during a whole-cell patch-clamp experiment (Fig. 7.2a) and the simplified circuit derived from it (Fig. 7.2b). The constant-current source (I-clamp) is a high voltage battery Ecc in series with a high resistance Rcc. The I-clamp also contains an ideal voltmeter (with an internal resistance even $\gg Rcc$) to record the membrane potential Vm. One of the outputs of the I-clamp is connected to the inside of the cell membrane via the Ag/AgCl electrode (spiral symbol) in the patch pipet and the pipet solution, which reaches into the cell.

Access resistances, from the electrode, through the pipet to the inside of the membrane, have been neglected here! The other I-clamp output is connected to the grounded bath electrode (Ag/AgCl spiral), which is in contact with the outside of the cell membrane via the low resistance extracellular solution. The membrane consists of a membrane capacity Cm, parallel to a conductive pathway consisting of a membrane resistance Rm in series with an intrinsic membrane potential Em. The voltage across Cm, Vcm, is equal to Vm.

Equations

According to Kirchof's current law, for the incoming current Icc, which distributes between Rm and Cm (see Fig. 7.2b), we may write:

$$Icc = Irm + Icm \tag{14}$$

For the capacitive current Icm, we may write:

$$Icm = Cm.dVm/dt \tag{15}$$

and Ohm's law says that

$$Irm = (Vm - Em)/Rm \tag{16}$$

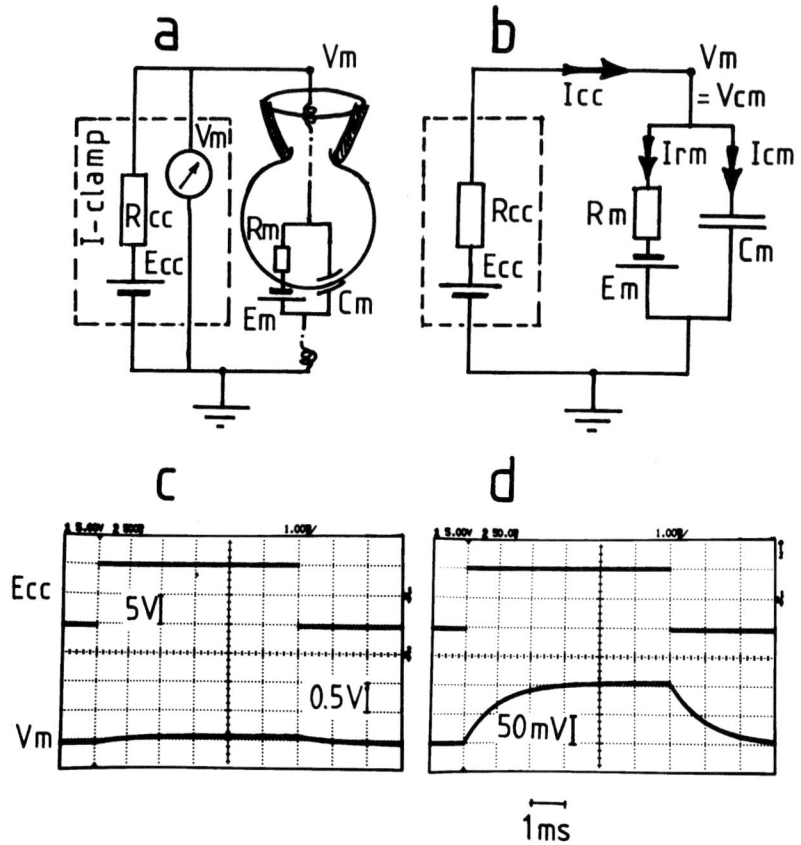

Fig. 7.2a–d. Membrane potential response to current-clamp stimulation. The whole-cell measurement configuration is shown in **a**, the electric circuit derived from **a** is given in **b** and example records are displayed in **c** and **d**. Component values are defined in the description of Experiments 2 and 3. Time base in **c** and **d** is 1 ms/div. *Upper sweeps* in **c** and **d** represent Ecc (5 V/div), *lower sweeps* represent Vm (0.5 V/ div in **c**, 50 mV/div in **d**). Ground potentials are indicated at the right sides of the frames by *small arrows*. Ground potential of the lower high-gain record (of Vm) in **d** is off screen. This record shows the exponential time course of the depolarization caused by the current step.

Assuming that $Rcc \gg Rm$, substitution of Eqs. (15) and (16) into Eq. (14) results in

$$Icc = Ecc/Rcc = (Vm - Em)/Rm + Cm.dVm/dt \qquad (17)$$

The solution of this differential equation for $Vm(t)$ is

$$Vm(t) = Vms + (Vm(0) - Vms).\exp(-t/tm) \qquad (18)$$

in which the stationary Vm (Vms) is

$$Vms = Em + Icc.Rm \qquad (19)$$

and

$$tm = Rm.Cm \qquad (20)$$

Experiment 2: Time Constant and Stationary Vm Change Under Ideal Current-Clamp (Rcc = 100 · Rm)

1. Make the circuit of Fig. 7.2b and take as values: $Rm = 1\,K$, $Em = -1.5\,V$, $Cm = 1\,\mu F$, $Ecc = 10\,V$ (ca. 100 Hz square wave amplitude) and $Rcc = 100\,K$.

2. Monitor Ecc and Vm ($= Vcm$) (see Fig. 7.2c, d) and determine Vms and tm.
   Are Vms and tm as expected from Eqs. (19) and (20), respectively?

3. Check Eqs. (19) and (20) further with different Rm and Cm.
   In the jargon of electrophysiologists, Vm changes in the direction of zero are called **depolarizations**, the opposite changes are called **hyperpolarizations**.

Experiment 3: Less Ideal Current-Clamp Conditions

1. Make $Ecc = 0\,V$ in the previous experiment and determine Vm under this condition.

2. Make $Rcc =$ infinite (remove the component) and observe that Vm hardly changes (in the order of 1 %, as may be expected from the ratio Rm/Rcc). This implies that the current-clamp condition hardly influenced Vm, as required for an ideal current-clamp experiment.
   Rm also did not influence Icc very much, since $Icc = Ecc/(Rcc+Rm)$ is very close to Ecc/Rcc, approximately 1 % smaller.

3. Repeat Experiment 1 with two other values of Rcc, 10 K and 1 K, and observe that these Rcc values cause larger Icc values but also spoil the current-clamp condition, in particular $Rcc = 1\,K$. These Rcc values depolarize Vm (10 K appr. 10 %, 1 K appr. 50 %, as can be expected!) and Rm also affects Icc. Equations (17)-(20) will no longer apply. If

the student is interested, he or she may derive more general equations, including also nonideal current-clamp conditions.

Comments   An excitable membrane only shows single exponential responses to constant-current steps if these responses are very small. With larger responses, the membrane may reach the firing threshold. In good current-clamp experiments this firing threshold can be measured in order to investigate the effect of certain conditions (e.g., pharmacological agents) on this threshold (cf. Wang et al. in this volume).

An additionally instructive experiment might be to simulate the firing of an action potential manually upon reaching threshold! The membrane time constant should then be slow enough (0.1–0.5 s) to be able to react upon watching V_m crossing an imaginary threshold. The action potential mechanism may be the same as the one in Fig. 7.1f (switchable resistors). One may also shape the form of the impulse by using two 10 K potentiometers, one for the Na^+ and one for the K^+ conductance.

Essentially the same electrical equivalent circuit as given in Fig. 7.2b can be used to measure pipet resistance (R_{pip}) and capacitance (C_{pip}) of the patch pipet before it is sealed to the cell. The current-resisting pipet tip in the bathing solution can be represented by a resistance (R_{pip}), while the glass wall separating the pipet fluid from the bathing fluid acts as a capacitance (C_{pip}, cf. Fig. 7.3) for electrical current flowing through the pipet to the bath. Equations (14)–(20) apply after substituting R_m for R_{pip}, C_m for C_{pip}, E_m for E_{pip} (a diffusion or liquid junction potential between pipet solution and bath solution) and V_m for V_{pip}. In the next section we discuss R_{pip} and C_{pip} measurements by voltage-clamp stimulation.

7.2
Voltage-Clamp/Current-Recording Exercises

Measuring Pipet and Seal Resistance by Voltage-Clamp

Any patch-clamp experiment begins with measuring the resistance of the patch pipet (R_{pip}) to see whether the pipet is good enough to approach the cell of interest for a gigaseal attempt. If R_{pip} is good (a few MΩ for a typical whole-cell experiment), the pipet is pressed slightly against the cell and the pipet mouth is sealed to the encircled patch of membrane by applying slight suction to the inside of the pipet. Sealing is successful if the seal resistance (Rs) against current flow from the pipet to the bath through the circular sealing space between the pipet tip and the cell membrane (see Fig. 7.3a) becomes 1 GΩ or larger. In order to establish successful sealing, Rs is monitored during the sealing procedure and the patch-clamp experiment is usually only continued if $Rs > 1$G Ω.

In the present exercise we simulate these resistance measurements of $Rpip$ and Rs as an introduction to the whole-cell voltage-clamp experiment in the next section. The $Rpip$ and Rs values used here are 100–1000 times smaller than in a real patch-clamp experiment.

Electrophysiologists usually measure resistance (unit: Ω) in voltage recording (**current-clamp**) experiments by dividing the measured dV by the applied (current-clamped) dI. In current-recording (**voltage-clamp**) experiments conductance is measured (unit: Siemens $= 1/\Omega$) by dividing the measured dI by the applied (voltage-clamped) dV. Thus, in current-clamp experiments on cells one speaks of membrane resistance and in voltage-clamp experiments one speaks of membrane conductance. An exception to this convention seems to be the routine voltage-clamp measurement of $Rpip$ and Rs which precede a typical voltage-clamp patch-clamp experiment, as presented here.

The basic idea of voltage-clamp measurements is to apply a voltage to the object of study (a cell, a pipet, etc.) from an ideal voltage source, i.e., one having such a low internal resistance (Rvc in Fig. 7.3) compared to the external resistance of the object (e.g., $Rpip$ in Fig. 7.3) that no significant voltage drop occurs across the internal resistance during application of the voltage. This means that the current measured in the voltage-clamp is solely determined by the applied potential (Evc) and the external resistance. Thus, current changes at a given voltage then faithfully reflect changes in the conductance of the object (more conductance $\rightarrow$ more current). We apply this concept in the present experiment for the measurement of $Rpip$ and Rs.

Further definition of voltage-clamp

Note. Although $Rpip$ and Rs are supposed to be constant during an experiment, they may very well change, in particular Rs, and should therefore be checked regularly.

Figure 7.3a shows the measurement configuration during $Rpip$ and Rs measurements. When the pipet tip is in the bath solution, but not in touch with the cell, current easily flows through the pipet to the bath. This is the case in the upper position of the pipet drawn in the figure; thus, we measure only the resistance of the pipet tip ($Rpip$) since the

Circuit

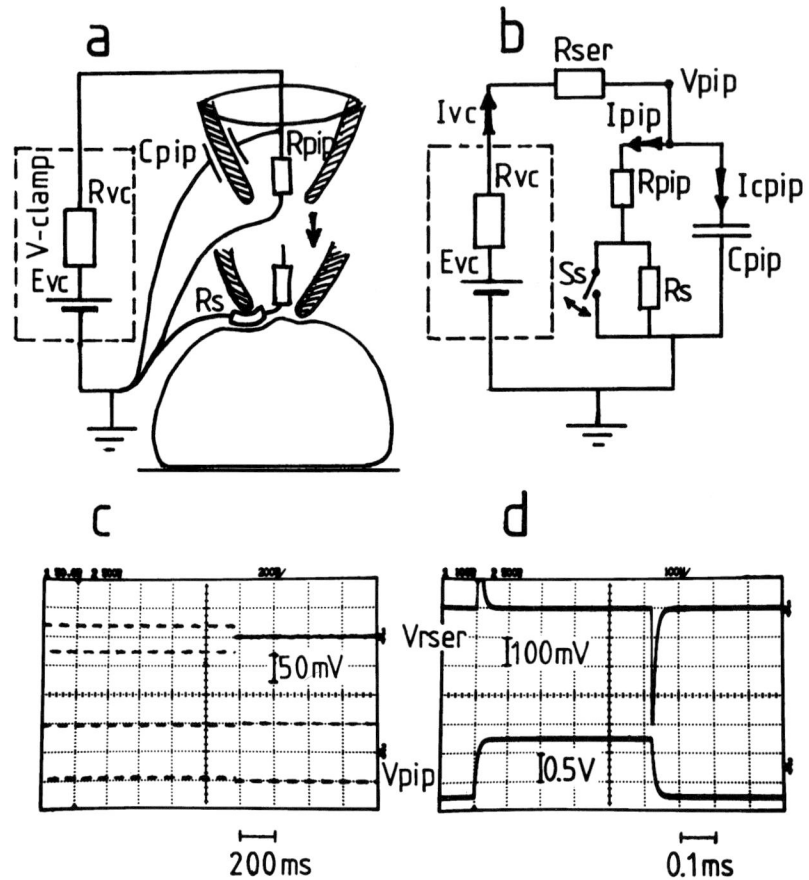

Fig. 7.3a–d. Measuring pipet and seal resistance by voltage-clamp (see Experiments 4 and 5). Measurement configuration in **a**, derived circuit in **b** and example records in **c** ($Vrser$, 50 mV/div; $Vpip$, 0.5 V/div) and **d** ($Vrser$, 100 mV/div; $Vpip$, 0.5 V/div). Frame **c** shows the abrupt sealing of the pipet on a slow time scale (0.2 s/div). Frame **d** shows the exponential time course of Ivc transients (*upper record*, monitored as $Vrser$) and of $Vpip$ (*lower record*) on a fast time scale (1 ms/div) after sealing.

resistance of the bath connection to the grounded reference electrode can be neglected. Upon aproaching and sealing the pipet against the cell membrane (see arrow), Rs is inserted between $Rpip$ and the bath. The pipet not only impedes current flow in the measurement circuit by the presence of $Rpip$, it also affects current flow through a capacitive current pathway parallel to $Rpip$, represented by the capacity $Cpip$ of the glass wall of the immersed pipet tip. The voltage-clamp block in the diagram contains the ideal voltage source Evc in series with a very small internal resistance Rvc. The latter resistance is required for the current measurement inside the voltage-clamp instrument.

Figure 7.3b is a schematic redrawing of the circuit identified in Fig. 7.3a. Resistor $Rser$ between Rvc and $Rpip$ has been added to the circuit in order to be able to measure the voltage-clamp current ($Ivc = Vrser/Rser$), since Rvc inside our voltage-clamp is not accessible for measurements. $Rser$ may also be seen as a representation of another series resistance in the circuit, e.g., the resistance of the Ag/AgCl electrode. The seal switch Ss serves to simulate the sealing procedure. When Ss is closed, Rs is short-circuited and $Rpip$ is the only current-limiting resistor in the circuit, which represents the measurement circuit before the pipet is touching the cell. Opening of Ss represents the sudden introduction of Rs into the circuit upon touching and sealing the cell. Sealing of a real cell may occur rather quickly (within a few seconds), but usually takes one or a few minutes. The currents through the different branches are indicated as Ivc, $Irpip$ and $Icpip$. The resistance for current through the sealed patch (with patch resistance Rpa) and the attached cell is neglected in this experiment, since often $Rpa \gg Rs$. Otherwise, the measured "Rs" is the equivalent resistance of Rs and Rpa in parallel.

Note. We do neglect pipet potentials, $Epip$. These potentials are usually (but not always!) small (a few mV) and do not affect our resistance measurements, but they should not be forgotten in a real experiment.

The relationship between the currents in the circuit of Fig. 7.3b **Equations** (Kirchof's current law) is

$$Ivc = Irpip + Icpip \tag{21}$$

Ohm's law applied to the currents results in

$$(Evc - Vpip)/Ro = Vpip/Rpip + Cpip.dVpip/dt \tag{22a}$$

or

$$(Evc - Vpip)/Ro = Vpip/(Rpip + Rs) + Cpip.dVpip/dt \tag{22b}$$

with

$$Ro = Rvc + Rser \tag{23}$$

for Ss closed (22a) or open (22b), respectively.

For stationary conditions ($dVpip/dt = 0$) we obtain for the stationary pipet potential, $Vpips$, for the conditions Ss closed or open

$$Vpips = Rpip.Evc /(Rpip+Ro) \tag{24a}$$

and

$$Vpips = (Rpip+Rs).Evc /(Rpip+Rs+Ro) \tag{24b}$$

respectively. In both cases $Vpips \sim Evc$, if $Rs \gg Rpip \gg Ro$, as is usually the case. Thus, by measuring stationary Ivc and dividing Evc by this Ivc, one can determine $Rpip$ and $Rpip+Rs \sim Rs$.

However, before becoming stationary, Ivc shows a sharp overshooting current transient, which is the current through $Cpip$, charging up the pipet to bring $Vpip$ to Evc.

To describe the time behavior of this current, we first have to solve the differential Eq. (22) for $Vpip(t)$. The solution is

$$Vpip(t) = Vpips + (Vpip(0) - Vpips).\exp(-t/tpip) \tag{25}$$

with

$$tpip = Rop.Cpip \text{ for } Rs \text{ shunted} \tag{26a}$$

or

$$tpip = Rops.Cpip \text{ for } Rs \text{ inserted} \tag{26b}$$

with

$$Rop = Rpip.Ro /(Rpip+Ro) \sim Ro \tag{27a}$$

or

$$Rops = (Rpip+Rs) .Ro/(Rpip+Rs+Ro) \sim Ro \tag{27b}$$

Thus, the time constant $tpip$ for charging up the pipet to $Vpip = Evc$ is for practical purposes

$$tpip = Ro.Cpip \tag{28}$$

independent of the position of the pipet, free in the bath or sealed to the cell. This constant is usually very small ($\sim 1\,\mu s$) compared to the time constant for charging up a membrane during a voltage-clamp experiment ($\sim 1\,ms$, see next paragraph).

The current $Ivc(t)$ can now easily be described by

$$Ivc(t) = [Evc - Vpip(t)]/Ro \tag{29}$$

Thus, $Ivc(t)$ overshoots its stationary value with the same time constant ($tpip$) as $Vpip(t)$ increases to its stationary value $Vpips = Evc$ (cf. Fig. 7.3d).

Experiment 4: Measuring Pipet Resistance and Capacitance

Make the circuit of Fig. 7.3b on the circuit board and take the following values for the components: $Rpip = 1\,K$, $Rs = 100\,K$, $Cpip = 0.1\,\mu F$, $Rser = 50\,\Omega$ and Evc a square voltage wave of 1 V peak-peak (pp) at a frequency of appr. 100 Hz. $Rvc = 50\,\Omega$. Keep switch Ss closed to short-circuit Rs.

Follow the next steps in this experiment:

1. Monitor $Vrser$ with a differential amplifier and measure the peak of the voltage change of the initial fast peak transient (pk $dVrser$), the subsequent stationary potential change (stat. $dVrser$) and the time constant of the decay of the initial transient.

2. Calculate the peak current change of Ivc ($= Irser = Vrser/50\,\Omega$), pk $dIvc$, and the stationary current change, stat. $dIvc$, and determine from these currents and the applied voltages Ro and $Rpip$, respectively.

3. Observe $Vpip(t)$ and measure the time constant of the initial exponential rise and compare it with the time constant of $Vrser(t)$. Measure also the change in $Vpips$, $dVpips$, and compare it with $dEvc$.

4. Recognize the effect of $Cpip$ on $tpip$ (cf. Eq. 28) by repeating the experiment with $Cpip = 0.01\,\mu F$ and $Cpip = 1\,\mu F$.

5. Calculate $Cpip$ from the measured Ro, $Rpip$ and $tpip$ using Eqs. (26a) and (28).

Experiment 5: Measuring Seal Resistance

1. Stimulate the pipet with voltage pulses while monitoring $Vpip$ and $Vrser$ on the oscilloscope, as in Experiment 4, and then open switch Ss to insert Rs in the $Rpip$ branch in order to simulate (abrupt) sealing (see Fig. 7.3c).

2. Observe that the peak and time constant of the initial capacitive current transient (cf. Fig. 7.3d) hardly change upon sealing, but that stationary $Vrser$ suddenly drops.

3. Calculate Rs and explain why the transient is not affected.
   If the calculated Rs is smaller than expected, then uncouple $Vpip$ from the oscilloscope to see whether the $Vpip$ measurement affects the current measurement.

Comments   The initial peak transient of Ipip upon voltage-step stimulation is never measured while preparing for a seal attempt in patch-clamp experiments. This capacitive current is so fast that it is usually filtered to a smaller and slower peak transient by the filters in the patch-clamp, even at the most favorable filter settings. Thus, an attempt to measure Rvc would lead to an overestimation of Rvc. Another reason that the initial current peak is smaller than expected for a given Rvc and voltage step is that the applied voltage step is not always exactly square. Often, the applied voltage step is rounded off a bit to prevent a too large initial current spike, since such a spike may bring the amplifier for a short time into saturation and spoil the rapid current measurement (cf. chapter by Sigworth in Sakmann and Neher 1995).

It is of interest to mention that even an overestimated Rvc would never cause a problem in whole-cell voltage-clamp experiments. We will see that the major problem interfering with ideal voltage-clamping of a cell membrane is the resistance of the access to the cell interior through Rpip and through the hole in the cell membrane (see chapter by Sigworth in Sakmann and Neher 1995).

Nevertheless, in whole-cell experiments on very small cells or vesicles it may be still important to reduce the duration of the capacitive transient by reducing the pipet capacitance. One way to do this is to use a low level bath for the cell, so that the pipet tip is less deeply immersed in the bathing solution and Cpip is smaller.

Voltage-Clamping an ERC Membrane

Voltage- and current-clamp compared   There has always been a great difference in attitude between membrane potential and membrane current recording electrophysiologists. The first group, the "current-clampers", usually biologists, were more classically oriented, more interested in the functional properties of excitable cells. The second group were the "voltage-clampers", usually biophysicists, who were more interested in the physical mechanisms of excitability. Although the inventors/developers of the voltage-clamp technique (see the Hodgkin and Huxley papers of 1952, discussed in Chapter 2 of Hille 1992) designed their voltage-clamp methodology to explain the current-clamp behavior of excitable cells (the ionic mechanism of the action potential) and easily switched from current-clamp thinking to voltage-clamp thinking, many following generations of current- and voltage-clampers did not understand each other very well and were unable to easily transfer voltage-clamp properties of nerve and muscle cells to current-clamp behavior or vice versa. Since it is important to master both ways of experimenting and thinking today, the present chapter offers both current-clamp and voltage-clamp exercises.

In current-clamp, the applied membrane current is under control, but excitability (action potential firing) is not. In voltage-clamp, the applied voltage and membrane excitability are under control. Thus, conceptually, current- and voltage-clamp are very different experimental conditions. Therefore, it is the more surprising that, electrotechnically speaking, the only difference between current- and voltage-clamp is the difference in the relative size of the voltage source and the source resistance between current-clamp ($Ecc \gg Vm$ and $Rcc \gg Rm$, see Fig. 7.2) and voltage-clamp (Evc appr. Vm and $Rvc \ll Rm$, see Fig. 7.4) conditions. The following voltage-clamp exercises serve to familiarize the student with the principle of voltage-clamp stimulation of cells.

Figure 7.4a shows a schematic diagram of a whole-cell voltage-clamp **Circuit** measurement on a spherical cell with the electrical circuit of the measurement configuration drawn in the figure. The voltage-clamp is an ideal voltage source Evc with a very small internal resistance Rvc ($\ll Rm$), required for current measurement inside the amplifier. Vvc is the voltage at the output of the voltage-clamp. The patch pipet adds series resistance ($Rser$) to the circuit between the voltage-clamp output and the inner side of the cell membrane. In practice, $Rser$ is two to three times the pipet resistance. The cell membrane is modeled by a simple ERC circuit consisting of a resistance branch parallel to a capacitive branch. The resistance branch consists of a membrane resistance Rm in series with an "intrinsic" membrane potential Em. Cm is the membrane capacitance.

The current in the voltage-clamp branch is called Ivc, the current in the resistive branch is Irm and the capacitive current is Icm. The capacitance of the pipet tip in the bathing solution is assumed to be negligible (zero) and the resistance of the gigaseal of the pipet tip to the cell membrane is assumed to be infinitely large. The bathing solution of the cell is grounded. The voltage-clamp circuit, redrawn in Fig. 7.4b, is essentially the same as the current-clamp circuit in Fig. 7.2b. Only the relative sizes of the different components are different: $Rcc \gg Rm \gg Rvc$ and $Ecc \gg Em$, Evc. Therefore, we can use the same equations for the passive responses of a cell membrane, as we will see below, but not for the active responses, such as action potential firing and conductance activation!

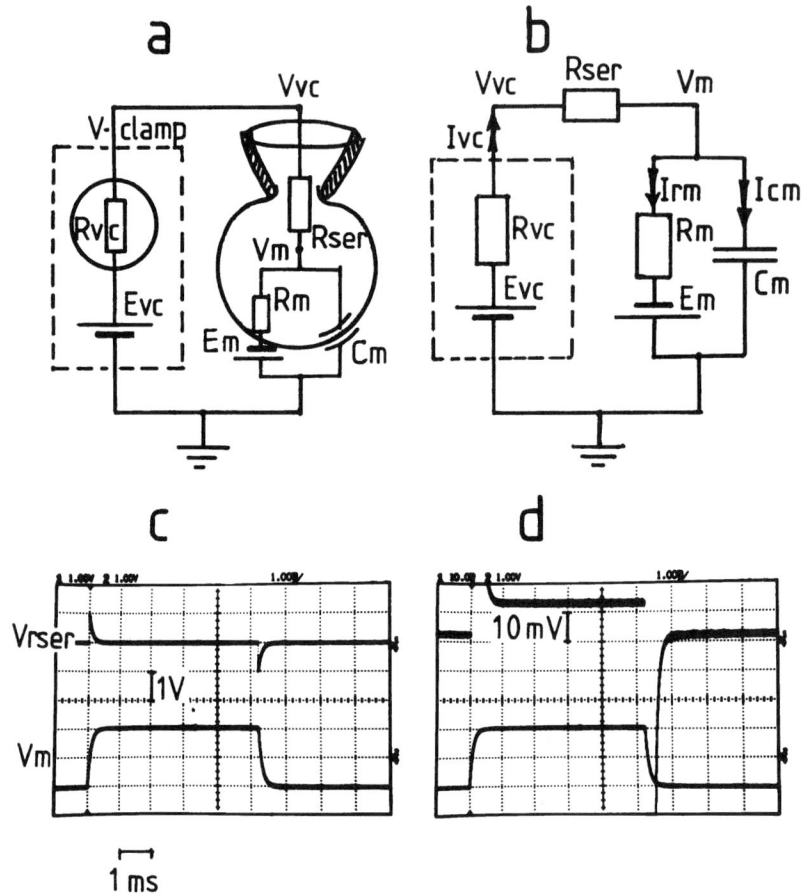

Fig. 7.4a–d. Voltage-clamping an ERC membrane (see Experiments 6 and 7). The whole-cell measurement configuration is shown in **a**, the derived circuit in **b** and example records in **c** and **d** on a fast time scale (1 ms/div). The *top records* in **c** (low gain, 1 V/div.) and **d** (high gain, 10 mV/div) are of Vrser (as a monitor of *Ivc*), while the *bottom records* show Vm (1 V/div). Ground potentials are indicated. Because of the absence of voltage-dependent conductances, the records are of a cell with only a leak conductance associated with a leak potential $El = Em$. The capacitive transients in *Ivc* have the same time constant as the Vm transients.

Equations   Kirchof's current law allows us to write for the voltage-clamp current *Ivc* in the circuit of Fig. 7.4b

$$Ivc = Irm + Icm \tag{30}$$

After substitution of Ohm's law for *Ivc* and *Irm* and the capacitive current equation for *Icm* we obtain

$$(Evc-Vm)/Ro = (Vm-Em)/ \, Rm \, + \, Cm.dVm/dt \qquad (31)$$

with

$$Ro = Rvc + Rser \qquad (32)$$

For stationary conditions $(dVm/dt = 0)$ the stationary value of Vm, Vms, derived from Eq. (31), is

$$Vms = (Rm.Evc+Ro.Em)/(Rm+Ro) \qquad (33)$$

This equation, which is essentially the same expression as Eq. (12), implies that Vms is very close to Evc under good voltage-clamp conditions $(Rm \gg Ro)$.

The time behavior of Vm, $Vm(t)$, can also be derived from Eq. (31). After rearranging terms we find an expression for differential Eq. (31) which better illustrates the relationship of Eq. (31) with Vms (Eq. 33):

$$dVm/dt = (-1/tmo).[Vm-Vms] \qquad (34)$$

with

$$tmo = Rmo.Cm \qquad (35)$$

and

$$Rmo = Rm.Ro/(Rm+Ro) \qquad (36)$$

and with Vms given by Eq. (33). Thus, the rate of change of Vm, dVm/dt, is proportional to the difference between the actual Vm and Vms and is positive when Vm is below Vms and negative when it is above Vms. Furthermore, the smaller tmo is, the larger dVm/dt at a given $[Vm-Vms]$.

The resistive part of tmo, Rmo, is mainly determined by Ro, since Eq. (36) implies that Rmo is very close to Ro, if $Rm \gg Ro$, as usually is the case and as required for ideal voltage-clamp conditions.

When Eq. (34) is solved for $Vm(t)$ we obtain

$$Vm(t) = Vms + (Vm(0)-Vms).exp(-t/tmo) \qquad (37)$$

Again, the time constant of this exponential function is mainly determined by the total output resistance $(Ro = Rvc+Rser)$ of the voltage-clamp before it connects to the membrane.

Equation (37) describes how soon $Vm(t)$ approaches the applied voltage Evc upon applying a step in Evc. This is important to know in order to decide whether this is rapid enough to record currents from rapidly voltage-activated conductances. However, it is also of importance to have an expression for the recorded **voltage-clamp current** flowing in the circuit upon applying a step in Evc. This current Ivc is the total membrane current, Im, and is equal to

$$Ivc = [Evc - Vm(t)]/Ro \tag{38}$$

This equation describes the spiking exponential shape of the voltage-clamp current upon steps in Evc, which is in contrast to the monotonic shape of $Vm(t)$. The rate of change of Ivc is again determined by tmo, as for $Vm(t)$, since $Vm(t)$ is the only part of Eq. (38) which is time-dependent. Thus, again, Ro rather than Rm determines the rate of current decline. And finally, Eqs. (37) and (38) both state that Vm has not yet approached Evc, as long as the Icc transient has not yet leveled off.

Experiment 6: *Vm and Im Transients During Voltage-Clamp Step Stimulation*

1. Make the circuit of Fig. 7.4b on the circuit board. Take $Rser = 50\,\Omega$, $Cm = 1\,\mu\text{F}$, $Rm = 10\,\text{K}$ and $Em = -1.5\,\text{V}$. We take again $Rvc = 50\,\Omega$, a frequently specified value for function generators. Evc usually includes the range of physiological Vm values (-100 to $+100\text{mV}$), which in our equivalent circuits corresponds to $-2\,\text{V}$ to $+2\,\text{V}$. Use for Evc an approximately 100 Hz square voltage block of 2 V peak to peak.

2. Monitor $Vm(t)$ and $Vrser(t)$ at approximately 1 V/div and 1 ms/div. $Vrser(t)$ monitors $Ivc = Im$, since $Ivc = Vrser/Rser = Vrvc/Rvc$, with $Rser = Rvc = 50\,\Omega$ in our case. Be aware of the fact that $Vrvc = Evc - Vvc$ and $Vrser = Vvc - Vm$. A patch-clamp amplifier essentially provides $Ivc = Vrvc/Rvc$, but we cannot measure $Vrvc$ in the function generator. Therefore, we derive Ivc from $Vrser/Rser$ by measuring $Vrser$ with a differential amplifier.

3. Observe the spiking exponential capacitive current $Im(t)$ and the monotonic exponential membrane potential change $Vm(t)$ (cf Fig. 7.4c). Note that they have the same time constant of approximately $Ro \cdot Cm = 0.1\,\text{ms}$ and that the peak value of $Vrser = 1\,\text{V}$ (cf. Fig. 7.4c), half the total voltage step applied (2 V). Explain why.
Calculate the peak current ($Vrser/Rser$) and explain why it has that value.

4. Try different $Rser$, Cm and Rm values to establish that Ro and Cm mainly determine the time constants of $Im(t)$ and $Vm(t)$.

5. Increase the recording sensitivity to approximately 10 mV/div and measure the stationary current at $Vm = +1$ V, -1 V and at 0 V (cf. Fig. 7.4d). Notice that there is also membrane current at $Vm = 0$ V, originating from Em, which can be checked by making $Em = 0$ V. Calculate the measured currents from the circuit components and check whether the measured values agree with the calculated values.

Experiment 7: Current-Voltage ($Im - Vm$) Relationships

1. Make $Em = 0$ V (remove the battery and connect Rm directly to ground) and measure stationary Im, Ims, at $Vm = -2$, -1, 0, $+1$ and $+2$ V. Then plot Ims as a function of Vm. The resulting $Im - Vm$ curve stands for a whole-cell I–V curve for a cell without membrane potential, e.g., for a cell with equal intra- and extracellular solutions or for a cell in which the membrane potential-generating conductances have been blocked and in which only a leakage conductance remains. The slope is the membrane conductance, the inverse of $Rm = 10$ K (100 μS). The intersection of the I–V curve with the voltage axis, the reversal potential Er, is $Er = Em = 0$ V.

2. Do the same experiment as in step 1, but now with $Em = -1.5$ V, as in Experiment 6. The resulting conductance may stand for a potassium conductance $Gk = 1/Rk = 1/10$ K with a Nernst potential $Ek = -1.5$ V determining the resting membrane potential. Take notice of the fact that the I–V curve reverses polarity at $Er = Ek$.

3. Do the same experiment as in step 2, but now with $Em = +1.5$ V, as if a sodium conductance Gna, with $Ena = +1.5$ V, determines Em. Observe that the I–V curve now reverses polarity at $Er = Ena$. The basic observation in this experiment is that the membrane potential or the dominating conductance determines the reversal potential, e.g., the position of the I–V curve along the voltage axis. In real experiments the I–V curve of for example a K^+ conductance can be shifted along the Vm axis by changing extracellular $[K^+]$ in order to prove K^+ selectivity.

Comments

The present experiments were an introduction into voltage-clamp measurements on a whole-cell membrane with only a linear conductance in series with a diffusion potential and with a parallel membrane capacitance. $Rser$ represented the sum of the pipet resistance ($Rpip$) and the access resistance through the broken patch to the inside of the whole-cell membrane. *Rpip* was set smaller than in Experiments 4 and 5 in order to obtain a good *Rm/Ro* ratio (100), but *Cpip* was assumed to be zero. A better simulation of the real measurement condition would be obtained by assuming *Cpip* = 0.1*Cm* = 0.1 μF, so that a two-component capacitive transient is obtained, a fast pipet and a slower membrane capacitive current. The larger *Rser*, the better the distinction between the fast and slow components, but this is at the expense of ideal voltage-clamp conditions. (Try *Rser* = 550 Ω consisting of *Rser1* = 50 Ω, the *Rser* already present for the *Ivc* measurement, and *Rser2* = 500 Ω, placed after *Rser1* and before the membrane. Then connect the point between *Rser1* and *Rser2* to ground via *Cpip* = 0.1 μF).

No voltage-activated conductances became active upon the voltage steps, so that the experiments were useful for obseving the baseline behavior of a voltage-clamped cell and for determining the quality of the measurement conditions for real voltage-clamp experiments on voltage-activated or agonist-activated conductances. In the next experiment we will consider a voltage-activated conductance (or channel) activated upon a step depolarization of the membrane.

Depolarization-Activated Channels and Conductances

The components of the circuits described in this course are all linear components, i.e., the resistors (or conductances) and capacitors are all constant and independent of the membrane potential *Vm*. Membrane electrophysiology is such an interesting science because the different ionic membrane conductances are so often voltage-dependent. This property is important for control of the resting membrane potential, but it also provides membranes with mechanisms to produce signals of great functional interest, such as action potentials.

From nonlinear electronic components it is possible to build voltage-dependent conductances with properties very similar to those of ionic conductances in excitable membranes (Ypey et al. 1980). In this course we do not use such electronic conductances because we would rather go to the living cell to study voltage-dependent conductances. However, we will still carry out a very instructive experiment in which we simulate the behavior of voltage-activated channels and conductances manually! We will give voltage-clamp pulses to a whole-cell with a very slow capacitive current transient (time constant appr. 0.1 s), so that there is time to manually activate a single channel just after the transient by inserting a resistor in the membrane circuit.

A channel (or conductance) activated by making the normally negative membrane potential more positive (depolarization) is called a **depolarization-activated** channel (or conductance). Other voltage-activated channels are **hyperpolarization-activated** channels, i.e., activated by making the negative membrane potential even more negative (cf. Ypey et al. 1992).

Definitions: voltage activation

We will also manually activate a whole-cell conductance by decreasing a potentiometer resistance in the membrane circuit upon a voltage-clamp step. In this case we can recognize the importance of fast capacitive current transients in order to obtain good separation of the membrane charging process from the activation process and to obtain a rapid (high-frequency) response of the voltage-clamp current. The experiment can also be used to understand how the polarity of a current response depends on the applied voltage and the reversal potentials involved. Finally, we will learn from the experiment that a high-conductance activation may affect the voltage-clamp conditions.

The circuits of Fig. 7.5 are simple modifications of the circuit of the whole-cell configuration of Fig. 7.4. Rl represents a small leakage conductance (high leakage resistance) of the whole cells in Fig. 7.5a, b. In Fig. 7.5a, Rch is the resistance of a single channel in the whole-cell membrane in series with a diffusion potential or Nernst potential Ech. The switch Sch represents the voltage-dependent (manually operated) opening and closing mechanism of the channel. In Fig. 7.5b, Rg represents the manually controlled variable resistance of the voltage-activated ionic conductance and Eg the diffusion or Nernst potential of that conductance. All other component symbols are as in Fig. 7.4.

Circuit

If we neglect the leakage current Il, all the equations of Sect. 7.2.2 apply to Fig. 7.5a, when Rch and Ech are substituted for Rm and Em, respectively, and to Fig. 7.5b, when Rg and Eg are substituted for Rm and Em, respectively. The assumption $Il = 0$ is not unreasonable in the experiment below, in which in Fig. 7.5a $Rl = 10 \cdot Rch$, while in Fig. 7.5b $Rl = 10$ to $100 \cdot Rg$, depending on the value of Rg.

Equations

Thus, for Fig. 7.5a, b we obtain for the stationary condition

$$Vms = (Rch.Evc+Ro.Ech)/(Rch + Ro) \tag{39a}$$

and

$$Vms = (Rg.Evc+Ro.Eg)/(Rg + Ro) \tag{39b}$$

with

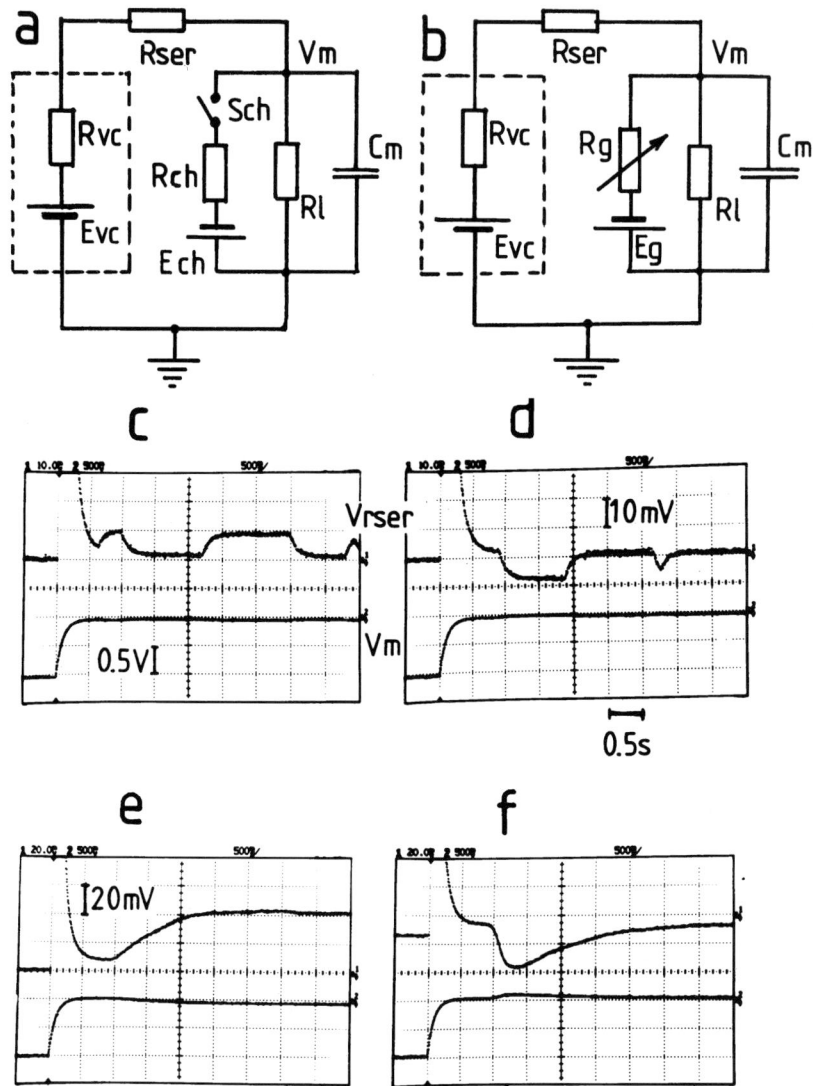

Fig. 7.5a–f. Depolarization activated channels (**a, c, d**) and conductances (**b, e, f**) (see Experiments 8 and 9). Frame **c** simulates depolarization activated K$^+$ channels, frame **d** depolarization activated Na$^+$ channels, frames **e** and **f** the corresponding conductances. Time base of all records is 0.5 s/div. *Bottom records*, from −1 V to 0 V, are of *Vm* (all frames). *Top records* of all frames represent current records (monitored as *Vrser*: 10 mV/div in **c, d**; 20 mV/div in **e, f**). Notice the small effect of conductance activation on the applied *Vm*, thus on the quality of voltage clamp.

$$Ro = Rvc + Rser \tag{40}$$

and $Ro \ll Rch$, Rg so that usually $Vms \sim Evc$.

For the time behavior of Vm upon a sudden switch-on of Rch or Rg, or upon a voltage step with Rch or Rg already on, we can derive

$$Vm(t) = Vms + (Vm(0) - Vms).\exp{-(t/tcho)} \tag{41a}$$

$$Vm(t) = Vms + (Vm(0) - Vms).\exp{-(t/tgo)} \tag{41b}$$

with

$$tcho = Rcho.Cm \; (\sim Ro.Cm) \tag{42a}$$

and

$$tgo = Rgo.Cm \; (\sim Ro.Cm) \tag{42b}$$

and

$$Rcho = Rch.Ro/(Rch + Ro) \; (\sim Ro) \tag{43a}$$

and

$$Rgo = Rg.Ro/(Rg + Ro) \; (\sim Ro) \tag{43b}$$

The membrane current $Im = Ivc$ upon voltage-clamp steps or upon sudden activation of Rch or Rg is described by

$$Ivc = [Evc - Vm(t)]/Ro \tag{44}$$

All these equations certainly apply if Rl is made infinitely large (component Rl removed).

Experiment 8: Single-Channel Current Upon Channel Activation

1. Make the circuit of Fig. 7.5a and take $Rser = Rvc = 50\,\Omega$, $Rch = 10\,\text{K}\Omega$, $Ech = -1.5\,\text{V}$ (resembling an Ek), $Rl = 100\,\text{K}\Omega$, $Cm = 1\,\text{mF}$ and $Evc = 1\,\text{V}$ block pulse pp, applied at about 0.2 Hz from a holding voltage of $-1\,\text{V}$.

   Trigger the scope at the positive step of the block pulse and display $Vrser$ (via differential amplifier to oscilloscope at 5 or 10 mV/div) as an indicator of Ivc and watch Vm (oscilloscope at 0.5 V/div) to monitor the clamped potential. Take a slow sweep speed of about 0.5 s/div (cf. Fig. 7.5c).

   Switch Sch on after the capacitive current transient upon the voltage step to 0 V and switch Sch off again before the step back to $-1\,\text{V}$.

Is the time constant of the channel-on current and that of the channel-off current predicted by Eq. (42a)?

Is the stationary channel current as expected from Ech and Rch?

Try to get the channel-on event during the capacitive transient in order to see how the capacitive transient masks the onset of the channel current.

Why is the channel current outward, both at $Vm = 0$ V and at -1.0 V?

2. Repeat the experiment with $Ech = +1.5$ V, simulating a depolarization-activated sodium channel, and observe that the channel currents now are inward (cf. Fig. 7.5d).

Experiment 9: Depolarization – Activated Conductance

1. Make the circuit of Fig. 7.5b with the component values as in Fig. 7.5a for experiment 8, except for Rg, which should be a potentiometer of 10 K in series with a 1 K resistance to limit the minimal resistance Rg to 1 K. Take $Eg = -1.5$ V, simulating an Ek.

   Repeat Experiment 8, but now by activating the conductance $Gg = 1/Rg$ upon a depolarization from -1 V to 0 V.

   Try to simulate a K^+ current with a rapid rise and a slower decay by turning the potentiometer in the right direction and with the appropriate speed.

   Try to activate the current during the decay of the capacitive current and then after this transient is over (cf. Fig. 7.5e). This will demonstrate again the importance of the separation in time of the capacitive and ionic membrane currents.

   Watch Vm during "K^+ conductance activation" and observe that the fully activated conductance ($Gg = 1/Rg = 1/1$ K) affects Vm, and thus makes the voltage-clamp conditions less ideal.

2. Repeat this experiment with $Eg = +1.5$ V to simulate the shape of inward Na^+ currents (see Fig. 7.5f).

Comments   The above experiments were probably an instructive confrontation with the measurement of voltage-activated channel activity and membrane conductance changes, since the experimenter was the gating mechanism of the channels studied. One significant observation was that current flow due to channel or conductance activation after the capacitive transient is over is still delayed (distorted) by the charging process of the membrane capacity. Thus, even if the capacitive transient does not mask the onset of the channel current(s), the recorded current is not a faithful reflection of the rapid event of channel opening or of conductance activation, since the current response is filtered by the components of the measurement configuration (mainly Ro and Cm). This is not always recognized by patch-clampers.

Experiment 8 is more representative for depolarization-activated channel activity in outside-out patches than for whole-cell current activation. The experiments described here showed that a depolarization-activated conductance causes more cell-outward current if it becomes activated above its reversal potential. Therefore, such a conductance is sometimes called an **outward rectifier.**

Simulation experiments on hyperpolarization-activated channels and conductances (cf. Ypey et. al. 1992) can be carried out in the same way as above if negative going steps are applied. Such conductances give rise to increasing inward currents if activation occurs below the reversal potential of the conductance. Such conductances are called **inward rectifiers.**

Comments

General Conclusion and Follow-Up

You now have obtained sufficient understanding of the principles of current-clamp and voltage-clamp measurements and have developed skills required to do actual patch-clamp experiments.

A useful first step to start these experiments is to repeat the exercises of this course with the use of a real-life patch-clamp amplifier. This will allow you to get to know the new features of this amplifier, such as capacitive transient cancelation. Although this is not covered here, you will understand what type of transient is canceled and why.

Another feature of certain patch-clamp amplifiers is the option to "electronically remove" a considerable part of the total series (or access) resistance to the inside of the cell during whole-cell measurements. We have represented this access resistance in the exercises by $Rser$ or Ro ($= Rvc+Rser$). You will now understand why it is useful to have this option.

The representation of a voltage-clamp patch-clamp amplifier by an Evc in series with a low current-measuring resistor Rvc has its limitations. To really understand the operation principle of the current-to-voltage converter and all its features (low impedance current measurement, transient cancelation, series resistance compensation, filtering), it is essential to consider the properties of operational amplifiers (see Chapter 4 in Sakmann and Neher 1995; Chapter 11 in Standen et al. 1987; Chapter 3 in Sherman-Gold 1993). However, the approach followed here is very useful for practical purposes. It explains the essence of the voltage-clamp technique, although this technique could only be developed due to operational amplifier technology.

Despite this course, you will quite often run into problems, both experimental and theoretical, and not understand what you are measuring or you will not be able to interpret your measurements. In those circumstances it is recommended to reduce the problem, or some aspect of it, to an "electrical equivalent circuit problem" and solve that problem, which will certainly contribute to solving the original problem.

For example, controls of patch-clamp measurements on an RC cicuit may help to decide whether an unwanted voltage offset during the experiments is localized inside or outside the amplifier.

References

Hamill OP, Marty A, Neher E, Sakmann B, Sigworth FJ (1981) Improved patch-clamp techniques for high resolution current recording from cells and cell-free membrane patches. (Appendix in Sakmann and Neher 1995). Pfluegers Arch. 391: 85–100

Hille B (1992) Ionic channels of excitable membranes. Sinauer, Sunderland, MA

Neher E (1992) Ion channels for communication between and within cells. Science 256: 498–502

Neher E, Sakmann B (1992) The patch-clamp technique. Sci Am 266: 28–33

Ravesloot JR, Van Putten MJAM, Jalink K, Ypey DL (1994) Analysis of decaying unitary currents in on-cell patches of cells with a high membrane resistance. Am J Physiol 266: C853-C869

Sakmann B (1992) Elementary steps in synaptic transmission revealed by currents through single channels. Neuron 8: 613–629

Sakmann B, Neher E (1995) Single-channel recording. Plenum, New York

Sherman-Gold R (1993) The axon guide for electrophysiology and biophysics. Axon Instruments, Foster City, CA

Standen NB, Gray PTA, Whitaker MJ (1987) Microelectrode techniques. The Plymouth workshop handbook. The Company of Biologists, Cambridge

Verveen AA, DeFelice LJ (1974) Membrane noise. Progr Biophys Mol Biol 28: 189–265

Ypey DL, Weidema AF, Hold KM, VanderLaarse A, Ravesloot JH, VanderPlas A, Nijweide PJ (1992) Voltage, calcium and stretch activated ionic channels and intracellular calcium in bone cells. J Bone Miner Res 7: S377-S387

Ypey DL, VanMeerwijk WPM, Ince C, Groos G (1980) Mutual entrainment of two pacemaker cells. A study with an electronic parallel conductance model. J Theor Biol 86: 731–755

Appendix

Results, Questions and Answers

Experiment 1: RC Behavior of a 2-Channel Membrane

Description of the step responses in the protocol (see Fig. 7.1c):
- Upon step 1 Vm slowly ($tm = 1$ s) moves to $Ek = -1.5$ V.
- Upon step 2 Vm does not or hardly change.
- Upon step 3 Vm moves rapidly ($tm = 200$ ms) to $Vm = 0$.
- Upon step 4 Vm moves rapidly ($tm = 170$ ms) to $Vm = -250$ mV.
- Upon step 5 Vm moves slowly ($tm = 1$ s) to $Vm = -1.5$ V.
- Upon step 6 Vm moves to $Vm = -250$ mV with the same tm as in step 4.
- Upon step 7 Vm moves to $Vm = 0$ V with the same tm as in step 3.
- Without Cm all the changes in steps 1–7 are virtually instantaneous.
- With smaller Cm the changes are faster, with larger Cm they are slower.

Questions

a. Explain the exponential time courses of the charging and discharging processes in steps 1 and 3.

b. Explain why Vm remains in principle constant in step 2, but why it may still very, very slowly decline.

c. Are the observed Vms values in steps 1 and 4 consistent with Eq. (6)?

d. Are the measured values of the time constants in steps 1–7 predicted by Eq. (8)?

e. If Sk is on and Sl is turned on too for a short time, the Vm response looks very much like a postsynaptic potential (try it! cf. Fig. 7.1d). Explain this resemblance.

f. Substitute the leakage branch of the circuit for a Na^+ branch with $El{\rightarrow}Ena = +1.5$ V, $Rl{\rightarrow}Rna = 100\ \Omega$ and $Sl{\rightarrow}Sna$. Switching on Sna for a short time simulates an action potential with only two ion channels! (cf. Fig. 7.1e).
Imagine, would this be possible in a real cell?

g. This experiment may also be seen as a simulation of a macroscopic action potential with a (rapid) sodium activation and inactivation, but without repolarizing K^+ activation. How could one add faster repolarization to the circuit? (cf. Fig. 7.1f).

Answers

a. At $t = 0$, i.e., just after switching on Sk, the voltage across Rk is Ek, since Vm is zero. This results in a sudden initial current Ek/Rk. As soon as Ik flows, it accumulates charge on Cm, so that the voltage across Rk and the current through Rk declines. Thus, the longer the charging proceeds, the smaller the current becomes, until it is zero at $Vm = Ek$ (no voltage drop across Rk anymore). This is, in fact, what Eq. (7) says!

A similar way of reasoning can be used for the other transients in the experiment.

b. In step 2 Vm is supposed to remain constant, since there is no discharging resistance in the circuit on the circuit board. However, there is still a very very slow discharge across the "hidden" input resistance (1 M) of the oscilloscope input ($tm = 1\,M\Omega \cdot 1\,mF = 1000\,s$)!

c. Equation (6) predicts for step 1 ($Rl \gg Rk$, since the leak channel is closed) $Vms = Ek$, consistent with the observation.
    For step 4 $Vms = 0.2\,K \cdot -1.5\,V/(1\,K\ +0.2\,K) = -0.25\,V$, consistent with the observation.

d. Equations (8) and (9) predict for step 1 ($Rl \gg Rk$) $tm = Rk \cdot Cm = 1\,s$, as observed. These two equations predict for step 3 $tm = 0.2\,K \cdot 1mF = 200ms$, as observed, and for step 4, ($Rkl = 0.17\,K$) $tm = 0.17\,K \cdot 1\,mF = 170\,ms$, as observed.
    The tm values in steps 5,6, and 7 are equal to those in steps 1, 4 and 3, respectively.

e. Excitatory transmitter operated conductances, such as the acetylcholine receptor in skeletal muscle, also have a rapid turn-on and a slower turn-off time course towards a receptor-associated diffusion potential close to 0 mV (cf. Fig. 7.1d).

f. In an actual excitable cell we would see a similar time course (fast-on, slower-off, cf. Fig. 7.1e), but not a reversal of the potential, since Na^+ channels do not have a 10x as large single-channel conductance, but approximately the same conductance. Thus, we would need a couple of open channels for potential reversal.

g. Rapid repolarization by K^+ activation could be added to the circuit by inserting a 100 Ω resistor parallel to Rk during the repolarization process (cf. Fig. 7.1f). This would simulate the activation of extra K^+ channels during the repolarization process.

Experiment 2: tm and Vms Under Ideal I-Clamp

An $Ecc = 10\,V$ pp square wave around 0 V was applied to $Rcc = 100\,K$, which is supposed to result in a 0.1 mA current block.
    The resting Vm at $Ecc = 0\,V$ was $-1.53\,V$. With $Ecc = 10\,V$, Vms alternated between the depolarized value of $-1.48\,V$ and the hyperpolarized value of $-1.58\,V$ (Fig. 7.2c). Thus, the ΔVms was 100 mV (Fig. 7.2d), as expected from a 0.1 mA current through a $Rm = 1\,K$ resistance (Eq. 35).
    The time constant tm was 1 ms (Fig. 7.2d), consistent with Eq. (20) ($1\,K \cdot 1\,\mu F$).
    ΔVms was proportional to Rm, tm was proportional to Rm and Cm.

Experiment 3: Less Ideal Current-Clamp

At $Vcc = 0\,V$, the next resting Vm values were measured at various Rcc values:

Rcc	infinite	100 K	10 K	1 K
Vm	$-1.55\,V$	$-1.53\,V$	$-1.4\,V$	$-0.8\,V$

The 10 % loading of Vm by $Rcc = 10$ K may just be acceptable in certain experiments, if one needs bigger currents than provided by $Rcc = 100$ K, but it would be better to choose a 10x stronger voltage source.

The more general equations, including the role of all resistors in the circuit, thus not only Rcc but also Rm, are in fact given as eqs. 30–38 for voltage-clamp stimulation of a membrane. To obtain the current-clamp equation one should replace Evc by Ecc, Rvc by Rcc and Ivc by Icc. This will become clear by comparing Fig. 7.2b with Fig. 7.4b.

Experiment 4: Measuring *Rpip* and *Cpip*

Questions

a. Are the measured results consistent with the equations?

b. What determines the overshoot potential?

c. What could be the reason that the peak $Vrser$ is somewhat smaller than expected?

Answers

a. The results are summarized as follows:

$dEvc$	pk d$Vrser$	stat . d$Vrser$	$t\ Vrser(t)$	d$Vpips$	$t\ Vpip(t)$
1 V	0.4 V	45m V	10 µs	0.9 V	10 µs
	pk dIvc	stat . dIvc			
	8 mA	0.9 mA			
	$Ro =$	$Rpip =$			
	1 V/8 mA	0.9 V/0.9 mA			
	125 Ω	1 K			

Thus, the time constants of $Vrser(t)$, $Ivc(t)$ and $Vpip(t)$ are equal, consistent with Eqs. (25) and (29).

d$Vpips$ is somewhat (10 %) smaller than dEvc, but this is consistent with Eq. (24a) and is due to non-ideal voltage-clamp ($Rpip = 10 \cdot Ro$).

b. According to Eq. (26a) and (28) $tpip$ is ten times smaller with $Cpip = 0.01$ µF and ten times greater with $Cpip = 1$ µF. With the greater $Cpip$, peak d$Vrser$ is close to 0.5 V, which indicates that the faster peak with 0.1 µF is filtered to smaller values by hidden filters in the measurement system. When this larger peak d$Vrser$ value is used in the calculation of Ro, we find $Ro = 100$ Ω, consistent with the nominal values of $Rser$ (50 Ω) and Rvc (50 Ω) . Thus, the underestimated peak d$Vrser$ caused an overestimated $Ro = 125$ Ω

c. The peak d$Vrser$ is equal to $Rser \cdot dEvc/(Rser+Rvc)$, since at $t = 0$, when still d$Vpip = 0$, Rvc and $Rser$ are the only resistors over which Evc is divided. Since $Rvc = Rser$, peak d$Vrser$ is expexted to be 0.5 V.

Experiment 5: Measuring *Rs*

Questions   What happens with *Vpips* upon sealing? And why?

If the calculated *Rs* is smaller than expected, then uncouple *Vpip* from the oscilloscope to see whether the *Vpip* measurement affects the current measurement. If so, what could be the reason?

Answers   Peak d*Vrser* (cf. Fig. 7.3d) is not expected to depend on sealing, since this value is equal to $Rser \cdot Evc/(Rser+Rvc)$, as explained in the previous experiment. The time constant of *Vrser*(t) (see Fig. 7.3d) is hardly changed by pipet sealing to the cell, as expected from Eqs. (27a, b).

Upon sealing, *Vpips* becomes equal to *Evc* (see Fig. 7.3c), since the voltage-clamp conditions become much better, consistent with Eqs. (24a, b).

Stationary d*Vrser* = 0.6 mV, thus stat. d*Ivc* = stat. d*Vrser*/50 Ω = 12 μA.

Then $Rs = Evc/12\,μA = 83\,K$, which is smaller than the nominal value of 100 K. This is due to the fact that the 1 M input resistance of the oscilloscope input, on which *Vpip* is measured, sligly shunts the $Rs = 100\,K$. Indeed, when *Vpip* is uncoupled from the oscilloscope, stat. d*Vrser*~0.5 mV and the calculated *Rs* is closer to 100 K. If it is still a bit more than 0.5 V, one may do a control for leak currents in the circuit board by removing *Rpip* and *Rs*.

Experiment 6: *Vm*(*t*) and *Im*(*t*) upon Voltage Steps

– According to Eqs. (36)–(38) *tmo* is mainly determined by *Ro*, if *Ro*≪*Rm*, as is the case.

Vrser(0) is 1 V (see Fig. 7.4c), since at $t = 0$ *Evc*−*Vm* = 2 V, which is equally divided over *Rvc* and *Rser*, because *Rvc* = *Rser*.

The peak current at $t = 0$ is (*Evc*−*Vm*)/(*Rvc*+*Rser*) = 20 mA, but only flows at $t = 0$ since, as soon as *Im* flows, *Cm* is charged up, so that the absolute value of (*Evc*−*Vm*) becomes smaller. This is as predicted by Eqs. (35) and (36).

– Observed and calculated *Im* values:

Vm	−1	0	+1 V
Measured (*Vrser*/*Rser*)	+0.05 mA	+0.16 mA	+0.26 mA
Calculated (*Vm*−*Em*)/*Rm*	+0.06 mA	+0.16 mA	+0.26 mA

Small reading errors may cause deviations.

Em was measured to be −1.55 V.

Experiment 7: *Im*−*Vm* Curves

The values from Experiment 6, step 5, already provide part of the I–V curve of Experiment 7, step 2. Extrapolation or an extra measurement at *Evc* < −1.55 V shows that $Er = Em = Ek = -1.55\,V$.

The rest of the experiment is straightforward and is already described with the excercise.

Experiment 8: Single-Channel Current Upon Channel Activation

- According to Eq. (42a) $tcho = Rcho \cdot Cm \sim Ro \cdot Cm = 100\,\Omega \cdot 10^{-3}\,F = 0.1\,s$.
  Indeed, both $Vm(t)$ and $Ivc(t)$ change with a time constant of 0.1 s upon channel on-switching (Fig. 7.5c, d).
  With $Ech = -1.55\,V$ (to be precise), the measured channel current at 0 V is $dVrser/Rser = 8\,mV/50\,\Omega = 0.16\,mA$.
  The expected current is $(Vm - Ek)/Rch = 1.55V/10^4\,\Omega = 0.155\,mA$, which is in good agreement with the measured value. Rl contributes to the total membrane current, but does not affect the current through Rch, which we measure here (from $dVrser$, not from $Vrser$!), since "all" channel current is short-circuited to $Ro \ll Rl$.
  The channel current at $Vm = 0\,V$ is outward (cf. Fig. 7.5c), as Ech is the only electromotive force in the channel circuit and causes cell-outward current of positive charges (K^+ ions) through Rch. At $Vm = -1\,V$, the current is also outward, despite the opposing effect of Evc, which has opposite polarity in the channel current circuit. However, Ech is "stronger" than Evc.
  The right polarity of the current is also defined by: $Ich = Gch \cdot (Vm - Ech)$, implying that outward current is positive.
- With $Ech = Ena = +1.5\,V$ (see Fig. 7.5d) the way of reasoning is the same.

Experiment 9: Depolarization-Activated Conductance

The polarity of the depolarization-activated currents will be the same as in Experiment 8. However, the differences are that the currents are now larger (minimal $Rg = 1\,K$!) and the current shape is not square as for single channels, but as determined by the way the potentiometer is operated upon occurrence of the depolarizing voltage step.

Chapter 8

Improved Electrophysiological Measurements by Series Resistance Compensation

A.C.G. Van Ginneken and A.O. Verkerk

Background

In studying the electrical behavior of ion channels and ion pumps of living cells, more insight is gained into the function of these cells. This is especially true for excitable cells, in which the transfer of ions plays a key role in the mechanism of impulse formation.

In the heart, various types of excitable cells are present. After the introduction of the voltage clamp technique by Hodgkin and Huxley (1952) using nervous tissue, this technique was also applied to cardiac tissue in an attempt to understand the mechanisms behind the various types of activity. Until the 1970s, however, it was only possible to perform voltage clamp experiments on multicellular preparations using the double micro-electrode voltage clamp, which severely limited the precision of the measurements. A large improvement was made when it became possible to isolate single, calcium-tolerant, intact cardiac cells. On these cells the whole cell variant of the patch clamp technique (Hamill 1981) could be applied. Since then much progress has been made in understanding how the electrical impulse is generated and conducted from cell to cell.

Despite improvements in the electrophysiological techniques which can be applied to cardiac cells, there are still several factors remaining which affect the accuracy of measurements of membrane voltages and currents. This chapter focuses on one of those factors, the disturbing effect of series resistance. Although the effect will be illustrated by experiments on heart cells, the results are applicable to every cell type on which whole cell measurements are performed.

With the patch clamp technique (Neher and Sakmann 1976) one can explore the electrical properties of an isolated piece of cell membrane ("patch clamp") or of the membrane of a whole cell ("whole cell clamp"). In the whole cell clamp configuration, two kinds of measurements can be made:

- Voltage clamp measurements, in which the cell membrane is clamped to a steady voltage or to a stepwise change in voltage, while the changes in membrane current are recorded

- Current clamp measurements, in which the current is clamped at a constant value (usually at zero) or is stepwise changed while changes in membrane voltage are recorded. The latter type of measurements is used for recording action potentials from excitable cells.

Commonly, a single pipette is used in the whole cell variant of the patch clamp technique. Therefore, there is a single common pathway for recording membrane potential and for injection of current. In this common pathway a series resistance (R_s, also called access resistance, R_a) is present. This series resistance is formed by the resistance of the narrow pipette tip, but intracellular organelles or membrane fragments clogging the tip may also contribute to it. When current is injected, a voltage drop will be present across this resistance. Accordingly, there will be a difference between the actual membrane voltage, V_m, of the cell and the voltage present at the wire connecting the patch pipette to the headstage (pipette voltage, V_{pip}). V_{pip} is the voltage measured in current clamp and is also the voltage which is controlled in voltage clamp. Both in current clamp and in voltage clamp, therefore, the actual membrane voltage V_m will deviate from the pipette voltage V_{pip} as soon as current flows through the pipette. The deviation will be proportional to the amplitude of current through the pipette (I_{pip}) and the size of the series resistance R_s:

$$V_m - V_{pip} = I_{pip}.R_s \qquad (1)$$

In voltage clamp experiments this deviation in membrane voltage will result in a time- and voltage-dependent distortion of the recorded currents.

For better control of the actual membrane voltage, most clamp amplifiers have a built-in series resistance compensation circuit. Series resistance compensation can be accomplished by subtracting a voltage, proportional to I_{pip}, from the voltage of the pipette:

$$V_{pip} = V_{com} - I_{pip}.K \qquad (2)$$

In Eq. (2) V_{com} is the desired command voltage and K is a constant, which can be set by a dial on the clamp amplifier. Substituting V_{pip} in Eq. (1) by the expression in Eq. (2) and rearranging results in:

$$V_m = V_{com} - I_{pip}.(K-R_s) \tag{3}$$

From Eq. (3) it is clear that when K equals R_s, V_m will also be equal to V_{com}. In voltage clamp mode, however, this ideal situation never will be reached, because the compensation is a kind of positive feedback. When this positive feedback is too strong, the circuit will oscillate and the cell will be destroyed. The frequency of the oscillations and the degree at which the series resistance can be compensated before the oscillations start depend largely on the frequency response of the clamp amplifier. Most clamp amplifiers therefore have one or more transient cancellation controls or capacity compensation controls which should carefully be adjusted. In practice, R_s can be compensated for by 80%–90% in voltage clamp mode.

In current clamp mode I_{pip} is independent of V_m and V_{pip}. Therefore R_s compensation in current clamp usually is more stable than in voltage clamp and R_s therefore can be completely compensated.

Materials

Equipment   **Electronics.** A custom-made clamp amplifier is used which can be switched between current clamp mode for measuring membrane voltage and action potentials, and voltage clamp mode for the measurement of membrane currents. The amplifier has provisions to compensate manually for pipette capacitance, voltage offsets and series resistance, just as is the case for most commercially available clamp amplifiers. (See list of suppliers in this book.)

The headstage of the clamp amplifier is separate from the main unit so that it can be mounted close to the cell chamber. In the headstage an airtight pipette holder is fitted which connects the input of the headstage to the solution in the pipette, usually via an Ag/AgCl pellet to reduce offset voltages. Pressure or suction can be applied to the lumen of the pipette holder via a connection tube.

Microscope and manipulators. For visual control during manipulation of cells and pipettes an inverted microscope is used (Nikon Diaphot). A cell chamber is mounted on the movable stage of the microscope. The bottom of this chamber rests on a translucent heating plate (Van Ginneken and Giles 1985) which is used to maintain the temperature of the chamber at 35°–37°C. The temperature of the bathing solution is monitored continuously by a thermistor probe. Solution flows into the cell chamber by gravity; the solution level in the cell chamber is kept constant by a suction tube which removes excess solution.

Just above the movable table a fixed stage, holding a micromanipulator (M103, Narishige Scientific Instrument Lab., Japan), is mounted. The headstage of the clamp amplifier is mounted on the manipulator. In this way the headstage, together with pipette holder and pipette, can be moved independently of the cell chamber.

Pulling pipettes. Patch pipettes are pulled on a laboratory-made one-stage pipette puller using borosilicate glass tubing with an outer diameter of 1 mm. The outer diameter used depends on the type of pipette holder that is used. A glass fiber inside the lumen makes filling the pipette easier.

Prior to use, the tips of the pipettes are heat-polished by means of a heating filament of platinum-iridium wire of 50 μm diameter mounted in the focus of a microscope so that the pipette tip can be independently manipulated near to it.

Stimulation and data acquisition. During the experiments membrane voltage and membrane current are stored on a video cassette recorder (Sony Betamax) via a pulse code modulation system (Sony PCM 501). After the experiments these signals are played back and filtered at 1 kHz. The signals are digitized by a 12-bit analog-to-digital converter (National Instruments, NB-MIO-16) at a sample frequency of 2 kHz, and stored on the hard disk of a personal computer (Apple Macintosh Quadra 650). Data are analyzed by a laboratory-designed data acquisition and analysis program. Stimulus protocols and voltage clamp protocols are generated by the same computer during the experiment.

Solutions and drugs

During the experiments the cells are superfused with Tyrode's solution containing in mM: NaCl 140, KCl 5.4, CaCl$_2$ 1.8, MgCl$_2$ 1.0, HEPES 5.0, and glucose 5.5; pH is adjusted to 7.4 with NaOH. The pipette solution contains in mM: KCl 140, HEPES 10; pH is adjusted to 7.2 with KOH.

Procedure

Cell Isolation

Single ventricular myocytes of the rabbit heart are isolated by the same procedures as described in Chap. 10. The cells are stored at room temperature (20°–22 °C) and are used within 8 h.

Not all cells survive this isolation procedure, but there is a clear distinction between dead and intact cells. Dead cells are rounded up; intact myocytes have a smooth surface and show clear cross-striations. These cells are selected for electrophysiological measurements.

Pulling and Filling the Pipettes

Heating and pulling force of the puller are adjusted until the tips of the pipettes have a diameter of 2–4 μm. After pulling, the tips are heat polished. Longer and thus more heavily heat-polished tips may seal more easily, but may result in large access resistances.

The pipette solution is filtered through a syringe-mounted Millipore filter of 0.22 μm pore (Millipore S.A Molsheim, France) before use to ensure that the pipette tip remains clean.

Making a Giga-ohm Seal

Put an aliquot of cells in the cell chamber. Allow the cells to settle for 5 min before superfusion with Tyrode's solution is started. Place a reference ("ground") electrode in the bath solution. This reference electrode preferably consists of an Ag/AgCl pellet connected via a KCl bridge to the bath solution.

Next the pipette is placed in the bath solution. While the pipette is in the bath, a slight positive pressure of approx. 50 mm Hg is applied to the lumen of the pipette to avoid flow of extracellular solution into the pipette, which would dilute the pipette solution at the tip and might contaminate the tip thereby making seals more difficult. The pressure can be applied by mouth.

The resistance of the pipette is determined either by making small voltage steps in voltage clamp mode while current is recorded or, in current clamp mode by injection of current pulses of known amplitude while the voltage displacement is recorded. In either mode Ohm's law is used to calculate the resistance of the pipette, by dividing the voltage displacement by the current change:

$$R_{pip} = \Delta V / \Delta I \tag{4}$$

Prior to making the seal, the measured voltage is set at 0 mV by adjusting the offset voltage or junction potential control. By doing this, the liquid junction potential between pipette and bath solution is also balanced. It should be noted that the junction potential may change when the pipette tip is brought into contact with a solution with another ionic composition, as is the case with the intracellular solution. Depending on the composition of the pipette solution, one may expect changes in liquid junction potential in the order of 15 mV (Barry and Lynch 1991).

When seals are made in voltage clamp mode, one should furthermore carefully adjust the voltage so that current is as close as possible to zero. In that way polarization of Ag/AgCl junctions is avoided. Some clamp amplifiers have a "search" mode, which keeps the long-term average of the current at zero.

Next the pipette tip is brought into contact with the cell. Once firm contact is established, a negative pressure of 50–100 mm Hg is applied. When the circumstances are correct, a giga-ohm seal will develop. In voltage clamp this is reflected by a decrease in the current step, while in current clamp formation of a seal is accompanied by an increase in voltage. A resistance can again be obtained by dividing the change in voltage by the change in current, as in Eq. (4). This calculated resistance is formed by the seal resistance R_{seal}, which is parallel to the resistance of the membrane patch if facing the electrode tip and in series with the pipette resistance R_{pip}:

$$R = R_{pip} + (R_{seal} \cdot R_{patch})/(R_{seal} + R_{patch}) \tag{5}$$

Usually, R_{seal} is much larger than R_{pip}. Also, R_{patch} is usually considered to be much larger than R_{seal}. For practical purposes therefore, the calculated resistance can be considered to be equal to R_{seal}. A seal resistance of several giga-ohms is considered to be acceptable.

From Seal to Cell

Once a seal has formed, one should wait several seconds to allow diffusion of the remainder of the extracellular solution in the tip into the pipette solution. In the mean time the frequency response of the measurements can be improved by compensating for stray capacitance between pipette and bath solution. This is achieved by adjusting the "fast transient" or the "pipette capacitance" compensation until the fast transient is removed (in voltage clamp) or, in current clamp, until the changes in voltage due to the stepwise change in current are fast and without overshoot.

Next, access to the cell interior is obtained by application of a suction pulse to the pipette lumen. When in current clamp mode access to the cell interior is obtained, one will observe a sudden decrease in voltage displacement during the current pulses while the measured voltage will drop to a negative value of −70 to −90 mV. In voltage clamp, one will observe a shift of the membrane current in the outward direction, while the change in current during the voltage step is also increased.

Determining Series Resistance in Current Clamp and Voltage Clamp

Current Clamp   The size of the series resistance is determined in current clamp mode. After having gained access to the cell, switch to current clamp mode. Inject current pulses of 2–5 ms with a rate of 1 Hz. Increase the amplitude of the pulses until action potentials are generated. Action potentials of normal amplitude and duration indicate that the condition of the cell is good. Record on a fast time scale (10–20 ms full scale) the changes in voltage during the current pulses. A fast, stepwise changing component will be superimposed on a slower one. The fast change reflects the instantaneous change in voltage across the series resistance R_s, the slower component reflects the charging of the membrane by the injected current. By dividing the amplitude of the fast stepwise change by the current amplitude, the value of R_s can be calculated:

$$R_{series} = \Delta V / \Delta I \tag{6}$$

Voltage clamp   A value for R_s can also be obtained in voltage clamp. Immediately after a step in voltage, ΔV, membrane voltage V_m will not yet be changed, since it takes some time to charge the membrane capacitance. Then ΔV will also be the voltage across R_s, and R_s can be calculated again by:

$$R_{series} = \Delta V / \Delta I_0 \tag{7}$$

where I_0 is the current amplitude immediately after the step. Note, however, that the membrane will be charged quickly and that current therefore also quickly diminishes. The value of the peak current is easily underestimated when the current is low-pass filtered or digitized with a too low sampling frequency. When the current is fitted with a decaying exponential function, it is, however, possible to extrapolate the fit to the time of the step in order to obtain a reasonable value for for I_0.

Effect of Series Resistance on Current-Voltage Relations

After having determined the series resistance, turn off series resistance compensation, switch to voltage clamp and make a series of voltage clamp steps of 200 ms to various voltages. Record current and voltage. Next, adjust series resistance and repeat the protocol for each step the amplitude of the inward transients and the amplitude of the current at the end of the steps. Construct current-voltage relations (I–V relations) by plotting the measured current amplitudes against the voltage of the steps. Compare the I–V relations obtained with and without compensation.

Results

Estimation of Series Resistance in Current Clamp

Figure 8.1A shows membrane voltage measurements recorded in current clamp. Suprathreshold current pulses of 2.5 nA are applied via the pipette to a rabbit ventricular cell and the voltage response is recorded without and with 100 % series resistance compensation. Without compensation a fast step is seen, superimposed on a slower changing voltage response (closed symbols). The fast step is due to the voltage drop across the series resistance. The series resistance was 100 % compensated (open symbols), whereafter only the slow membrane response remains. Of the action potentials, elicited by the current pulses, only the upstroke and the first 12 ms are shown. From the uncompensated steplike response, a series resistance of 15 MΩ is calculated.

Estimation of Series Resistance in Voltage Clamp

Series resistance and its compensation are also evaluated in voltage clamp. A voltage step from −40 mV to 0–30 mV is made (Fig. 8.1B, trace *a*) while current is recorded. Without series resistance compensation (Fig. 1B, trace *b*) a relatively long-lasting transient with moderate amplitude is seen. From the current immediately after the step (marked ΔI in Fig. 8.1B) a series resistance of 18 MΩ was calculated. In the current trace, recorded with an 80 % compensation of series resistance (Fig. 8.1B trace *c*), there is a faster and larger transient, indicating that the membrane capacitance is charged considerably faster than in trace *b*. When series resistance is compensated to 90 % (Fig. 8.1B trace *d*) , an increasing tendency to oscillate is seen.

A

a ΔI $\downarrow$ ⎡‾‾⎤ 2.5 nA

b -

ΔV ○ ● 50 mV

5 ms

B

a ΔV $\downarrow$

b ΔI $\downarrow$ 1 nA

50 ms

c

d

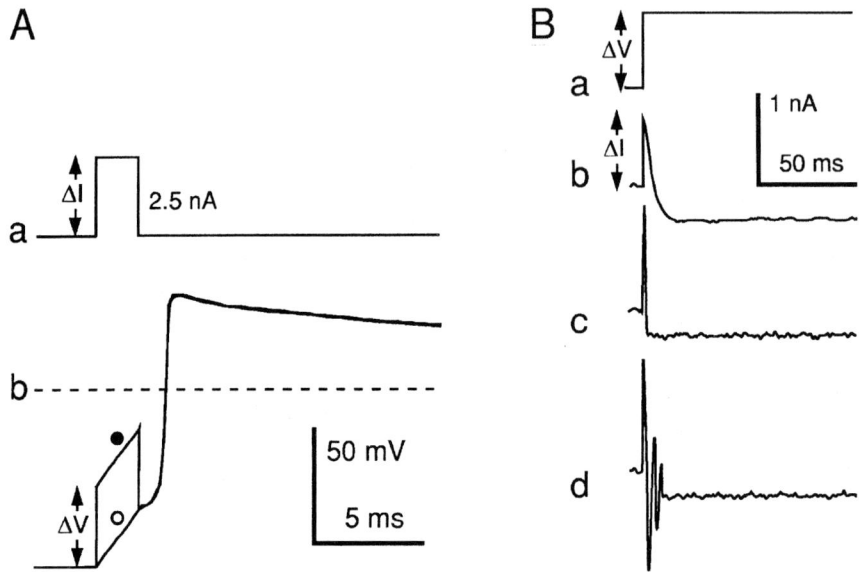

Fig. 8.1A, B. Effect of series resistance compensation in current clamp and voltage clamp. **A** Current clamp. A current step of 2 ms and 3.6 nA (ΔI) is applied to a ventricular cell via a patch pipette in whole cell mode (*a*). Membrane responses without (*closed circles*) and with (*open circles*) 100 % series resistance compensation. Without compensation a step in voltage (ΔV) is present when current is applied (*b*). **B** Voltage clamp. (*a*) The command voltage. A voltage step from a holding voltage of -40 mV to -30 mV (ΔV) is applied. (*b*) Membrane current without series resistance compensation. Note the transient peak current, which charges membrane capacitance. From the initial current series resistance can be calculated. (*c*) Membrane current with 80 % compensation. Note the increase in peak current. (*d*) Membrane current with 90 % compensation shows a dampened oscillation, indicating a tendency to oscillate

Effects of Series Resistance on Time- and Voltage Dependence of Current

In voltage clamp, 200 ms steps are applied to various voltages ranging from -100 mV to $+50$ mV in 10 mV increments. A holding voltage of -40 mV was used to inactivate the fast sodium current. Upon depolarizing steps a transient inward current was activated, which very likely is the calcium current $I_{Ca,L}$. Without series resistance compensation (Fig. 8.2A, traces in *b*), $I_{Ca,L}$ is activated rather slowly and reaches its maximum value at -20 mV (Fig. 8.2B, closed circles), while with compensation (Fig. 8.2A, traces in *c*), the activation is faster and maximum is reached at 0 mV (Fig. 8.2B, open circles). When the current traces in Fig. 8.2A *a* and *b* are compared, it is clear that the time course of the

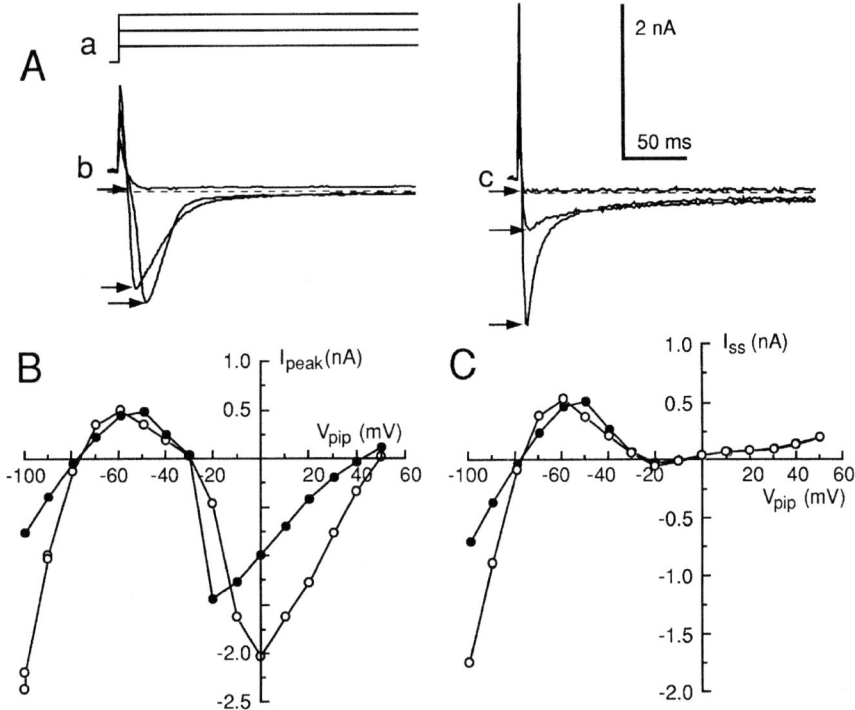

Fig. 8.2A–C. A Effect of series resistance compensation on time- and voltage dependence of currents. In voltage clamp, 200 ms steps are applied from a holding voltage of -40 mV to voltages ranging from -100 mV to $+50$ mV with 10 mV increments. (*a*) Only the voltage steps to -30, -20 and -10 mV are shown. (*b*) Current traces during the steps in (*a*), recorded without series resistance compensation; (*c*) same as (*b*), but recorded with 80 % compensation. inward peaks in (*b*) and (*c*) are caused by calcium current, $I_{Ca,L}$. See text for further explanation. **B** Peak inward currents plotted versus voltage, without (*closed circles*) and with (*open circles*) 80 % series resistance compensation. Peak inward currents (*arrows* in **A**) were plotted versus pipette voltage. Note the more negative voltage of the maximum peak current (-20 mV, respectively, 0 mV) when series resistance is compensated. **C** Currents at the end of the 200 ms voltage steps plotted versus voltage, recorded without (*closed circles*) and with (*open circles*) 80 % series resistance compensation. Note the increase in slope of the current-voltage relationship when series resistance is compensated

currents is affected by the series resistance. Also the voltage dependence of currents is affected by series resistance. The more steep voltage dependence of activation of $I_{Ca,L}$ between -30 and -20 mV, seen without compensation, is an artifact caused by poor control of the actual membrane voltage. The maximum at -20 mV is recorded at a V_m which is more positive than V_{pip}. The controlled voltage-comparable artifact is seen in Fig. 8.2C, where current at the end of the 200 ms steps is plotted versus voltage, again without (closed circles) and with 80 % series resistance compensation. I–V relations are dominated by the characteristics of the inward rectifying K$^+$-current (I_{K1}). They show a steep slope negative to −80 mV, while an outward "hump" in current between −80 mV and −20 mV is present. At more positive values I_{K1} hardly carries any current due to its inward rectifying properties. Note that the presence of a series resistance decreases the slope of the curve negative to −80 mV, while the negative slope between −50 mV and −20 mV is increased.

From these experiments it can be concluded that the presence of a series resistance results in a distortion of the time course of currents and in a distortion of current-voltage relationships. The distortions are proportional to the size of the series resistance and will therefore be minimized when series resistance is compensated. Since series resistance cannot be completely compensated in voltage clamp, these distortions will always be present to some extent.

References

Barry PH and Lunch JW (1991): Liquid junction potentials and small cell effects in patch-clamp analysis. J Membrane Biol 121:101–117

Neher E, Sakmann B (1976) Single channel currents recorded from membrane of denervated frog muscle fibres. Nature 260: 779–802

Hamill OP, Marty A, Neher E, Sakmann B, Sigworth FJ (1981) Improved patch-clamp techniques for high-resolution current recording from cells and cell-free membranes patches. Plugers Arch 391: 85–100

Hodgkin AL, Huxley AF (1952) A quantative description of membrane current and its application to conduction and excitation in nerve. J Physiol 117: 500–544

Van Ginneken ACG, Giles WR (1985) A transient outward current in isolated cells from the crista terminalis of rabbit heart. J Physiol 368: 243–264

Preparation of Patchable Plant Cell Protoplasts and a Procedure for the Improvement of Gigaseal Formation

Sake A. Vogelzang and Margaretha Blom-Zandstra

Background

Patch-Clamping Plant Protoplasts

The patch-clamp technique is generally used to study ion transport characteristics of plasma membranes or organelle membranes of individual cells. Application of this technique to plant cells is much more complicated than to animal cells (Hamill et al. 1981) due to the presence of a cell wall, which has to be removed prior to the patch-clamp experiment. After removal of the cell wall a glass pipette is placed against the (plasma) membrane of the isolated protoplasts. Suction applied to the pipette interior is required for the formation of a very tight seal between the membrane and the glass tip, thus realizing a very high electrical resistance (gigaseal). In general, obtaining a gigaseal is more difficult for plant cells than for animal cells. The physical nature of gigaseal formation is still unknown. However, the seal success rate can be increased if a number of factors in the isolation and sealing procedure are optimized (Elzenga et al. 1991; Vogelzang and Prins 1992; Blom-Zandstra et al. 1995)

Here we present a general isolation procedure and discuss the factors involved in the seal formation.

Outline

The following short protocol will be used:

Cut a small piece of plant material (e.g., root, leaf or pitch) and incubate it in an enzyme mixture (check osmolarity) in a small petridish at room temperature. Cells are slightly plasmolysed and the cell wall is consumed. After 30–120 min the root or leaf should almost fall apart. Then, a pair of tweezers is used to pick up the flabby material and dip it in the measuring chamber several times. With the microscope one can observe whether protoplasts fall to the bottom of the dish. After carefully rinsing the cells with an enzyme-free bath solution the patch-clamp experiment can be performed.

Materials

- Enzyme solution: 1 % (w/v) cellulase (R10 Onozuka, Serva) and 0.2 % (w/v) macerozyme (R10 Onozuka, Serva) in 10 mM MES (pH 5.5 with KOH), 2 mM $MgCl_2$, 2 mM $CaCl_2$, 10 mM sucrose, the osmotic value is adjusted with mannitol or sorbitol (see below). Make a 10 ml stock solution of the enzyme medium and store it in 0.5 ml dishes at $-20\,°C$. These dishes can be used for a period of 6 weeks or more.

Note. This mixture is useful for most tissues, however an adjustment of the enzyme composition may be required for some plant tissues, e.g., for root hair cells, cellulase RS is used instead of cellulase R10. Also, 1 % macerozym R10, 0.02 % (w/v) pectolyase, and 1 % (w/v) cellulysin are added (Gassmann and Schroeder 1994).

Note. Addition of sucrose helps to sustain the viability of the protoplasts

Note. Addition of magnesium and calcium are important for membrane stability.
- Bath solution: 10 mM MES (pH 5.5 with TRIS), 10–150 mM KCl (concentration depends on the desired K^+ equilibrium potential), 10 mM sucrose, 2 mM $MgCl_2$, 2 mM $CaCl_2$, replenished with mannitol (or sorbitol) to an osmolarity of 25 mOsmol/kg under the plasmolytic osmotic value (see below).
- Pipette solution suitable for whole cell recording: 1 mM HEPES (pH 7.2 with BTP), 10–150 mM KCl (concentration depends on the desired K^+ equilibrium potential), 2 mM $MgCl_2$, 2 mM K_4BAPTA (Sigma), 5 mM MgATP, 0.1 mM $CaCl_2$, replenished with mannitol (or sorbitol) to an osmolarity 10 mOsmol/kg greater than the bath solution (see below).

Note. Addition of MgATP to the pipette solution helps to maintain the stability of the protoplast in the whole cell configuration.

Note. Use of K_4BAPTA as Ca^{2+} chelator is recommended. The Ca^{2+}/Mg^{2+} binding ratio is higher for K_4BAPTA than for EGTA or EDTA.

The protoplast isolation procedure can be used for a diversity of plant species (both monocotyledonous and dicotyledonous) and different plant tissues (leaf, root or pitch). Root material preferably should be obtained from plants grown on hydroculture. In addition, it requires less than 100 µg plant material to obtain enough protoplasts for patch-clamp studies. Use of young parts of plants is preferred. In general, stress (water stress, nutrient deficiency, use of herbicides etc.) during plant growth should be avoided.

Plant material

Electrodes can be prepared with different types of borosilicate glass: thick- and thin-walled, with or without filament. Thin-walled pipettes are more easily pulled and shaped than thick-walled pipettes, but have a higher degree of background noise. The presence of a filament in the pipette is very helpful because it prevents the formation of air bubbles in the tip of the pipette and because it makes filling the pipette easier. We recommend the use of Sylgard to reduce background noise and capacitance, especially when thin-walled glass is used.

Patch-clamp pipettes

Tip. We suggest using GC150–15 (Clark Electromedical Instruments, Reading, UK) for the use of thick walled pipettes without filament or TW150F-4 (World Precision Instruments, Florida USA) for the use of thin-walled pipettes with filament.

Procedure

Isolation of Protoplasts

Here, a protoplast isolation procedure is described in which centrifugation steps are omitted to restrict mechanical stress. Both mechanical stress factors and washing of the protoplasts are minimized. Protoplasts can be isolated from different monocotyledonous and dicotyledonous plant species and different tissues (leaves and roots). The duration of the isolation procedure is flexible: either overnight ("long and cold") or 30–90 min ("short and warm"). This procedure requires very little plant material.

For a correct choice of the osmotic value of the solutions, the plasmolytic threshold level of the tissue has to be determined. This can be done by a rough estimate of the osmotic values of expressed cell sap using an osmometer.

Determination of osmotic value

Another possibility with intact plant material is to determine the osmotic value at which 50% of the cells plasmolyse. Use the resulting value to adjust the enzyme, bath and pipette solutions with mannitol or sorbitol.

Isolation procedure

1. The abaxial surface of leaf material must gently be ground with carborundum (SiC) to remove the lower epidermis. Root or pitch tissue does not require SiC treatment.

2. Plant material is cut in pieces of ca. 5 mm × 5 mm (leaf) or 5–10 mm (root) or slices of 1 mm thickness (pitch). The material is then placed in the enzyme medium (16 h, 12 °C or 30–90 min at room temperature).

3. After the incubation period, pieces of tissue are picked up with a pair of tweezers, transferred directly to the patch-clamp dish and gently shaken through the bath solution (0.5–2 ml) to release protoplasts. In case of root material, the release of protoplasts generally requires additional mechanical pressure applied with a pair of tweezers. Small spots are visible on the bottom of the dish, containing at least 20–200 protoplasts. It takes 2 min for the protoplasts to adhere to the bottom of the dish.

4. Gently substitute half of the bath solution with fresh bath solution (four times) to wash the protoplasts and remove debris.

Note. Plant tissue, left over from the incubation but not used for the isolation of protoplasts immediately, can be held in the enzyme mixture and stored on ice. From this plant material, fresh viable protoplasts can be obtained over a period of at least 8 h.

Optimizing Success of the Seal

Here we discuss some factors involved in the formation of gigaseals. The probability of obtaining a gigaseal can vary strongly within a batch of isolated protoplasts. Clearly, if the seal success rate of a batch is low one can either isolate a new batch or compensate for the low success rate by increasing the number of sealing attempts. In addition, the seal success probability can be optimized if special attention is paid to the choice of the pipette, its shape, the angle of contact between pipette and membrane, the choice of the type of protoplast and the suction protocol.

Different types of glass capillaries are commercially available (see below). The pipettes are made with a pipette puller in two or three steps.

1. The diameter must be ca. 1 µm, corresponding to a pipette resistance of 5–10 MΩ (at the given bath and pipette ion concentrations). Especially for single channel recordings, the noise and capacity of the pipettes have to be minimized. Therefore, the part of the pipette in contact with the bath solution has to be coated with Sylgard starting close to the tip (closer than 250 µm) and continuing up the pipette well past the beginning of its taper.

2. A rigid metal wire (diameter 0.5 mm) is used to carefully apply Sylgard to the tip of the pipette. The wire is bent 180 °C at its end so that it can hold a small droplet. Make sure that the Sylgard is uniformly distributed, e.g., by rotating the pipette around its axis during the coating procedure. Finally, the tip is fire polished. Extreme fire polishing can significantly reduce the tip diameter and therefore increase the pipette resistance.

3. Fill the pipette with solution and remove any air bubbles that may occur in the tip. Often the same setup used for fire polishing the tip is used for coating the pipette with Sylgard.

   Tip 3. For the whole cell configuration, a large tip diameter is recommended (corresponding to a pipette resistance of 5 MΩ). This also has the advantage that the exchange of intracellular solution with the pipette solution is maximal. In addition, the access resistance is low permitting a reliable clamp of the voltage in the whole cell configuration. However, the seal success probability in general is reduced if large tip pipettes are used.

Contact between the pipette and the protoplasts is only possible with cells that are adhered to the bottom of the dish. Therefore, the glass should be clean (use new glass or wash the dish with methanol). Adherence of the protoplasts can be tested by gently moving the dish while observing their response under the microscope. The protoplast selected for the patch-clamp experiment (diameter 20–30 µm) should be free from other cells and of debris. In general, the highest seal success rate can be obtained by using viable protoplasts, which can be identified by cytoplasmic streaming.

Gigaseal formation

1. Apply positive pressure to the pipette interior (10 mbar) before the pipette is placed in the bath solution. When the pipette is brought into the solution, the pipette offset must be adjusted to zero. The subsequent procedure is depicted in Fig. 9.1.

2. Place the tip of the pipette approximately 100 μm above the protoplast. Focus on the top of the protoplast to determine level 1 (100 %), then focus on the centre of the protoplast to determine level 3 (0 %). The height for perpendicular contact between pipette and protoplast is at 70 % (level two), assuming that the pipette approaches the protoplast at an angle of 45°.

3. Focus on that level and place the pipette in position A shown in Fig. 9.2. Make sure that the tip is in focus and exactly placed on the line. Pull the pipette away from the line (situation B) and move the protoplast along the line in the patch-clamp position (situation C) and finally, place the pipette against the protoplast (situation D). The contact causes a slight inward dent of the membrane. At the same time the

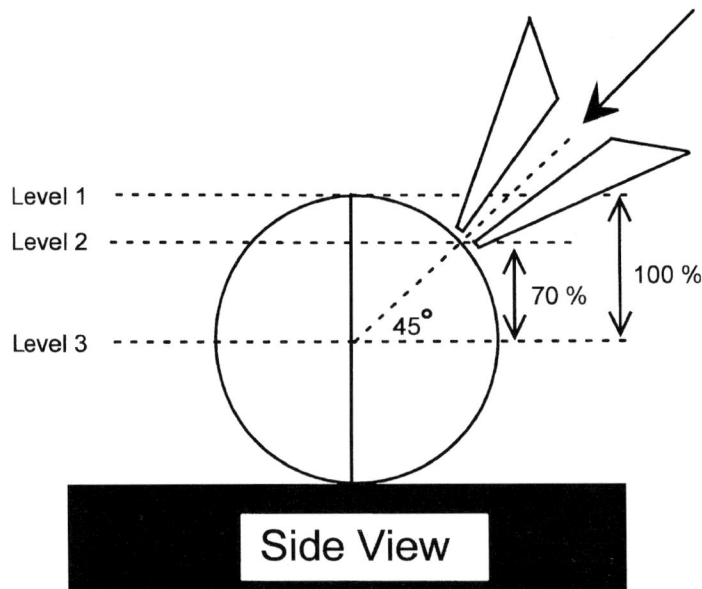

Fig. 1. Side view of a protoplast, adhered to the bottom of the dish, and the tip of a pipette approaching the cell at an angle of 45°. The *dotted lines* represent the three levels on which the experimenter should focus. The lower level (*level 3*) and the upper level (*level 1*) correspond to a level of 0 % and 100 %, respectively. At *level 2* (70 %) the tip and the plasma membrane should make (perpendicular) contact

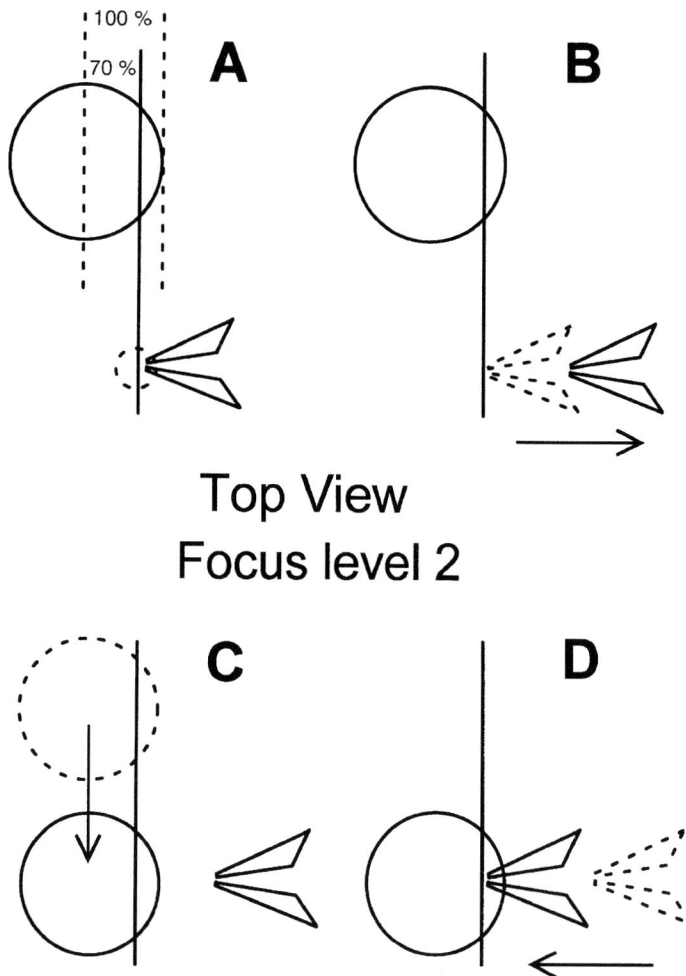

Top View
Focus level 2

Fig. 2A–D. Top view of four different configurations in which the protoplast and the pipette are observed at focus level 2 (Fig. 1). The tip of the pipette is moved downward in the bath solution until it is in focus. **A** The tip is then placed at the straight line under the protoplast. This 70 % line is used to estimate at which point the tip should make contact with the plasma membrane. **B** The tip is moved to the right and **C** the protoplast is placed in position by moving the dish in the downward direction. **D** Subsequently, the pipette is moved back in its original position and touches the plasma membrane

pipette resistance will increase. This can be monitored on the oscilloscope by measuring the decrease of the current in response to 5 or 10 mV step-changes. The increase of the pipette resistance can also be observed by a rise of the baseline if a negative pipette potential (-5 to -30 mV) is applied.

4. After making contact the pressure inside the pipette should be reversed, resulting in a value of -5 to -20 mbar. Subsequently the pressure is slowly increased (more negative) in response to changes in pipette resistance until a gigaseal is formed. This process can easily take 5–10 min. However, on some occasions a gigaseal is formed within seconds. A critical point is reached when the pipette resistance is approximately 100 MΩ. The suction pressure can now be increased more rapidly and the pipette resistance should increase to 10 GΩ or more. In general, a suction pressure of -30 to -50 mbar is sufficient.

5. The formation of a gigaseal requires a certain amount of pressure (P_{seal}) imposed to the pipette interior. The resulting suction force depends on the suction pressure and the diameter of the tip of the pipette. Clearly, it is the amount of suction force rather than the amount of suction pressure that determines the intensity of contact between glass and membrane. Consequently, a decrease of the diameter of the tip will result in an increase of P_{seal}.

Comments and Troubleshooting

- Keep the sealing protocol simple and fast.

- Whole cell measurements must be preceded by the formation of a gigaseal. During the suction protocol it may happen that a whole cell is obtained before the pipette resistance has reached the gigaohm range. This may cause an unreliable clamp of the voltage and the seal resistance may become voltage-dependent. If the pipette resistance is low, it is advised not to waste time: prepare a new pipette and try again.

- Wash the bath solution at regular intervals (60 min) to compensate for evaporation of the water, which causes an increase of its osmotic value.

- If the pipette resistance if low (5 MΩ) and the membrane is flabby then the suction protocol may be accompanied by the formation of vesicles in the pipette. Do not waste time on such an experiment, even if a gigaseal is obtained the configuration will remain obscure and the data useless.

- An increase of the pipette resistance and the calcium concentration (both of bath or pipette solution) may improve the seal success rate.

Patch-Clamp Configurations

There are four different configurations: cell-attached patch (CAP), inside out patch (IOP), outside out patch (OOP) and whole cell (WC). The calcium concentration is critical for the obtainment of a particular configuration. A high calcium concentration in the pipette (>1 mM) favors the stability of the membrane while a low concentration (0.1 mM) facilitates the formation of a whole cell because the patch is easily ruptured by suction. Although, all configurations start with the cell attached configuration (CAP), we advise a low calcium concentration in the pipette if a WC configuration is required. To compensate for the instability of the membrane the calcium concentration in the bath solution can be increased (>1 mM).

- If a WC is required, we advise rupturing the membrane within 1 min after gigaseal formation. Attempts to rupture the patch at a later time are usually very difficult. Apparently the patch has become very stable.

- Rupture of the plasma membrane is more successful with application of suction than with application of an electrical high-voltage pulse. The latter often causes loss of the cell or of the gigaseal. This is in contrast to electrical high-voltage pulses applied to vacuoles, which often result in a successful rupture of the patch.

- As gigaseal formation with plant protoplasts can be very difficult for inexperienced experimenters, we recommend that they practice first on material that is known for its excellent seal success rate, e.g., Chara droplets (Lühring 1986). Helpful information regarding general problems with the patch-clamp technique can be obtained from: "Introduction to single channel recording by microelectrode techniques" by Ogden and Stanfield (1988).

References

Blom-Zandstra M, Koot H, Hattum J van and Vogelzang SA (1995) Isolation of protoplasts for patch-clamp experiments: an improved method requiring minimal amounts of adult leaf or root tissue from monocotyledonous or dicotyledonous plants. Protoplasma 185: 1–6

Elzenga JTM, Keller CP, Van Volkenburgh E (1991) Patch-clamping protoplasts from vascular plants. Method for quick isolation of protoplasts having a high success rate of gigaseal formation. Plant Physiol 97: 1573–1575

Gassmann W, Schroeder JI (1994) Inward-rectifying K^+ channels in root hairs of wheat. A mechanism for aluminum-sensitive low-affinity K^+ uptake and membrane potential control. Plant Physiol. 105: 1399–1408

Hamill O, Marty A, Neher E, Sakmann B, Sigworth FJ (1981) Improved patch-clamp techniques for high-resolution current recording from cells and cell free membrane patches. Pflügers Archiv 391: 85–100

Lühring H (1986) Recording of single K^+ channels in the membrane of cytoplasmic drop *Chara australis*. Protoplasma 133: 19–28

Ogden DC, Stanfield PR (1988) Introduction to single channel recording. In: Standen NB, Gray PTA, Whitaker MJ (eds) Microelectrode techniques. The Plymouth workshop handbook, IIth Impression. The Company of Biologists, Cambridge, pp 63–81

Vogelzang SA, Prins HBA (1992) Plasmalemma patch-clamp experiments in plant root cells: procedure for fast isolation of protoplasts with minimal exposure to cell wall degrading enzymes. Protoplasma 171: 104–109

Ion Channel
and Membrane Potential Measurements Using
the Patch-Clamp Technique

Single Channel Measurements

Recording and Analysis of ATP-Sensitive Potassium Channels in Inside-Out Patches of Rabbit Ventricular Myocytes

Marieke W. Veldkamp

Background

An important method to study the behavior of ion channels is the patch-clamp technique, developed by Neher and Sakmann (1976). This method allows clamping of a voltage difference across the cell membrane and, at the same time, measurement of the currents that flow through the ion channels as result of the applied voltage difference. With this technique one can either measure the sum of currents flowing through the channels in the entire cell membrane, referred to as whole-cell configuration (Fig. 10.1A), or one can measure the very small currents through single channels in a patch of cell membrane (Fig. 10.1B–D). In the cell-attached configuration this patch of membrane is still attached to the intact cell (Fig. 10.1B), in the inside-out configuration and outside-out configuration the patch of membrane is excised from the cell (Fig. 10.1C, D).

Correct kinetic analysis of ionic currents requires complete isolation of the current that is investigated. In the whole-cell configuration this sometimes proves difficult. The presence of other current types activated in the same voltage range cannot always be excluded and may distort the measurements. A frequently applied method to isolate a current component in the whole-cell configuration is the use of blocking agents. This requires, however, that the blocking action of the agent is specific for one channel type and voltage-independent. Unfortunately, both conditions can often not be satisfied.

At the single channel level, different types of channels can be distinguished from one another by their single channel conductance, channel open probability and some other properties. Moreover, the pipette solution can be composed in such a way that it favors the detection of just one type of single channel (current). Therefore, single channel measurements may provide more accurate analysis of channel kinetics. Another advantage of single channel measurements is that the ionic composition

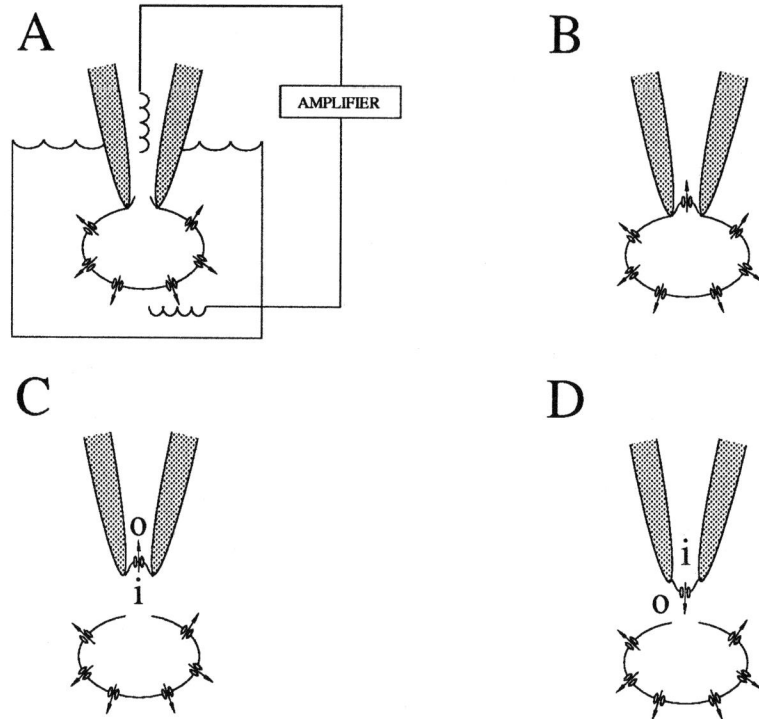

Fig. 10.1A–C. Four different configurations of the patch-clamp technique. **A** Whole-cell configuration. **B** Cell-attached patch configuration. **C** Inside-out patch configuration. **D** Outside-out configuration. The letters *i* and *o* refer to intracellular and extracellular side of the membrane, respectively

of intra- and extracellular solutions, in the inside-out and outside-out configuration, can be precisely defined and quickly and easily changed.

In this chapter the sequence of steps in a single channel experiment is described and basic single channel analysis will be presented. For this purpose I show activity of single ATP-regulated potassium (K.ATP) channels in inside-out patches of ventricular myocytes from rabbit heart. For more advanced single channel analysis, the reader is referred to the textbook by Sakmann and Neher(1983).

Materials

Solutions for cell isolation
– Normal Tyrode's solution (in mM): NaCl 140, KCl 5.4, $CaCl_2$ 1.8, $MgCl_2$ 1.0, glucose 5.5, HEPES 5.0, buffered with NaOH at pH 7.4.

- Nominally Ca^{2+}-free Tyrode's solution (in mM): NaCl 140, KCl 5.4, $MgCl_2$ 0.5, KH_2PO_4 1.2, glucose 5.5, HEPES 5.0, buffered with NaOH at pH 6.9.
- Kraftbrühe (KB) solution (in mM): KCl 85, K_2HPO_4 30, $MgSO_4$ 5.0, glucose 20, pyruvic acid 5.0, creatine 5.0, taurine 5.0, EGTA 0.5, β-hydroxybutyric acid 5.0, succinic acid 5.0, Na_2ATP 2.0, along with 50 g/liter polyvinyl pyrrolidone-40, buffered with KOH at pH 6.9.

Note. The solutions for cell isolation can be stored as stock solutions. It is advisable, however, to leave glucose out of the stock solution in view of fungal and bacterial contamination. Glucose can be added prior to use.

- Pipette solution: 140 mM KCl, 10 mM HEPES, pH adjusted to 7.4 with KOH. Pipette solutions can be stored in small quantities (5 ml) at $-20\,°C$.
- Bath solution 1: 140 mM KCl, 10 HEPES, pH adjusted to 7.2 with KOH.
- Bath solution 2: 140 mM KCl, 10 HEPES, 5 mM MgATP, pH adjusted to 7.2 with KOH.

Solutions for single channel measurements

Note. The small amplitude of single channel currents makes it often hard to distinguish them from noise. Therefore, to increase single channel currents, a high concentration of the permeant ion is often used. The absence (or low concentration) of other ions prevents the detection of currents through other channel types (for instance Na^+ or Ca^{2+} channels).

Note. To improve the success rate of seal formation, the pipette solution should be filtered with a 0.22 µm filter in order to remove small particles.

Patch pipettes are generally made of either soda glass, which is soft, or from borosilicate glass, which is hard. Both types of glass can be used for single channel recordings, although soft glass is noisier than hard glass. In our laboratory we use thin-walled borosilicate glass tubes with an inner diameter of 0.74 mm and an outer diameter of 0.97 mm. The lumen of the glass tube is provided with a filament which facilitates filling with pipette solution and prevents formation of air bubbles.

Patch-clamp pipettes

Note. To reduce noise and capacitance, the taper of the pipette near the tip may be coated with Sylgard (Dow Corning), a nonconductive material that has minimal slow charge movement. The coating of the pipette taper can best be done under low magnification (10x) while the pipette is fixed in a holder which can be rotated. An iron wire, which is hooked at the end, is dipped in Sylgard so that a droplet of Sylgard remains in the hook. The hook with the droplet is placed around the taper of the pipette and the pipette is then rotated so that Sylgard is fixed all round the taper. Subsequently, the pipette taper with Sylgard is briefly placed in a heating coil to harden the Sylgard.

Procedure

Preparation of Single Ventricular Myocytes

Single ventricular myocytes are isolated from rabbit hearts by an enzymatic dissociation procedure (Veldkamp et al. 1993). Briefly, New Zealand white rabbits (1–2 kg) are anesthetized with 1 ml/kg bodyweight Hypnorm i.m. (10 mg/ml fluanisone +0.315 mg/ml fentanyl citrate; Janssen Pharmaceutics, Tilburg, The Netherlands). The heart is quickly taken out, mounted on a Langendorff perfusion apparatus, and retrogradely perfused with solutions in the following sequence:

1. Normal Tyrode's solution is used for 5 min to wash out the blood.

2. Nominally Ca^{2+}-free Tyrode's solution is used for 10 min.

3. The enzyme solution is used for 20 min and consists of nominally Ca^{2+}-free Tyrode's solution to which 1.43 mg/ml collagenase (type B, Boehringer Mannheim, Germany) is added. During the last 5 min of this period 0.1 mg/ml protease (type XIV, Sigma, St. Louis, USA) is added. The enzyme solution is recirculated.

4. KB solution is used to wash out the enzyme (5 min). All solutions are saturated with O_2. Temperature is maintained at 37 °C. The left and right ventricular wall are cut into pieces and gently agitated in a small beaker with KB solution to obtain single cells. These are kept in KB solution and stored at 4 °C for later use on the same day.

Note. Experiments should be performed only on myocytes that are rod-shaped and show clear cross-striation (see Fig. 10.2).

Note. The use of laboratory animals is strictly regulated. An experimental animal committee should agree with the protocols and number of animals that will be used. Furthermore, if you are not used to operating on animals, do not try it yourself but ask an experienced person to do so.

Note. To keep the period that the heart is not perfused as short as possible, the heart should be taken out quickly in the following way. After

Fig. 10.2. Photograph of a rod-shaped, ventricular myocyte with clear cross-striations

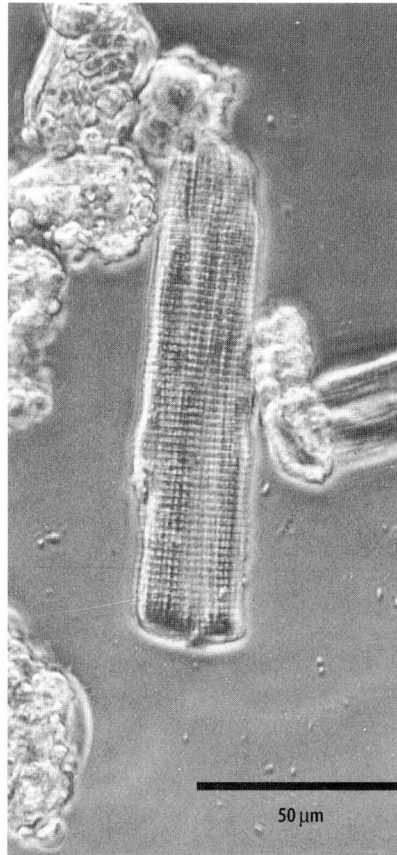

50 µm

opening the thorax, the heart is lifted by the apex with a pair of twee-zers and the vena cava inferior is squeezed with a clip at the height of the diaphragm. Next the vena cava inferior is cut with a pair of scis-sors and subsequently, at the rostral side, also the aorta and the vena cava superior are cut. Now the heart can be taken out. Do not take too much time to trim the heart while still inside the animal. Removal of the thymus, lungs and esophagus can be done when the heart is mounted on the Langendorff perfusion apparatus.

Preparation of Patch Pipettes

Pipettes can be pulled in one or more stages. In our laboratory pipettes are pulled with a vertical one-stage patch pipette puller and have a tip diameter of about $1-2\,\mu$m. After pulling, the tip of the pipet is heat polished. Heating should be such that there is a clear deformation of the tip. Next the pipette is filled with pipette solution and placed in the bath solution in the recording chamber/tissue bath. Under these conditions the pipette resistance is $3-5\,\text{M}\Omega$

Gigaseal Formation

The pipette is placed in the bath solution while applying a positive pressure to the patch pipette. The positive pressure causes an outward flow of pipette solution, preventing the attachment of small particles to the tip of the pipette, which may hamper the formation of a seal. Next, it is important to first adjust the voltage difference between the patch pipette and bath to 0 mV. Subsequently, the pipette is carefully maneuvred into the close vicinity of the cell and the positive pressure is released. The tip of the pipette is gently placed on the surface of the cell membrane so that a small indentation arises. A giga-ohm seal, the tight connection between the tip of the pipette and the cell membrane, can now be established by applying a slight negative pressure to the patch pipette. Sometimes the contact between the tip of the pipette and the cell membrane is already sufficient for immediate seal formation and application of negative pressure is not necessary.

Monitoring gigaseal formation The formation of a gigaseal can be monitored in the voltage-clamp mode by measuring the changes in current in response to 1 mV rectangular pulses. When the patch pipette is placed in the bath solution the resistance between pipette and bath (solution) is very low and consequently there are large current changes in response to the 1 mV pulses. Upon placement of the pipette on the cell membrane the resistance between the pipette and the bath (solution) is slightly increased, seen as an attenuation of the current changes in response to the voltage pulses. The moment a gigaseal is formed, the resistance between pipette and bath (solution) becomes very high, resulting in a large drop in current change.

Obtaining the Inside-Out Configuration

Upon formation of a gigaseal, the cell-attached patch configuration is established. To obtain the inside-out configuration the pipette is lifted.

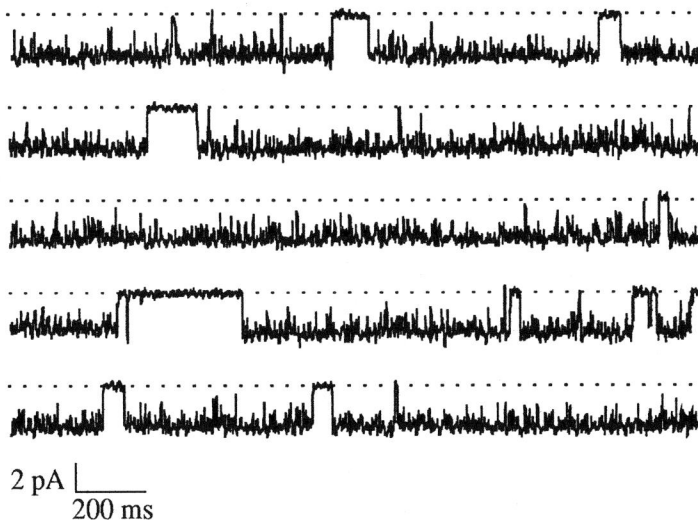

Fig. 10.3. Example of K.ATP channel activity recorded at −40 mV from an inside-out patch. *Dotted line* indicates the level at which all channels are closed. Channel openings are seen as downward deflections

Since cardiac ventricular myocytes are not tightly adhered to the bottom, they will stick to the tip of the pipette. By tapping gently on the stage of the microscope, and sometimes somewhat more vigorously, the cell is released from the pipette leaving behind a patch of membrane. The inside-out patch configuration is now established. K.ATP channel activity can be observed when the pipette is clamped at, e.g., −40 mV. Figure 10.3 shows examples of current traces with K.ATP channel activity characterized by openings occurring in bursts, which are separated by closures of relatively short duration.

Note. The inside-out configuration can be obtained in several ways. Instead of tapping against the stage of the microscope to release the cell from the pipette, one can also lift the pipette for a very short period out of the bath solution. Although this is an appropriate way to obtain an inside-out patch, it seems not to be that successful with cardiac ventricular myocytes. For cells that are tightly adhered to the bottom it suffices to simply lift the patch electrode.

Note. It may happen that a vesicle is formed instead of an inside-out patch. One should take into account that this is the case when:

- Channel activity is absent after formation of an inside-out patch. The density of ATP-regulated potassium (K.ATP) channels in rabbit ventricular myocytes is very high and normally every inside-out patch contains one or more K.ATP channels.
- The shape of the unitary current is decaying instead of rectangular. A vesicle can be converted to an inside-out patch by lifting the pipette for a very short period out of the bath solution.

Note. The transmembrane potential is defined as the potential at the intracellular side minus the potential at the extracellular side. Consequently, the patch membrane potential in the inside-out configuration is equal, but opposite, to the patch pipette potential.

Comments

ATP-Regulated Potassium Channels

The K.ATP channel was first identified in heart by Noma (1983). The dominant characteristic of the K.ATP channel is that it is closed by the application of ATP to the intracellular side of the channel. The concentration at which half-maximal inhibition occurs lies at approx. 0.1 mM ATP. At low concentrations of ATP channel activity is characterized by bursts of channel openings separated by relatively short closed periods. Since in heart cells the physiological ATP concentration is in the millimolar range, the channels are normally closed. When cardiac metabolism is compromised and the intracellular ATP concentration ($[ATP]_i$) is reduced, these channels may be activated. The interested reader is referred to three excellent reviews about the properties and functional role of K.ATP channels in heart and other tissues(Ashcroft 1988; Nichols and Lederer 1991; de Weille 1992)

K.ATP channels have a number of specific properties by which they can be identified. In the following, it will be explained how these properties can be determined. This will be illustrated with the help of some of the figures.

As mentioned before an important characteristic of the K.ATP channel is that it is closed by the application of ATP to the intracellular side of the channel. Figure 10.4 shows a typical example of such an experiment on an inside-out patch containing 12 K.ATP channels. In the absence of ATP the channels have a high open probability, that is, a major part of time the channel is in the open state. Shortly after ATP is applied to the bath solution, the channels rapidly close. Channel activity reappears some time after washout of ATP.

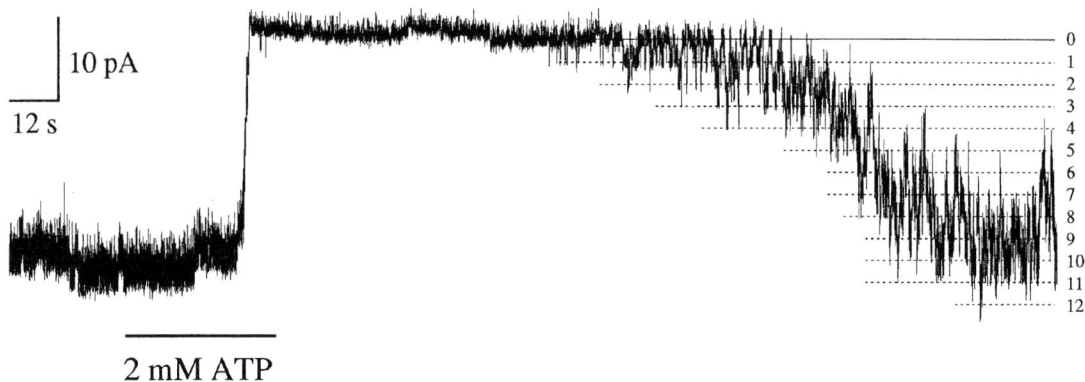

Fig. 10.4. K.ATP channel inhibition by intracellular ATP. K.ATP channel activity recorded from an inside-out patch at a membrane potential of +40 mV. The patch contained at least 12 channels. Channel openings are seen as downward deflections. The current level at which all channels are closed is indicated by a *solid line*, the unitary current levels of the K.ATP channel are indicated by *dotted lines*

- It may happen that channel activity gradually declines after excision of the patch and eventually completely disappears, a process called "run down." This process is enhanced when the intracellular side of the membrane is exposed to divalent cations. Reactivation of channel opening can be induced by short exposure of the intracellular side of the channel to MgATP. Shortly after washout of MgATP spontaneous channel openings will reappear.

- The conventional way to display current is such that inward currents are shown as downward deflections and outward current are shown as upward deflections. Inward current is defined as positive current flowing from the intracellular side to the extracellular side of the cell membrane, and outward current is defined as positive current flowing from the extracellular side to the intracellular side of the cell membrane.

Single channel conductance

The single channel conductance of the K.ATP channel is relatively large compared to other K^+ channels in the heart and is dependent on the extracellular K^+ concentration. When exposed to symmetrical concentrations of 150 mM K^+, the channel has a conductance of about 70 pS. The single channel conductance can be determined by measuring the single channel current amplitude as a function of membrane potential (Fig. 10.5A). Single channel current amplitudes can be obtained from amplitude histograms which are constructed of several current traces (Fig. 10.5B, C). In an ampli-

A

+40 mV

+30 mV

+20 mV

0 mV

-20 mV

-30 mV

-40 mV

-50 mV

-60 mV

5 pA

250 ms

B

V_m = -20 mV
i = -1.46 pA
P_O = 0.11

O C

N
4000
3000
2000
1000
0

-12 pA   -10   -8   -6   -4   -2   0

C

V_m = -50 mV
i = -3.53 pA
P_O = 0.42

O   C

N
1600
1200
800
400
0

12 pA   -10   -8   -6   -4   -2   0

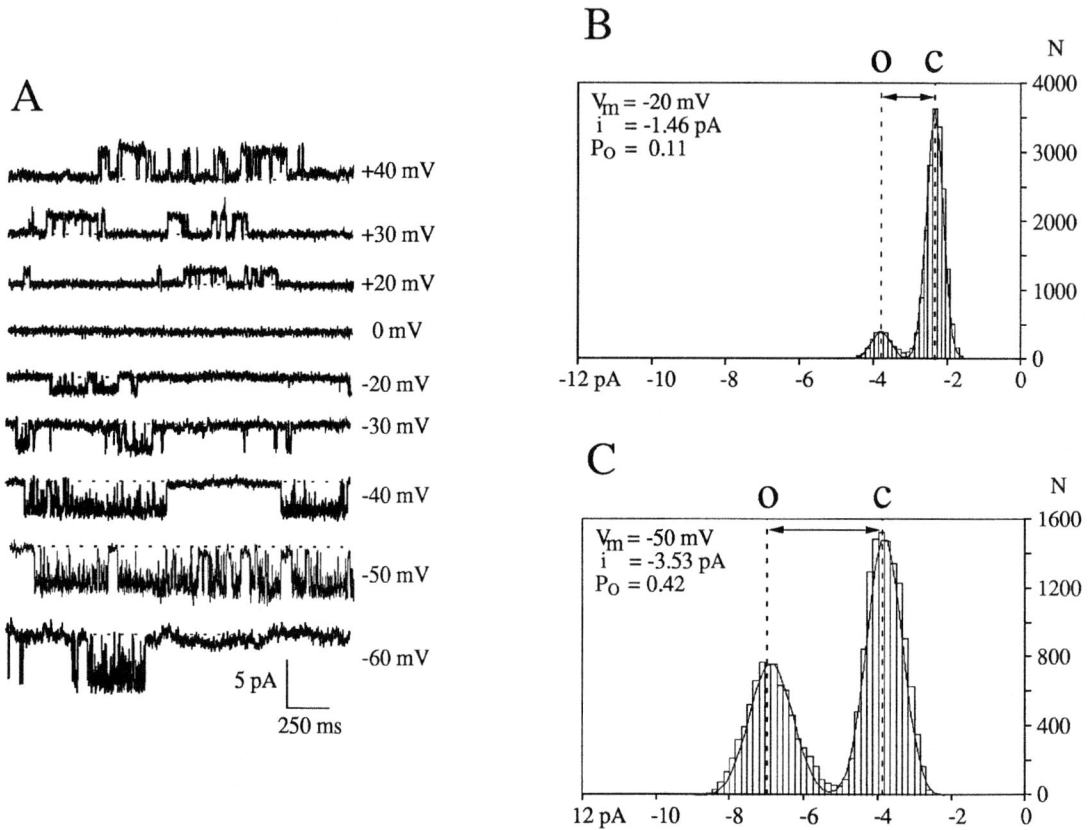

Fig. 10.5A–C. Determination of the single channel current amplitude. **A** Single channel currents recorded from an inside-out patch at various potentials. The level at which the channel is closed is indicated by a *dotted line*. **B, C** Amplitude histograms constructed from five current traces of 2000 ms in duration. The current traces were recorded at a membrane potential of −20 mV (**B**) and −50 mV (**C**). The letters *c* and *o* refer to the closed and open state of the channel, respectively

tude histogram, the number of observations that fall in "bins" of specified width is represented. The peaks in an amplitude histogram can be fitted with gaussian functions of which the mean is taken as the current amplitude. Note in Fig. 10.5B and C that the current level at which the channel is closed, is not equal to 0 pA, meaning that some current is still flowing from the patch pipette to the bath (leakage current). Therefore, to obtain the correct single channel current amplitude one should subtract the closed channel current level (c) from the open channel current level(o).

Next, a current-voltage (I–V) relationship is constructed to obtain a value for the single channel conductance (Fig. 10.6). The I–V relationship of the K.ATP channel is shown for an extracellular concentration of 10 mM (filled triangles) and 140 mM (filled circles), respectively. The data points are fitted with a linear function of the form:

$$i = g(V_m - V_{rev}) \tag{1}$$

where i is the single channel current amplitude, g is the single channel conductance, V_m is the trans-patch membrane potential, and V_{rev} is the reversal potential (see below).

The single channel conductance of the K.ATP channel was 29 pS and 66 pS for 10 mM and 140 mM extracellular K^+, respectively.

- The value found for the single channel conductance may vary with species and type of tissue (e.g., atrial, ventricular).

- If the open probability is very high, so that the channel seldomly fully closes, it will be difficult to determine correctly the single channel current amplitude from the amplitude histograms. In this case it might be helpful to add low concentrations of ATP (e.g., 0.1 mM). This will decrease the open probability of the channel and increase the duration of closures.

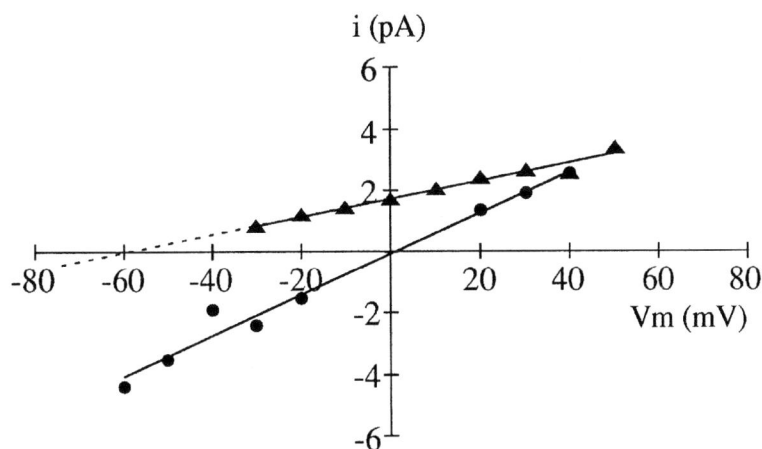

Fig. 10.6. Current-voltage relationship of the K.ATP channel at 10 (*filled triangles*) and 140 mM (*filled circles*) extracellular K^+. Data points were fitted by linear regression. The slope conductance was 67 pS and 29 pS for 140 and 10 mM K^+, respectively. The reversal potential was +1 mV and −60 mV for 140 and 10 mM K^+, respectively.

Reversal potential

The reversal potential is defined as the potential at which the current through a channel changes sign. That is, the voltage at which the inward current becomes outward or vice versa. The reversal potential of the K.ATP channel is very close to the equilibrium potential for K^+ (E_K), indicating that the channel is highly selective for K^+ ions. The value for the reversal potential can be obtained from the linear fit shown in the I–V relationship in Fig. 10.6. At an extracellular K^+ concentration of 10 mM, a V_{rev} of −60 mV is determined. This is close to the calculated E_K (−68 mV). When the extracellular K^+ concentration is increased to 140 mM, a V_{rev} of 0 mV is determined which is equal to the calculated E_K. In both experiments the intracellular K^+ concentration was 140 mM.

Open probability

The open probability is the fraction of time that an individual channel spends in the open state. Values for the open probability can be derived from amplitude histograms as shown in Fig. 10.5B, C in the following way. The area under each peak in the amplitude histogram represents the probability that 0, 1, 2, ..., n channels are simultaneously in the open state. When channels are identical and behave independently, these probabilities are binominally distributed, given by:

$$(P_o + P_c)^n = 1 \tag{2}$$

where P_o is the open probability, P_c is the closed probability, and n is the number of channels. The area under the peaks can be calculated from Gaussian functions fitted through these peaks. The open probability of the K.ATP channel is mainly determined by the intracellular ATP concentration. At zero ATP, immediately after patch excision, the open probability is normally very high, around 0.9. Increasing the ATP concentration decreases the open probability. The concentration at which the channel is half maximally inhibited lies around 0.1 mM (Fig. 10.7).

- The calculation of the open probability as described previously may only be applied when it concerns stationary channel kinetics, that is, the probabilities do not change with time. For analysis of nonstationary channel kinetics the reader is referred to Sakmann and Neher (1983).

- Other important parameters of channel kinetics are the mean open time and the mean closed time. The mean open time is the average duration of channel openings, and the mean closed time is the average duration of the channel closures. It must be realized that the open probability gives no information about the mean open time, as can be

Fig. 10.7. ATP dependence of K.ATP channel open probability. (From Noma 1983)

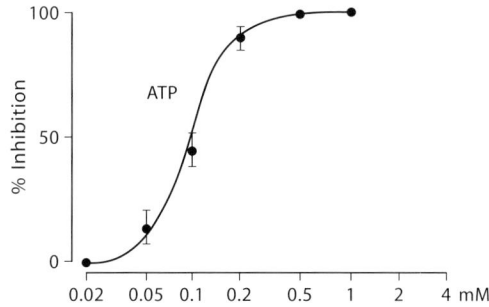

appreciated from an example in Fig. 10.8. Although the open probability for both channels is the same, 0.5, the mean open and mean closed times differ by a factor of 2.

- To get a reliable estimate of the open probability, a relatively large number of events (channel openings) is needed. Thus, the chosen sampling time depends on the open probability itself and mean open and mean closed times. For example, when the open probability is low a relative long sampling time is needed to get a large number of events. As another example, with the same open probability, a shorter sampling time is needed when the mean open time is short than when the mean open time is longer (see also Fig. 10.8).

Fig. 10.8. Difference between mean **A** open time and **B** open probability

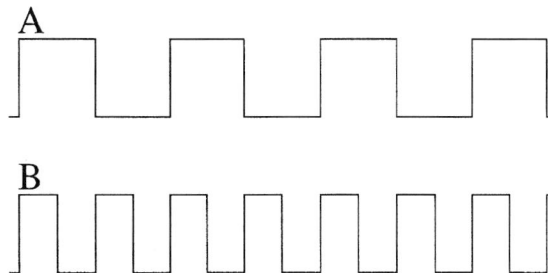

References and Further Reading

Ashcroft FM (1988) Adenosine 5'-triphosphate-sensitive potassium channels. Annu Rev Neurosci 11:97–118

Nichols CG, Lederer WJ (1991)Adenosine triphosphate-sensitive potassium channels in the cardiovascular system. Am J Physiol 261:H1675–1686

Noma A (1983) ATP-regulated K^+ channels in cardiac muscle. Nature 305:147–148

Sakmann B, Neher E (1976) Single channel currents recorded from membrane of denervated frog muscle fibers. Nature 260:779–802

Sakmann B, Neher E (1983) Single channel recording. Plenum, New York

Veldkamp MW, van Ginneken ACG, Bouman LN (1993) Single delayed rectifier channels in the membrane of rabbit ventricular myocytes. Circ Res 72:865–878

Weille de JR (1992) Modulation of ATP sensitive potassium channels. Cardiovasc Res 26:1017–1020

Chapter 11

Mechanotransduction and Mechanosensitive Ion Channels in Osteoblasts

RANDALL L. DUNCAN

Background

Mechanotransduction

Mechanotransduction is a process through which a cell can detect a physical stimulus, convert this stimulus into a biochemical signal and, ultimately, alter function at the tissue level. While balance, hearing and tactile sensation are common examples of mechanotransduction, cells in almost every tissue are capable of responding to their mechanical environment. For example, numerous studies have demonstrated that bone can alter its shape and mass in response to exogenous mechanical strain (Goodship et al. 1979; Rubin and Lanyon 1984; Turner et al. 1991; Chow et al. 1993). While the tissue response to mechanical strain has been well documented, the mechanisms through which a single cell can transduce a physical signal and initiate this response in bone remains unclear. The process of mechanotransduction has four distinct phases (Duncan and Turner 1995); these are described below.

Mechanocoupling. Mechanocoupling relates to the physical transduction of energy from a physical stimulus applied to the tissue to a form of stimulus which can be detected by cells. This phase would include the characteristics of the physical stimulus imposed such as magnitude, frequency and duration.

Biochemical coupling. Biochemical coupling, or the conversion of the mechanical stimulus outside the cell into a biochemical signal within the cell, refers to the initial mechanosensitive mechanism located in or near the cell membrane which makes this conversion.

Signal transmission. The third phase of mechanotransduction, transmission of signal, is the cascade of events, including second messengers, autocrine and paracrine factors, which arise from the initial biochemical signal and lead to stimulation of the effector cells.

Effector cell reponse. The effector cell response is the final cellular response which alters tissue form or function. For example, this phase is characterized in bone by mineralization of bone induced by the osteoblast. To study mechanotransduction in any cell, one must consider each of these phases and how they correlate with the final tissue response.

Mechanosensitive Ion Channels

Since they were first described in chick skeletal myocytes (Guhuray and Sachs 1984), mechanosensitive channels have been reported in a variety of cells from diverse tissues, ranging from bacteria to mammalian neural cells (for review, see Morris 1990). These channels form a large family of membrane proteins which are all gated by mechanical deformation and include both stretch-activated and stretch-inactivated channels. Mechanosensitive channels can be further subdivided, based on their activation properties, kinetic characteristics and ion selectivity. One of the most common types of mechanosensitive channel is the stretch-activated, cation-nonselective channel (SA-cat) which has been reported in a wide range of mammalian cells including osteoblasts (Duncan and Misler 1989; Davidson et al. 1990). Even though these channels are gated by the common stimulus of mechanical deformation, the mechanism of gating may not be similar among these channels. Although he demonstrated that SA-cat channels in skeletal myocytes respond to increased membrane tension, Sachs (1988) proposed that this membrane tension could be focused by the cytoskeleton. However, using a variety of cytoskeletal severing agents, he was unable to demonstrate that actin or tubulin filaments were involved in activation of these channels. Yet, we have recently found that altering the polymerization of the actin cytoskeleton produced a six- to tenfold increase in SA-cat channel activity in osteoblasts (Duncan et al. 1992). Thus, while all mechanosensitive channels are gated by mechanical deformation, the mechanisms of gating may be quite different.

Since mechanosensitive channels were first reported, the primary function of mechanotransduction has been proposed. However, correlating whole cell functions with single channel activity has been difficult. Partly, this difficulty arises with problems in eliciting whole cell currents with mechanical stimuli which induce single channel activity (Morris and Horn 1991). However, recent evidence has pointed to mechanically induced whole cell currents in yeast spheroplasts (Gustin 1991) and mammalian smooth muscle (Davis et al. 1992), as well as mechanically stimulated intracellular calcium increases in hepatocytes (Bear and Li 1991) and lung airway epithelia (Sanderson et al.1990). These observations would indicate that mechanosensitive channels may be important to the normal response of cells to their mechanical environment.

Mechanosensitive channels have been reported in rat (Duncan and Misler 1989) and human (Davidson et al. 1990) osteoblast-like osteosarcoma cell lines. Application of chronic, intermittent mechanical strain to the osteosarcoma line UMR-106.01 has been shown to increase single channel activity and single channel conductance of SA-cat channels as well as increase stretch sensitivity and induce spontaneous activity (Duncan and Hruska 1994). Interestingly, parathyroid hormone alters channel kinetics in much the same way as chronic intermittent mechanical strain (Duncan et al. 1992) indicating that the anabolic response of osteoblasts to both of these stimuli may be mediated through these channels.

Application of Mechanical Stimulus to Osteoblasts

An important consideration to make in the study of the effects of mechanical deformation on cells in culture is the type of mechanical perturbation to which the cells are subjected in vivo. Due to the curvature of weight-bearing bones, normal loading combines bending and compression, producing variable strains on bone surfaces. This type of global loading would produce two types of mechanical signals that bone cells would perceive, mechanical stretch (strain) and extracellular fluid flows driven across bone cells in response to bending of the bone. Many studies have demonstrated that mechanical strain stimulates osteoblasts in culture.

Mechanical strain. Changes in proliferation rates, cell alignment, second messenger levels, autocrine and paracrine secretions, and matrix protein expression and production are all modulated by application of mechanical strain to cultured osteoblasts and osteoblast-like clonal cell lines (for reviews, see Burger 1992; Duncan and Turner 1995). The magnitudes of the applied strain in these studies were typically greater than or equal to $10\,000\,\mu$strain, which is equivalent to 1% cellular deformation. During vigorous exercise, bone is only subjected to $500–2000\,\mu$strain (Burr et al. 1996). Thus, due to the tenfold higher strains of these in vitro studies, the physiologic validity of these studies becomes a concern.

Shear stress. Bending of bone also produces pressure gradients which drives extracellular fluid through canaliculae, creating shear stresses on

Principles and applications

the cell membrane (Weinbaum et al. 1994). In vitro studies have shown that osteoblasts rapidly respond to fluid shear with an increase in cAMP and inositol trisphosphate as well as prostaglandin E_2 (Reich and Frangos 1990; Reich et al. 1991). Additionally, cultured osteocytes have been shown to be more responsive to fluid shear than osteoblasts (Klein-Nulend et al. 1995), suggesting that osteocytes, which are aligned in canaliculae, may be the cells which initially sense mechanical loading and start the cascade of events resulting in bone modeling.

Loading systems. Whether one studies the effects of strain or fluid shear on the effect of mechanical loading, it is important to understand the types of mechanical signals that a loading apparatus applies to the cells. A primary concern with many commercially available loading systems is the nonuniform strains imposed across the growing surface. While this is a minor problem in patch studies in which the strain of a defined area of the culture plate can be determined and relocated with each patch, it is difficult to quantitate the magnitude of strain in studies which utilize cells from the entire plate. Additionally, many mechanical loading systems are unable to decrease the magnitude of strain to physiologic levels, which for bone is less than 3000 μstrain. Another characteristic of most loading devices of which one should be aware is the fluid movement that occurs when the growing substrate is bent, pulled or stretched through the culture medium. These fluid flows can subject the cells to an additional mechanical stimulus which could approach or exceed physiologic levels of in vivo fluid shear and may affect the response of the cells. Thus it is important to choose a loading system in which mechanical forces have been well characterized or that the experimenter is aware of the various forces imposed on the cells in culture.

In the study of mechanotransduction, it is also important to consider the parameters of strain which affect the bone response in vivo. Magnitude of the stimulus imposed is, of course, a primary focus of study, but duration, frequency and strain rate of the mechanical stimulus are all important factors which affect bone formation. Lanyon and Rubin (1984) found that dynamic strain produces a much greater response than static strain and that as few as four loading cycles/day can maintain bone mass in the turkey ulna model. Turner et al. (1994) have observed that the response of bone to mechanical bending is frequency-dependent as well, with the threshold for bone formation shown to be 0.5 Hz. Whether this is an effect of the frequency or the strain rate, i.e., the speed of the application of strain, is still under study. In vitro studies have yet to define the effects of these parameters on osteoblasts or osteocytes, however examination of the reports of in vitro studies to date indicate that these parameters of strain interact with each other. Thus, special care should be taken to define the optimal levels of each parameter to accurately study mechanosensitive pathways.

Patch Clamp Study of Mechanosensitive Ion Channels in Osteoblasts

Several problems exist in the study of mechanosensitive channels in the osteoblast. The first, and foremost, problem pertains to the application of mechanical stimulus to a patched cell. Since exquisite stability of the patched cell is required to maintain the patch, application of a stimulus which deforms the cell becomes problematic. Two methods which can provide a mechanical stimulus to the cell, yet maintain cell stability, are hypotonic swelling and fluid shear. Typically, reducing the bath osmolarity from 310 mosm to 220 mosm will activate SA-channels in previously mechanically stimulated osteoblasts (Duncan and Hruska 1994). However, the physiologic relevance of such a stimulus on an osteoblast comes into question. An answer to this problem is the addition of an apparatus to the patch clamp assembly which subjects the patched cell to fluid shear forces via a fluid-filled micropipette positioned near the patched cell and attached to a picopump. While this technique may perturb the cell to the point of occasionally losing a patch, the type of mechanical stimulation is much more relevant to the osteoblast in vivo.

Growth phase of the cells. A question which is essential to the study of osteoblasts in vitro is what point in the growth of the cells in culture is most similar to the state of the cells in vivo? For example, osteoblasts in a proliferative state respond to various stimuli, such as PTH, quite differently than cells that are growth arrested by cell to cell contact. We have previously demonstrated that SA-cat channels are activated by parathyroid hormone (PTH) in osteoblast-like cells which are growth inhibited by cell-to-cell contact, yet do not respond to PTH stimulation when osteoblasts are isolated on the culture plate (Duncan et al. 1992). Thus, most of our experiments have been done using contact-inhibited cells. Osteoblasts are large cells that form extended processes when isolated and are electrically coupled through gap junctions when in the contact-inhibited state. These morphological characteristics make measuring whole cell currents difficult, due to both the ability of the membrane to act as a capacitor and voltage clamp limitations. Rather than measure currents, one way to determine loading-stimulated changes in channel activity is to examine whole cell conductance using current clamp techniques.

Materials

Solutions and reagents

Bath solutions typically used are sodium Ringers' (NaRB) and potassium Ringers' (KRB).

- NaRB (in mM): 136 NaCl, 5.5KCl, 1 $MgCl_2$, 1 $CaCl_2$, 20 HEPES, pH to 7.3 with NaOH.
- KRB (in mM): 137 KCl, 1 $MgCl_2$, 1 $CaCl_2$, 20 HEPES, pH 7.3 with KOH. The pipette solutions for single channel and whole cell conductance measurements are abbreviated SCP (single channel pipette solution) and NPP (nystatin permeabilized pipette solution).
- SCP (in mM); 40 KCl, 107 K^+ acetate, 6.7 HEPES, 1.3 EGTA, pH 7.3 with KOH.
- NPP (in mM); 12 NaCl, 64 KCl, 28 K_2SO_4, 47 sucrose, 1 $MgCl_2$, 0.5 EGTA, 20 HEPES, pH 7.35 with KOH. Nystatin is added just prior to patching the cell at a concentration of 300 µg/ml. The SA-cat channel blocker gadolinium chloride ($GdCl_3$) and nystatin for whole cell conductance measurements are available from Sigma Chemical (St. Louis, MO).

Equipment

The patch clamp system is composed of a List EPC-7 patch clamp amplifier (Adams-List, Westbury, NY), an eight-pole Bessel filter (Frequency Devices, Haverhille, MA), an Axon Instruments Digidata 1200 analog to digital/digital to analog converter (Foster City, CA) controlled by a 486 DX microcomputer (Dell Computers, Austin, TX) with pClamp6 software (Axon Instruments).

Stretch to the patch membrane for single channel analysis is performed by a microprocessor-controlled solenoid switch attached to a vacuum source and interfaced with the patch clamp software (personal construction). Fluid shear to the single patched cell is controlled by a World Precision Instruments picospritzer (New Haven, CT) and interfaced with the patch clamp software.

Borosilicate glass pipettes (DiRuscio and Assoc. Manchester, MO) are pulled on a Narishige PP83 two-stage pipette puller (Fryer Instruments, Huntley, IL), fire-polished and coated with a microforge (Medical Systems, Great Neck, NY). Pipettes are coated with Sylgard (Dow Chemical).

Procedure

Pipette Preparation Protocol

1. Pipette glass is cleaned in acetone, blown dry with nitrogen and stored in a clean, dry glass tube until used.

2. Pipette glass is placed in the holder of a Narishige PP83 two-stage pipette puller and pulled to a point which has a resistance of $2-5\,M\Omega$. New pipettes are pulled each day to prevent clogging of the tip.

3. Pulled pipettes are placed in a holder on a microforge and the tip moved close to a heated filament for 10 s. Sylgard is applied to the shank and within 200 μm of the tip with a glass rod. Hot air then is blown across the tip to solidify the Sylgard.

4. Pipettes are stored in a clean, dry glass jar until used.

Single Channel Kinetic Studies

1. Osteoblasts or osteoblast-like cells are subjected to mechanical perturbation for the magnitude, duration and frequency desired by the experimenter.

2. The culture plate is rapidly removed from the loading device and placed into the patch chamber containing NaRB. Cells are rinsed once with NaRB.

3. The micropipette is filled with SCP solution, inserted into the pipette holder and head stage of the patch apparatus and positioned over the cell. A software sequence which pulses $\pm2\,mV$ through the pipette is initiated, which produces a square wave on the computer monitor oscilloscope window of pClamp6. As the pipette is lowered to the membrane of the cell, the square wave becomes smaller, indicating an increase in resistance. A gentle suction is then applied to the pipette through the sideport of the pipette holder. The membrane then seals to the glass of the pipette so that a resistance of 10–100 gigaohms is attained.

   Tip. Osteoblasts and osteoblast-like cells grown in culture are usually flattened, which makes attaining a patch seal difficult. To induce a morphological change in the osteoblast which will help to achieve more consistent patches, a calcium-free bath solution containing

5 mM EDTA can be introduced to the patch chamber for 15 min (Loza et al. 1994). After this time, the normal bath solution is reintroduced into the chamber and a patch seal may be attempted immediately. This produces a more rounded cell shape without altering cell attachment, which may be quite important to the functional expression of channels.

4. The presence of an SA-cat channel is determined using the microprocessor controlled suction generator which is attached to the sideport of the pipette holder and interfaced with the data acquisition software. A current-voltage sequence is initiated with clamp voltages (V_c) ranging from -80 to $+80$ mV at 20 mV intervals. These V_c s are held for 1 s prior to initiation of vacuum. The vacuum is held for 1 s, then released. The patch is allowed to stabilize following vacuum-induced stretch for 1 s. Thus, the V_c at each voltage interval is held a total of 3 s. Channel amplitudes are graphed against V_c to determine channel conductance. Typical channel conductance for the SCP solution is approx. 41 pS.

5. Single channel stretch sensitivity, activity (NP_o) and open times are determined using a computer sequence which applies a V_c of -50 mV with vacuum-induced stretch applied for 1 s at 2 s intervals. The magnitude of vacuum is changed using a vacuum regulator in line with the vacuum source. SA-cat channels are usually activated at 15–20 mm Hg vacuum with maximum activity observed at 40 mm Hg. Following chronic strain, sensitivity is increased so that the channels are activated at 10 mm Hg and maximum activity seen at 20–25 mm Hg. NP_o and channel open times are determined using pClamp data analysis software. A minimum of 500 channel events are required for accurate measurement of open time and currents should be sampled at 5–20 KHz.

6. Changes in single channel NP_o and conductance can be measured by hypotonically challenging the cell or introducing fluid shear to the cell. Hypotonic NaRB is made by reducing the NaCl concentration to 80 mM (220 mosm). Fluid shear is introduced by filling a pipette with a large tip diameter (1 MΩ), attaching it to the computer interfaced picospritzer, and pulsing fluid by the cell.

Whole Cell Conductance Measurements

1. Cells are mechanically loaded and placed in the chamber as described above.

2. Pipettes are prepared by front-filling the pipette with NPP solution without nystatin present. This is done by applying suction to the back of the pipette while the tip is in the filtered NPP solution. A 200 μm column of nystatin-free NPP solution from the tip should be attained for best results. The pipette is then backfilled with the NPP solution containing 300 μg/ml nystatin. This procedure is necessary since nystatin interfers with the ability of the pipette to seal to the membrane. A patch seal is then performed as described above. Once the seal has been made, the amplifier is turned to the current clamp mode and clamped at 0 so that as the nystatin starts to permeabilize the membrane, the membrane voltage will start to increase. Usually, osteoblasts and osteoblast-like cells have a membrane voltage of -30 to -50 mV. The degree of permeabilization can then be determined by using the amplifier to measure the access resistance. Nystatin will typically produce an access resistance of less than 50 MΩ

3. Whole cell conductance changes as a result of fluid shear or hypotonic challenge are measured by pulsing ± 50 pA through the pipette. The osteoblast has a resting membrane conductance between 10 and 20 nS. Following hypotonic challenge, membrane conductance increases to 30–60 nS. This increase can be blocked by the application of 10 μM GdCl$_3$. Similar changes are observed with application of fluid shear.

References

Bear CE, Li C (1991) Calcium-permeable channels in rat hepatoma cells are activated by extracellular nucleotides. Am J Physiol 261:C1018–C1024

Burger EH, Veldhuijzen JP (1993) Influence of mechanical factors on bone formation, resorption and growth, in vitro. In: Hall BK (ed) Bone (vol. 7) Bone Growth B, Boca Raton, FLA: CRC, pp 37–56

Burr DB, Milgrom C, Fyhrie D, Forwood M, Nyska M, Finestone A, Hoshaw S, Saiag E, Simkin A (1996) In vivo measurement of human tibial strains during vigorous activity. Bone 18:405–418

Chow JW, Jagger CJ, Chambers TJ (1993) Characterization of osteogenic response to mechanical stimulation in cancellous bone of rat caudal vertebrae. Am J Physiol 265:E340–E347

Davidson RM, Tatakis DW, Auerbach AL (1990) Multiple forms of mechanosensitive ion channels in osteoblast-like cells. Pflugers Arch 416:646–651

Davis MJ, Donovitz JA, Hood JD (1992) Stretch-activated single channel and whole cell currents in vascular smooth muscle cells. Am J Physiol 262:C1083–C1088

Duncan R, Misler S (1989) Voltage-activated and stretch-activated Ba^{2+} conducting channels in an osteoblast-like cell line (UMR-106). FEBS Lett 251:17–21

Duncan RL, Hruska KA, Misler S (1992) Parathyroid hormone activation of stretch-activated cation channels in osteosarcoma cells (UMR-106.01). FEBS Lett 307:219–223

Duncan RL, Hruska KA (1994) Chronic, intermittent loading alters mechanosensitive channel characteristics in osteoblast-like cells. Am J Physiol 267:F909–F916

Duncan RL, Turner CH (1995) Mechanotransduction and the functional responsse of bone to mechanical strain. Calcif Tissue Int 57:344–358

Goodship AE, Lanyon LE, McFie H (1979) Functional adaptation of bone to increased stress. J Bone Joint Surg 61A:539–546

Guharay F, Sachs F (1984) Stretch-activated single ion channel current in tissue cultured embryonic chick skeletal muscle. J Physiol (Lond) 352:685–701

Gustin MC (1991) Single-channel mechanosensitive currents. Science 253:800

Klein-Nulend J, Van der Plas A, Semeins CM, Ajubi NE, Frangos JA, Nijweide PJ, Burger EH (1995) Sensitivity of osteocytes to biomechanical stress in vitro. FASEB J 9:441–445

Lanyon LE, Rubin CT (1984). Static vs. dynamic loads as an influence on bone remodeling. J Biomech 17:897–905

Loza, J, Stephan E, Dolce C, Dziak R, Simasko S (1994) Calcium currents in osteoblastic cells: Dependence upon cellular growth stage. Calcif Tissue Int 55:128–133

Morris CE (1990) Mechanosensitive ion channels. J Membrane Biol 113:93–107

Morris CE, Horn R (1991) Failure to elicit neuronal macroscopic mechanosensitive currents anticipated by single-channel studies. Science 251:1246–1249

Reich KM, Gay CV, Frangos JA (1990) Fluid shear stress as a mediator of osteoblast cyclic adenosine monophosphate production. J Cell Physiol 143:100–104

Reich KM, Frangos JA (1991) Effect of flow on prostaglandin E_2 and inositol trisphosphate levels in osteoblasts. Am J Physiol 261:C428–C432

Rubin CT, Lanyon LE (1984) Regulation of bone formation by applied dynamic loads. J Bone Joint Surg 66A:397–402

Sachs F (1988) Mechanical transduction in biological systems. Crit Rev Biomed Eng 16:141–169

Sanderson MJ, Charles AC, Dirksen ER (1990) Mechanical stimulation and intercellular communication increases intracellular calcium in epithelial cells. Cell Regul 1:585–596

Turner CH, Akhter MP, Raab DM, Kimmel DB, Recker RR (1991) A noninvasive, in vivo model for studying strain adaptive bone modeling. Bone 12:73–79

Turner CH, Forwood MR, Rho IY, Yoshikawa T (1994) Mechanical loading thresholds for lamellar and woven bone formation. J Bone Min Res 9:87–97

Weinbaum S, Cowin SC, Zeng Y (1994) A model for the excitation of osteocytes by mechanical loading-induced bone fluid shear stresses. J Biomechanics 27: 339–360

Chapter 12

The Study of (Plant) Ion Channels Reconstituted in Planar Lipid Bilayers

Henk Miedema

Background

The development of the patch-clamp technique by Neher and Sakmann (1976) opened the door to the study of single ion channels. Initially performed on animal cells, 8 years later the technique was succesfully applied to plant cells as well (Moran et al. 1984). One reason for the delay in applying patch-clamp to plant cells is the presence of a cell wall which has to be removed before the plant protoplast can be patch-clamped (Takeda et al. 1985). Today, patch-clamp is an almost standard technique in laboratories around the world specializing in plant cell electrophysiology. Despite the loss of the cell wall, during a patch-clamp experiment the ion channel is studied in its more or less native membrane environment. An alternative strategy to study ion channels is a method based on the reconstitution of the isolated, purified channel in an artificial membrane system, such as a liposome or a planar lipid bilayer. Obviously, these techniques are especially useful in cases in which the particular membrane is inaccessible to the patch-clamp pipette (Miller 1983). The small size of liposomes and the high turnover rate of ion channels, however, limit the use of liposomes for studies of channel mediated transport (Miller 1983). The performance of electrophysiological measurements requires either enlargement of the liposomes to a size suitable for patch-clamping (Tank and Miller 1983) or the reconstitution of the channel into a so-called planar lipid bilayer (PLB). The PLB technique allows the study of ion channels on the single channel level under precise, well defined experimental conditions and has been appplied succesfully to study ion channels originating from animal cell membranes and endomembranes (see references in Labarca and Latorre 1992; Coronado et al. 1992). White and Tester (1992) were the first to reconstitute plant plasma membrane channels in a PLB. At the same time, Klughammer et al. (1992a,b) applied the PLB technique to study ion channels in the plant vacuolar membrane. Here we describe a methodology to set up such a

PLB system and show measurements obtained after incorporation of plant vacuolar channels in a PLB. The more interested reader is referred to the superb textbooks by Hanke and Schlue (1993) and Miller (1986) and to Labarca and Latorre (1992).

Principles and applications

A schematic diagram of a planar lipid bilayer (PLB) setup is shown in Fig. 12.1.

Basically, the system consists of two electrically isolated compartments filled with salt solution and separated by the lipid bilayer containing the reconstituted channel. By convention, one compartment is defined as the cis chamber and one as the trans chamber. Ag/AgCl wires are inserted in two separate compartments. Via a saltbridge the cis side is connected to the measuring electrode and to the headstage of the patch-clamp amplifier, while the trans side is electrically grounded. Photographs of the equipment currently in use by the author are shown in Fig. 12.2 (see legend for detailed description).

Figure 2C shows a homemade polycarbonate cuvette or cup and a PVC cuvette holder in more detail. At one side, the wall of the cuvette is partly flattened, resulting in a wall thickness of just 0.25 mm. At this position the wall contains a small hole (or aperture) with a diameter of 0.25 mm. The cup fits into the rear hole of the holder, thereby serving as the cis chamber, while the second hole serves as the opposite or trans chamber. By using some kind of a "brush," lipid is "painted" across the aperture. In a few seconds, the lipid film may thin out, a process thermodynamically controlled and resulting in the formation of a true lipid bilayer. In front, the cuvette holder contains a glass window, providing the possibility to optically follow bilayer formation using a modified microscope setup (for technical details of the PLB setup, see Hanke and Schlue 1993).

Measurement of Bilayer Capacitance

In addition to electric resistance, the quality of a bilayer is determined by its surface area and therefore its capacitance, as both are interrelated: the capacitance of a membrane is directly proportional to its surface area. A drop of lipid itself has almost no capacitance, only a true phospholipid bilayer has a significant capacitance. Therefore, instead of following the process of bilayer formation optically, one can monitor the increase in capacitance that occurs when the lipid film thins out. The capacitance is measured by applying a triangular voltage wave of appropriate size and duration. The current output signal will be a square wave whose ampli-

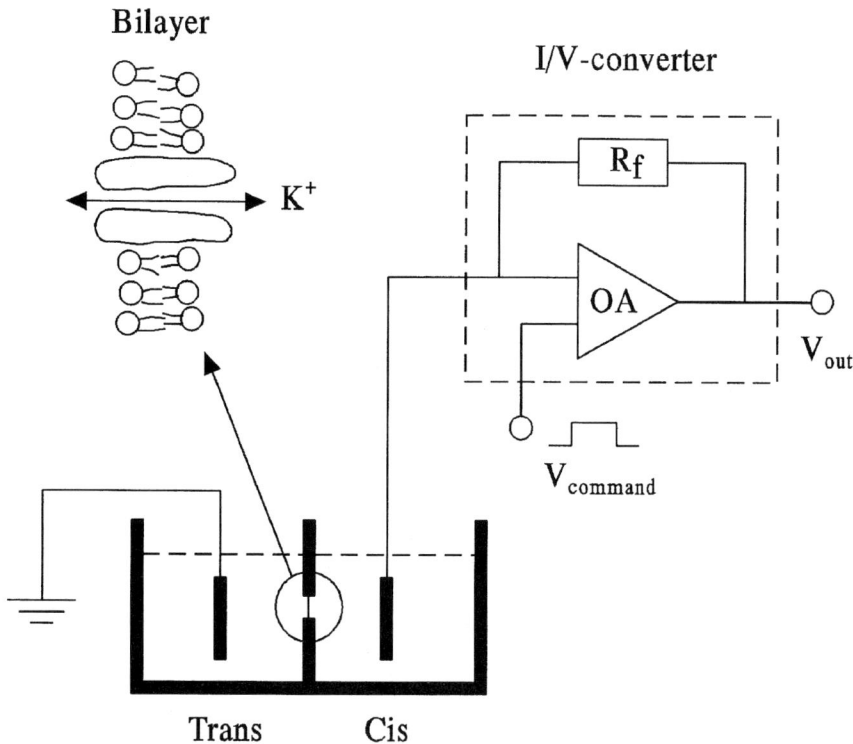

Fig. 12.1 Scheme of the planar lipid bilayer setup and the preamplifier in the head-stage of the patch-clamp amplifier. A lipid bilayer separates two electrically isolated compartments (cis and trans) of which one (cis) is connected to the measuring electrode and to the amplifier headstage and the other is electrically grounded (trans). The amplifier clamps the voltage difference (V) between the two compartments and across the bilayer on the holding or command potential ($V_{command}$). Current flow, arising from channel opening, changes V. However, the operational amplifier (OA) compares the value of V with the value of $V_{command}$ and compensates instantaneously for any difference by injecting current through the feedback resistor R_f. This "correcting" current is equal in magnitude but opposite in sign to the current passing the membrane. The electronic circuit functions like a current-to-voltage or I/V converter, i.e., the output voltage of the headstage (V_{out}) is directly proportional to the current across the bilayer. Typically R_f has a value of $1\,G\Omega$, then, an input of $1\,pA$ corresponds to an output of $1\,mV$

tude is a measure of the capacitance of the system. Prior to an actual experiment the system must be calibrated by connecting a capacitor of known value to the input of the headstage (see Fig. 12.3). For this purpose, Axon Instruments provides a convenient 100 pF bilayer model cell.

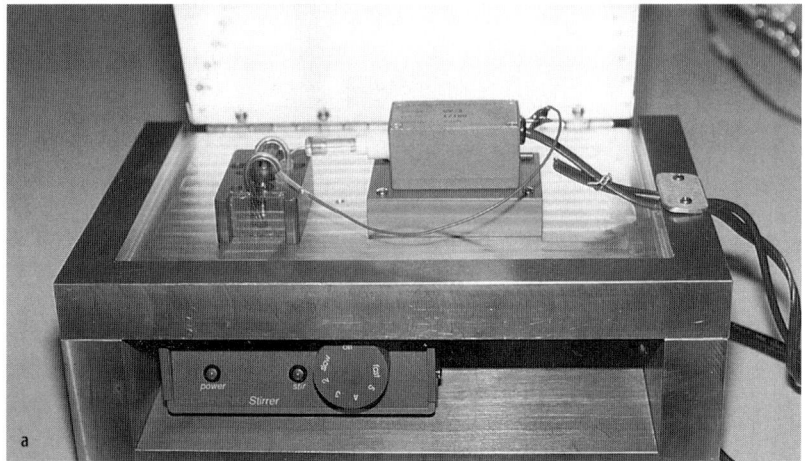

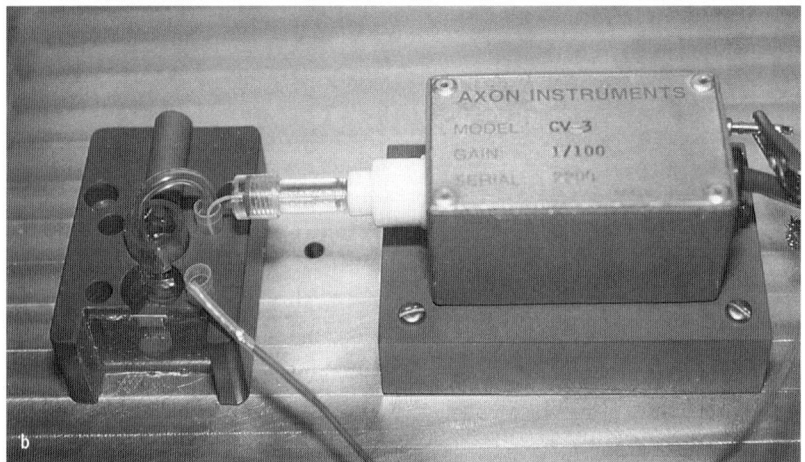

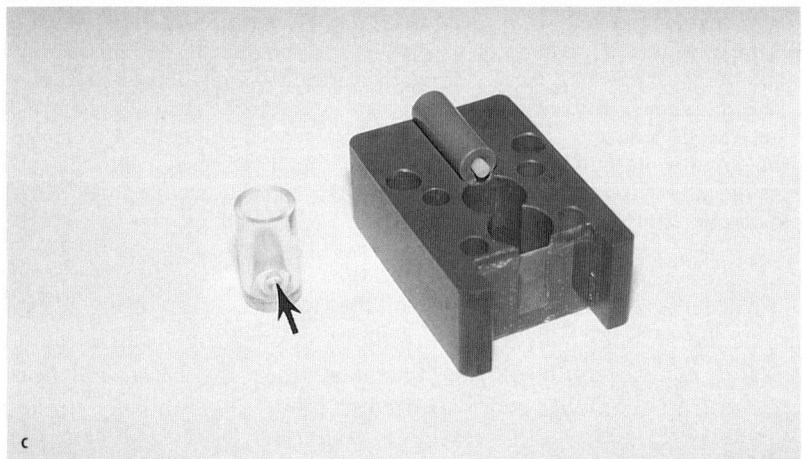

Bilayer Surface Area

The specific capacity of a decane containing bilayer is around $0.4\ \mu F/cm^2$ ($=0.004\ pF/\mu m^2$), about half the specific capacitance of a biological, i.e., solvent-free, bilayer ($0.8\ \mu F/cm^2$; see Hanke and Schlue 1993). A typical PLB has a capacitance of 200 pF, corresponding to a membrane surface area of $5\times10^4\ \mu m^2$ ($=0.05\ mm^2$). In contrast, the surface area of a membrane patch at the tip of a patch-clamp pipette is $5\ \mu m^2$, at most. Note that just part of the aperture, with a total surface area of $0.2\ mm^2$, is covered by a bilayer and that most part is covered by the so-called lipid annulus (see Miller 1986).

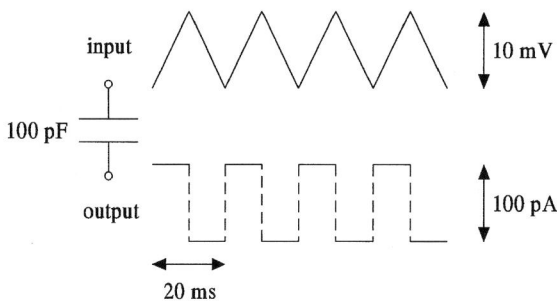

Fig. 12.3 The current output of a 100 pF capacitor (C) in response to an input of a triangular voltage wave (V) of 10 mV (peak-to-peak), applied at a frequency of 50 Hz. The current, I, given by $C\cdot dV/dt$, equals 100 pA. Keep in mind that the amplifier may attenuate the signal applied to the "external command" input of the amplifier. For instance, the Axopatch-1B attenuates the signal by a factor 50, implying the necessity of an input signal in the voltage range

Fig. 12.2A–C. Photographs of the bilayer setup in use by the author. **A** Headstage and cuvette holder are mounted on a homemade table. Note the stirring device underneath and, in the background, the lid attached to the table with a hinge and functioning as a shielding or Faraday cage. **B** Headstage and cuvette holder in more detail. Measuring and ground electrode are inserted into electrode cups and, via a saltbridge, connected to cis and trans compartment, respectively. For clarity, the saltbridge in the front chamber has been removed. **C** Close-up of cuvette and cuvette holder. The position of the 0.25 mm aperture in the wall of the cuvette is indicated by the *arrow*

Resolution

Although both patch-clamp and bilayer techniques allow single channel measurements, the resolution of the patch-clamp exceeds the resolution of the PLB technique. One reason is the strong background noise in a PLB experiment and the consequent strong filtering. Low pass filtering below 100 Hz is not an exception. The second reason is that the voltage step resolution is lowered by a slow capacitance current transient. Both, excessive noise and high capacitance are due to the relatively huge membrane surface area of the PLB. One way to reduce the background noise as well as the capacitance is to reduce the size of the aperture. However, there is a limit to decreasing the diameter: (1) no bilayer will be formed when the aperture is too small and, (2) the chance of channel incorporation is directly related to the surface area of the bilayer.

Protein Incorporation

Dependent on the hydrophillic properties of the channel, the protein can either be added directly to one of the aqueous compartments of the PLB setup or, prior to the actual PLB experiment, be reconstituted in a vesicle that in turn is fused with the PLB. Naturally, membrane proteins have a strong hydrophobic character and therefore vesicle fusion is the common way to incorporate the channel protein into the bilayer. Figure 12.4 shows a cartoon of vesicle attachment and subsequent fusion with the bilayer. Sticking of the vesicles to the phosphatidylserine-containing and therefore negatively charged bilayer is promoted by Ca^{2+}.

The process of fusion is still not fully understood but stirring is an absolute necessity. Swelling and fusion of the vesicle is accomplished by the flow of water from trans to cis across the PLB into the vesicle. The process of swelling seems to be controlled by the osmolality inside the vesicle and by the difference in osmolality between cis and trans compartments (see Miller 1986).

Sign Convention

As explained in Fig. 12.4, when added on the cis side, after fusion of a right-side out tonoplast microsome with the bilayer, the cytoplasmic face of the channel is adjacent to the cis side. Given that the measuring electrode is connected to the cis side, potentials refer to the potential at the

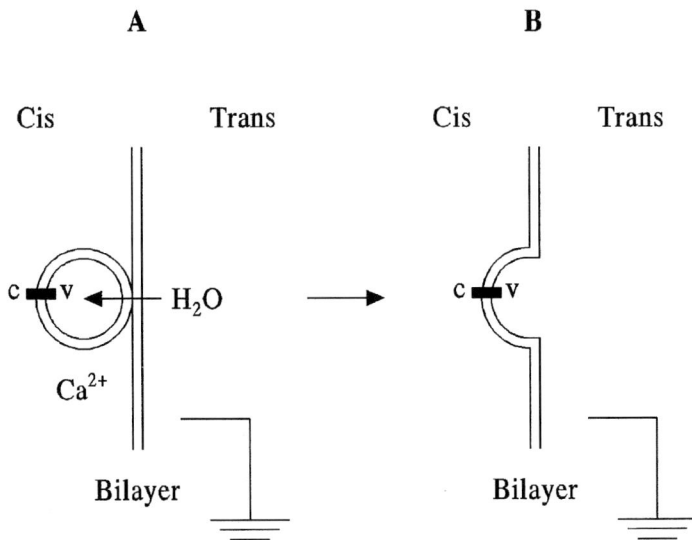

Fig. 12.4A, B. Sticking to (A) and fusion of (B) a vesicle with the bilayer. Vesicle attachment to the bilayer requires Ca^{2+} and, due to the hyperosmolality of the cis solution, a water flow from trans to cis swells the vesicle, finally resulting in a fusion with the bilayer. After adding right-side out tonoplast vesicles on the cis side and after vesicle fusion with the PLB, the cytoplasmic face of the channel protein (*c*) faces the cis compartment, while the vacuolar face (*v*) is exposed to the trans solution

cytosolic face with respect to the potential at the vacuolar face of the channel. The Axopatch patch-clamp amplifier defines a positive current as a flow of cations out of the pipette, i.e., in case of our bilayer setup, as a flow of cations from cis to trans. The resulting sign convention resembles the one commonly used in cell electrophysiology for (endo)membranes: a positive current refers to a flux of positive charge out of the cytoplasm. Note that according to this convention, the signs for potential and current inverse when fusing right-side out plasma membrane-enriched microsomes with the PLB.

Outline

The entire protocol of a PLB experiment is outlined in Fig. 12.5.

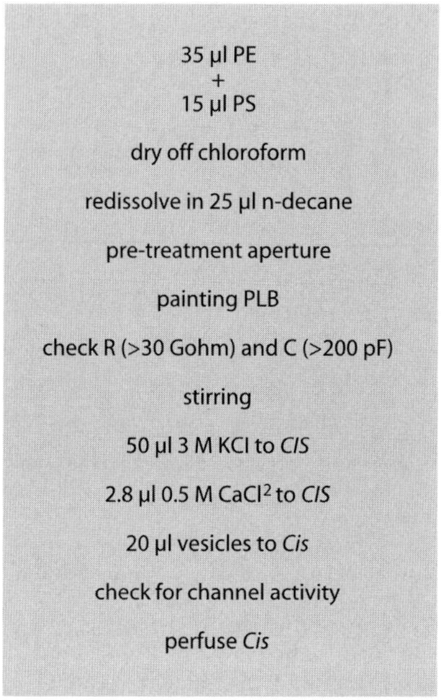

Fig. 12.5 Summary of the entire protocol for a bilayer experiment.

Abbreviations: *PE*, phosphatidylethanolamine; *PS*, phosphatidylserine; *PLB*, planar lipid bilayer; *R*, electrical resistance of PLB; *C*, electrical capacitance of PLB. Stock solutions of PE and PS contain 10 mg lipid per ml chloroform. The PLB is painted on using a solution containing 20 mg lipid per ml decane

35 µl PE
+
15 µl PS

dry off chloroform

redissolve in 25 µl n-decane

pre-treatment aperture

painting PLB

check R (>30 Gohm) and C (>200 pF)

stirring

50 µl 3 M KCl to *CIS*

2.8 µl 0.5 M CaCl^2 to *CIS*

20 µl vesicles to *Cis*

check for channel activity

perfuse *Cis*

Materials

Preparation of Tonoplast Microsomes

A detailed description of the isolation of red beet tonoplast vesicles or microsomes is beyond the scope of this chapter. The isolation, however, of tonoplast (and plasma membrane) vesicles is among the standard techniques in plant cell transport studies. The procedure is based on membrane seperation by application of aqueous-polymer two-phase partioning using differential centrifugation (Rea and Poole 1985). Following this procedure, the sideness of the vesicles will be predominantly right-side out. Marker tests, performed after the isolation procedure, should establish that the purified membrane fraction indeed originates from the tonoplast.

Preparations Before the Actual Experiment

General precaution One of the most serious caveats in bilayer experiments is the fact that all kinds of contaminations can cause channel-like behavior. In this respect, traces of detergent, used for cleaning the setup, are particular infamous.

Therefore, it is almost a prerequisite "to be obsessive about cleanliness" (Ehrlich 1992). Do not try to save time on cleaning: it simply does not work.

1. Dissolve 25 mg phosphatidylethanolamine (PE) in 2.5 ml chloroform.

2. Dissolve 25 mg phosphatidylserine (PS) in 2.5 ml chloroform.

3. Store these lipid stock solutions at −80 °C.

Lipid stock solutions

After pipetting lipid out of the stock solutions, fill the vials with a gentle stream of nitrogen, close the vials tightly and put them back at −80 °C. Use glass vials, as chloroform destroys plastic (e.g. 5 ml Kimble-Accuform Micro Vials with chemically inert, PTFE-faced silicon septum, Fischer Scientific, USA).

1. Pipette 35 µl PE and 15 µl PS stock solution into a 1 ml glass vial.

2. Evaporate the chloroform with a gentle stream of nitrogen.

3. Place the vial under nitrogen for an additional 15 min.

4. Redissolve the lipids in 25 µl n-decane (Sigma).

Preparation of lipids

Decane and chloroform are difficult to pipette. Use Hamilton syringes to pipette accurately.

1. 100 mM KCl, 2 mM $MgCl_2$ and 10 mM MES buffer, adjusted to pH 6.5 with N-methyl-D-glucamine. Filter this solution using a 0.22 µm nylon filter (Micron Separations Inc., MA).

2. 3 M KCl + 1 % agar

3. 3 M KCl

4. 0.5 M $CaCl_2$

Experimental solutions

Prior to use, cups and holder are soaked overnight in a mild nonionic detergent, e.g., 2 % Triton-X100. Afterwards, rinse thoroughly with deionized water. As already mentioned, traces of detergent may result in a leaky, unstable bilayer and possibly in artificial channel activity. Prepare two or three cups at a time in order to be able to switch in case a cup does not yield satisfactory results.

1. Place cups and holder in running deionized water for at least 2 h.

Cleaning the experimental setup

2. Try to rinse the aperture by forcing flowing water through the small hole. Be careful, as a too high pressure may damage the aperture. Do this for several minutes to be sure all the detergent has been removed.

3. Use a stream of nitrogen to dry the cup, especially the aperture. The pretreatment with lipid requires an absolutely dry aperture.

Pretreatment of the aperture

Prior to painting the actual bilayer, the hole is pretreated with lipid. Such pretreatment significantly improves the chances of obtaining a high resistance, stable bilayer. Again, do this carefully and do not cut corners to save time. For brushing the lipid, we use a glass pasteur pipette, melted at the tip into an appropriate spherical shape. One must be able to clearly feel the hole with the brush. Make sure the tip of the brush is clean. Fire polishing of an already used brush will ensure that there is no chloroform or protein left from previous use.

1. Paint a drop of lipid onto the aperture and dry gently with a stream of nitrogen.

2. After drying, repeat this step one or two times. Prevent clogging of the hole with lipid by blowing through with nitrogen.

Assembling the setup

1. Place a cuvette in the rear position of the cuvette holder and place a small (1.5 mm) stirbar in this compartment.

2. Pipette 700 µl of 100 mM KCl solution (solution 1) into the cuvette, i.e., the cis compartment.

3. Pipette 1400 µl of solution 1 in the front, i.e., the trans compartment.

4. Put two small cups for inserting the electrodes in place and fill both with 3 M KCl.

5. Put the measuring electrode in the rear and the ground electrode in the front cup.

6. Fill two plastic tubes with 3 M KCL+1 % agar and put these saltbridges in place, i.e., one end in one of the electrode cups and the other end in the corresponding chamber. The saltbridges can be made ahead of time. However, to prevent artefacts, e.g., due to bacterial contamination, avoid using the same saltbridge twice.

7. Avoid air bubbles and make sure that the stirbar can rotate freely and does not jar the saltbridge.

Procedure

Painting a Bilayer

As is usual in patch-clamp experiments in following the process of seal- **Measurement**
ing, painting of the lipid film is followed by applying a voltage pulse and **of membrane**
watching the increase in electric resistance (R). According to Ohm's law **resistance**
and given a constant voltage (V), an increase in R will be accompanied
by a decrease in current (I): $\Delta I = V/\Delta R$. The voltage pulse (e.g., a 10 mV
pulse for 10 ms) can be applied by using the "seal-test" option on the
patch-clamp amplifier, by an external function generator or, alternatively, can be defined in pClamp 6.0.3 software. Due to the high capacitance of the PLB, the duration of the test pulse should not be too short,
otherwise the calculated R will be artificially low. Depending on whether
one can watch the screen of the oscilloscope or computer during the process of painting, an audio monitor can be very helpful in following the
resistance. With the Axopatch-1B, for instance, a decrease in volume
indicates an increase of R.

1. Set the GAIN of the amplifier to 1 and the HOLDING potential to
   0 mV.

2. Zero the current to compensate for electrode offsets by using the
   "junction null" of the amplifier.

3. Apply the seal-test pulse and turn the audio monitor on.

4. Dip the brush into the lipid-decane mixture and paint lipid on the
   aperture.

5. Repeat until the audio becomes totally silent, indicating a high resistance.

   Note. In practice, just a few trials should suffice, otherwise there will
   be too much lipid, which will prevent the forming of a bilayer anyway.
   Between paintings, clean the brush with a Kimwipe before dipping it
   again in the lipid solution. The lipid-decane solution should remain
   clear.

6. Turn the amplifier GAIN to 500. If the current deviates significantly
   from 0 pA, try to paint another bilayer (the reason is that the compensation for electrode offsets (step 2) is never 100 %. Then, at low resistance, a relatively small offset will result in a relatively large current).

7. Check the bilayer resistance by applying a HOLDING potential of −100 mV. The resistance should be at least 20 GΩ but resistances of 50–100 GΩ are preferable. Probably due to the slight curvature of the bilayer, the membrane is more stable at high negative than at high positive potentials.

8. Once the resistance is high enough, the capacitance must be tested. Set the GAIN to 1 again and the HOLDING potential to 0 mV. Apply the triangular wave and check the capacitance. A bilayer worth using for vesicle fusion will have a capacitance of at least 200 pF.

When the membrane breaks during this stage of the experiment, very often it is not necessary to add lipid again. Repainting by simply touching the aperture with the brush may suffice to obtain a new bilayer. Do not waste time and money by adding vesicles to a bad bilayer!

Channel Reconstitution

1. While gently stirring, add 50 µl 3 M KCl on the cis side, resulting in a 300/100 mM KCl (cis/trans) osmotic gradient. To prevent the buildup of a hydrostatic pressure, pipette the same amount of solution out of the cis chamber.

2. Pipette 2.8 µl 500 mM $CaCl_2$ into the cis chamber.

3. Before adding vesicles, check over the entire voltage range for channel activity. Any activity or channel-like activity at this stage of the experiment must be due to contamination or bilayer instabilities. In case of activity or membrane instabilities, break the PLB by firmly stirring and paint a new membrane. When activity is observed again, try another, clean, cup.

4. If the membrane is still "clean", add 20 µl tonoplast vesicles (2 mg protein/ml) to the cis side, corresponding to a final protein concentration of around 50 µg/ml.

5. Stop stirring every minute and check for channel activity at different HOLDING potentials.

6. As soon as channel activity is recorded, stop stirring immediately and perfuse the cis chamber with solution 1.

Results

One of the most prominent roles of the plant cell vacuole is its storage capacity. Therefore, it is not surprising that a variety of transport systems has been demonstrated in the vacuolar membrane (see Schulz-Lessdorf and Hedrich 1995). Examples of current recordings obtained with a PLB setup and with purified tonoplast microsomes isolated from *Beta vulgaris* (red beet) are shown in Fig. 12.6. After the addition of vesicles on the cis side, typically, within 20 min, channel activity could be observed. From a holding potential of 0 mV, the membrane was clamped at different potentials for a duration of 7 s. The conductance of 151 pS was calculated from the I–V curve in (D). As the observed reversal potential (voltage of zero current) of -26.1 mV is close to the theoretical Nernst or equilibrium potential of K^+ (-24.4 mV, corrected for ion activity), the channel shows a high selectivity for K^+ over Cl^+. See figure legend for details on experimental conditions.

Troubleshooting

- Before painting a PLB, the resistance is already in the gigaohm range: Aperture may be clogged by lipid; alternatively, check aperture and saltbridges for air bubbles.

- It seems impossible to obtain a gigaohm membrane: Check the setup for fluid leaks as a leak may electrically short-cut cis and trans compartments.

- Bilayer is consistently unstable: Including cholesterol may stabilize the membrane (Piñeros and Tester 1995). Otherwise, the lipid stock solutions may need to be replaced.
- No indication of any channel activity after adding vesicles: No channel activity may indicate that:
- fusion does not occur,
- fusion occurs but vesicles are empty, e.g., nonchannel containing vesicles are more likely to fuse with the PLB than channel containing ones or
- channel reconstitution occurs but the channel is inactive. Vesicle fusion is promoted by adding more vesicles, increasing the Ca^{2+} level, or increasing the osmotic gradient. When no channel activity appears within 15–25 min, one may consider one of these options. However, keep in mind that at high Ca^{2+} and high vesicle concentrations, vesi-

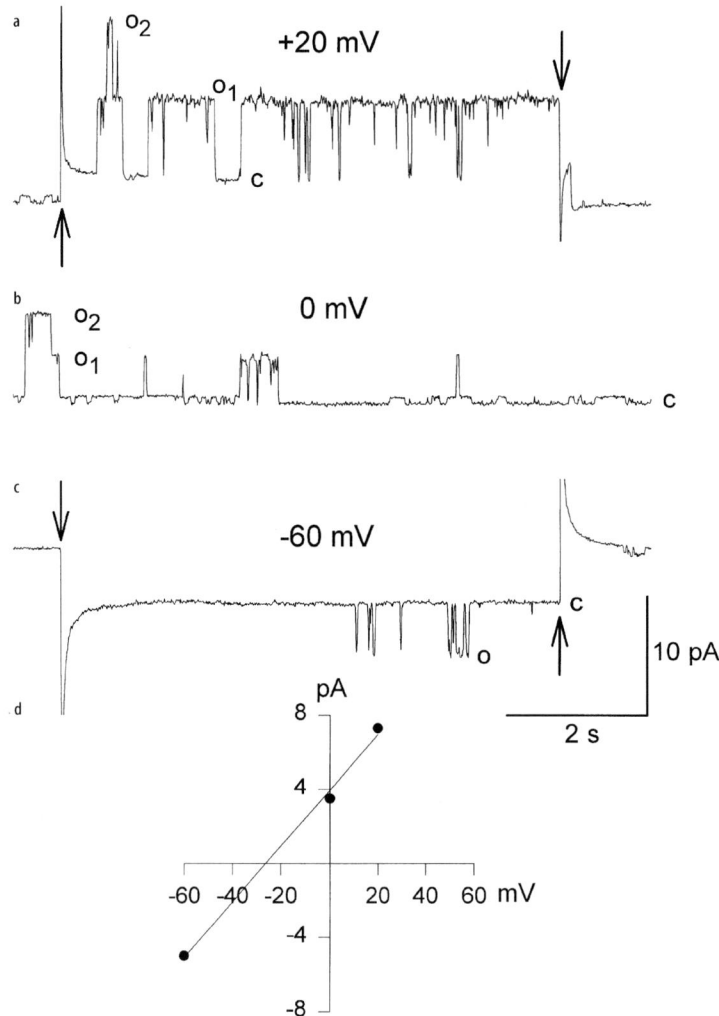

Fig. 12.6A–C. Examples of current recordings at different membrane potentials obtained after fusion of tonoplast-enriched-microsomes, isolated from red beet, with the PLB. The holding potential was 0 mV and the beginning and end of the voltage pulse of 7 s duration are indicated by the *left* and *right arrow*, respectively. Note in the current traces in (**A**) and (**B**) the presence of a low and a high conductance channel. Closed (*c*) and open (*o*) states of the large channel are indicated, o_1 and o_2 refer to the current level representing one and two open channels, respectively. The sampling rate was 1 kHz and the filter frequency was 40 Hz. (D) The current-voltage relationship (I–V curve) of the high conductance channel in **A–C**. The calculated conductance is 151 pS, recorded in 300/100 mM KCl (cis/trans). The observed reversal potential of –26.1 mV is close to the theoretical reversal potential of K^+, –24.4 mV, indicating the K^+ selectivity of the channel under these conditions

cles may aggregate, something one has to prevent. Therefore, taking another cup and painting another bilayer may be more promising.

- Vesicle fusion breaks the bilayer: For some preparations, it may be a problem to get the channel into the bilayer, while for others mass fusion may destroy the bilayer. Adjusting one of the factors listed above may prevent the fusion of too many vesicles.

Ideally, one should prevent the incorporation of more than one channel as multiple channel activity makes channel analysis on a more advanced level practically; if not totally; impossible. Therefore, as soon as channel activity is recorded, stop stirring immediately and perfuse the remaining vesicles out of the cis chamber.

Comments

Compared to the patch-clamp technique, the PLB technique has its own advantages but also its own limitations (see also White and Tester 1994; Miller 1983):

- Advantages of the PLB
- Membranes that are inaccessible for the patch-clamp pipette can be studied by the PLB technique.
- Some type of plant cells, e.g., root cells, are extremely difficult to patch-clamp. Reconstitution into a PLB may be an attractive alternative (see White and Tester 1992).
- By choosing the right solutions, channels of low abundance may be easier to detect in a PLB than with patch-clamping. Piñeros and Tester (1995) succesfully applied this strategy and recorded in monovalent cation-free, calcium-containing solutions an, until then, unknown Ca^{2+} selective channel in the plasma membrane of wheat root cells.
- Using the PLB technique, Klughammer et al. (1992a) found channels in the tonoplast of *Hordeum vulgare* that were not observed with conventional patch-clamp techniques. White et al. (1993) came to a similar conclusion after the fusion of *Acetabularia* plasma membrane-enriched microsomes with a PLB. In this respect, the PLB technique complements the patch-clamp technique by providing additional information on the presence of ion channels in (plant) cell membranes.
- As demonstrated by Klughammer et al. (1992b), the PLB technique can be used as a functional assay during channel purification.

- Compared to an artificial PLB, a native membrane is, even in the cell-free, excised patch configuration, a rather complex entity, which may complicate the interpretation of patch-clamp data.
- The PLB technique allows the study of ion channels under strictly defined conditions, not only with regard to the solutions on either side of the membrane but also with respect to the phospholipid composition of the membrane. Therefore, the PLB technique is a powerful tool for the biochemical and biophysical characterization of ion channels, e.g., with respect to the structure-function relationship. The first plant ion channels have been cloned succesfully and it is expected that in the near future the PLB technique will be applied more frequently in order to study plant ion channels on a more fundamental, biophysical level.

- Limitations of the PLB
- The highly reductionist character of the PLB technique mentioned above is, at the same time, the Achilles heel of the technique. Extrapolation of conclusions based on PLB measurements to the situation in vivo should be done carefully.
- Excessive noise and strong low-pass filtering make the PLB technique less suitable for the study of small conductances and for the more advanced analysis of channel gating kinetics.
- The effect of solvent in the bilayer, e.g., decane or hexane, on channel behavior is unclear. One may consider using solvent-free monolayers (see Miller 1986).
- Despite the use of highly purified membrane preparations, contamination by channels originating from other membranes remain a serious caveat in bilayer studies.

Acknowledgements. The recordings shown in this chapter were obtained in the laboratory of Dr. R. Poole at McGill University, Montreal. The author was supported during the writing of this chapter by NSF grant MCB-9316319 and USDA grant 92–37100–8333 to S.M. Assmann. The author thanks Drs. S.M. Assmann and Y.-R.J. Lee for helpful comments.

References

Coronado R, Kawano S, Lee CJ, Valdivia C, Valdivia HH (1992) Planar Bilayer recording of ryanodine receptors of sarcoplasmic reticulum. In: Rudy B, Iverson LE (eds) Methods in enzymology, Academic, San Diego, 207:699–707

Ehrlich BE (1992) Incorporation of ion channels in planar lipid bilayers. In: Fozzard et al. (eds) The heart and cardiovascular system. Raven, New York, pp 551–560

Hanke W, Schlue WR (1993) Planar lipid bilayers. Academic, London

Klughammer B, Benz R, Betz M, Thume M, Dietz K-J (1992a) Reconstitution of vacuolar ion channels into planar lipid bilayers. Biochem Biophys Acta 1104: 308–316

Klughammer B, Betz M, Benz R, Dietz K-J (1992b) Partial purification of a potassium channel with low permeability for sodium from tonoplast membranes of *Hordeum vulgare* cv. Gerbel. J Membrane Biol 128: 17–25

Labarca P, Latorre R (1992) Insertion of ion channels into planar lipid bilayers by vesicle fusion. In: Rudy B, Iverson LE (eds) Methods in enzymology, Academic, San Diego, 207:447–463

Miller C (1983) Reconstitution of ion channels in planar lipid bilayer membranes: a five-year progress report. Comments Mol Cell Biophys 1: 413–428

Miller C (1986) Ion channel reconstitution. Plenum, New York

Moran N, Ehrenstein G, Iwasa K, Bare C, Mischke D (1984) Ion channels in plasmalemma of wheat protoplasts. Science 226: 835–838

Neher E, Sakmann B (1976) Singel channel currents recorded from membrane of denervated frog muscle fibres. Nature 260: 779–802

Piñeros M, Tester M (1995) Characterization of a voltage dependent Ca^{2+} selective channel from wheat roots. Planta 195: 478–488

Rea PA, Poole RJ (1985) Proton translocating inorganic pyrophosphatase in red beet (*Beta vulgaris* L.) tonoplast vesicles. Plant Physiol 77: 46–52

Schulz-Lessdorf B, Hedrich R (1995) Protons and calcium modulate SV-type channels in the vacuolar-lysosomal compartment; channel interaction with calmodulin inhibitors. Planta 197: 655–671

Takeda K, Kurkdjian AC, Kado RT (1985) Ionic channels, ion transport and plant cell membranes: potential applications of the patch-clamp technique. Protoplasma 127: 147–162

Tank DW, Miller C (1983) Patch-clamped liposomes: Recording reconstituted ion channels. In: Sakmann B, Neher E (eds) Single channel recording. Plenum, New York, pp 91–106

White PJ, Tester, MA (1992) Potassium channels from the plasma membrane of rye roots characterized following incorporation into planar lipid bilayers. Planta 186: 188–202

White PJ, Tester MA (1994) Using planar lipid bilayers to study plant ion channels. Physiologia Plantarum 91: 770–774

White PJ, Smahel M, Thiel G (1993) Characterization of ion channels from *Acetabularia* plasma membrane in planar lipid bilayers. J Membrane Biol 133: 145–160

Suppliers

All the equipment used by the author is listed below. Standard equipment used in electrophysiology:

- IBM compatable 486-Computer (Gateway 2000, N. Sioux City, SD, USA)
- Patch-clamp amplifier (Axopatch-1B, Axon Instruments, Foster City, CA, USA)
- AD-Converter (DigiData 1200, Axon Instruments)
- pClamp software (version 6.0.3, Axon Instruments)
- External function generator (Model 420, Simpson, IL, USA)
- External 8-pole Butterworth low-pass filter (Model 900, Frequency Devices, Haverhill, MA, USA)
- Oscilloscope (VC-6020, Hitachi)
- Vibration free, air-lifted table (Newport Corporation, Foutain Valley, CA, USA)

Specific equipment for the bilayer setup:

- The quality of the phospholipids is essential. We purchase the lipids from: Avanti Polar Lipids, 700 Industrial Park Drive, Alabaster, Alabama 35007, tel: 1-800-227-0651, fax: 1-205-663-2494
- Cuvette holder (model BCH-13) and cuvette (model CUV-13) were from: Warner Instrument Corporation, 1125 Dixwell Ave Hamden, CT 06514, tel: 203-776-0664, fax: 203-776-1278.

  The aperture in the wall of the cuvette, e.g., its geometry, is critical and defines the quality of the bilayer. We succeeded in making our own cuvettes using polycarbonate, instead of polystyrene. The cups could be made in large quantity, thereby significantly lowering the cost. We did not see any difference between the Warner cups and these homemade ones.
- Finally, some kind of a construction table which is provisioned with a Faraday cage is needed to mount the headstage and the cuvette holder. A magnetic stirring device underneath the table drives the magnetic stirbar and, ideally, a perfusion system is set up for at least the cis chamber.

Chapter 13

Ion Channel Measurement on the Cell Nucleus

FEDERICA BERTASO, RAFFAELLA TONINI, AND MICHELE MAZZANTI

Background

The purpose of a biological membrane is to separate two different environments. The lipid bilayer can function as a semipermeable barrier against free diffusion using protein structures, ionic channels and transporters which regulate the movement of molecules through the membrane. This is known to be the case for the plasma membrane and for most membranes of intracellular organelles. The nucleus is a special intracellular component, not only because it contains the genetic material, but also because it is delimited by a double membrane, probably originating from the endoplasmic reticulum. Nucleocytoplasmic communication is achieved by unique protein complexes, the nuclear pores, which span the double membrane and modulate nuclear envelope permeability. During regular cell functioning the nuclear envelope represents a barrier only for solutes above 40 kDa; these must have a consensus sequence, a sort of molecular key, to cross the envelope through the pores. Small molecules and ions are freely diffusible in and out of the nucleus (Csermely et al. 1995), although under certain conditions the nuclear envelope is able to reduce its passive permeability. Loewenstein and colleagues demonstrated that the resistance of the envelope increases upon hormone stimulation (Ito and Loewenstein 1965). The same group, and later others, showed the presence of a nuclear resting potential (Kanno and Loewenstein 1963; Mazzanti et al. 1990; Matzke et al. 1990; Dale et al. 1994) that could be due to the presence of an ion selective membrane. More recently in situ patch-clamp experiments on *Xenopus* oocyte nuclei demonstrated that ATP is essential to observe large ionic conductances in the envelope (Mazzanti et al. 1994). However, there are still many conflicting results and opinions regarding nuclear Ca^{2+} permeability (Nicotera et al. 1989; Gerasimenko et al. 1994; Al-Mohanna et al. 1994; Stehno-Bittel et al. 1995).

Based on the results of the above mentioned investigations, we hypothesized that under particular cytoplasmic conditions, e.g., during stimulatory stress or in pathological situations, the nucleus is able to discriminate not only between large proteins and RNA but also between small particles and ions.

The patch-clamp technique, used for the study of ion permeability of cell membranes, was successfully applied to the nuclear envelope on isolated (Mazzanti et al. 1990; Matzke et al. 1990; Bustamante 1992) and in situ nuclei (Mazzanti et al. 1994). We believe that this technique will contribute to understanding nucleocytoplasmic permeability mechanisms not only for ions but, hopefully, also for proteins and RNA.

Experimental On-Nucleus Patch-Clamp Configurations

Two experimental conditions are used for single-channel recordings on the nuclear envelope. Both configurations offer advantages and disadvantages. The in situ procedure requires large transparent cells with a very tight plasma membrane (e.g., I stage *Xenopus* oocytes). With the pipette perforating the external membrane and crossing the cytosol before reaching the nuclear envelope, the possibility of obtaining gigaseals are limited; however, these experiments represent the situation most closely resembling physiological conditions. The nucleus remains in its place surrounded by nothing else other than cytoplasm. Despite these advantages of in situ measurements, the low success rate and the complex technical approach make experiments on isolated nuclei more feasible. Isolated nuclei can be easily obtained from different cell types and treated in electrophysiological experiments like cells. Of course, cell enucleation disrupts the physiological environment in which the nucleus operates. However, electron microscopy on isolated nuclei showed that the integrity of the nuclear envelope remained intact. In addition, using appropriate solutions it is possible to reestablish protein and RNA transport in vitro, demonstrating physical integrity of the system after the isolation trauma.

13.1 Isolation of Nuclei and Pronuclei

Materials

These solutions are needed for both hepatocyte and cultured cell isolated   **Solutions**
nuclei.
- 0.25 M sucrose TKM: 25 mM KCl, 50 mM MgCl2, 5 mM TRIS-HCl
  (pH 7.8), 250 mM sucrose
- 2.3 M sucrose TKM: 2.3 M TKM sucrose, 25 mM KCl, 50 mM MgCl2,
  5 mM TRIS-HCl (pH 7.8), 2.3 M sucrose
- High-K: 120 mM KCl, 2 mM MgCl2, 1.1 mM EGTA, 10 mM HEPES,
  0.1 mM CaCl2; pH 7.35 with KOH
- Bath solution (required for zygote pronuclei isolation): 120 mM KCl,
  2 mM MgCl2, 1.1 mM EGTA, 0.1 mM CaCl2, 5 mM glucose, 10 mM
  HEPES; pH 7.4 with KOH

Procedure

Isolation of Nuclei from Hepatocytes

1. Put on ice: **Hepatocytes**
- Two 60 mm glass tissue culture dishes with 10 ml of 0.25 M TKM solu-
  tion
- A Teflon/glass pestle with 0.025 µm clearance
- Two 15 ml polypropylene tubes
- Two 10 ml centrifuge glass tubes
- A syringe with its needle made of glass (from a Pasteur pipette), con-
  taining 3 ml of 2.3 M TKM solution

2. Euthanize one adult mouse by cerebral dislocation (or four/five neonatal mice by decapitation).

3. Extract the liver.

   Note. The next steps must be carried out at 4 °C.

4. Put the liver in the 0.25 M TKM dish.

5. Clean away blood and hairs, and transfer to the second 0.25 M TKM
   dish.

6. Cut the tissue into small pieces.

7. Homogenize in the pestle with 10–15 strokes at 15 000 rev/min.

8. Transfer to a 15 ml polypropylene tube and use the other tube to balance the centrifuge.

9. Spin for 5 min in a swing rotor at 1000 rpm (200 g at r max $= 16$ cm)

10. Dispose of the supernatant.

11. Resuspend pellet in 3 ml of 0.25 M TKM.

12. Add 6 ml of 2.3 M TKM.

13. Mix with Vortex at 15–20 Hz.

14. Transfer to a glass tube.

15. Underlay with 1 ml 2.3 M TKM by syringe.

16. Fill the other glass tube with 0.25 M TKM and use it to balance the centrifuge.

17. Spin for 48 min in a swing rotor at 4000 rpm (2900 g at r max $= 16$ cm).

18. Dispose of the supernatant.

19. Dilute the pellet in 3 ml of High-K solution.

20. Spin for 5 min in a swing rotor at 1000 rpm (200 g at r max $= 16$ cm)

21. Repeat the last three steps twice (washing in High-K).

22. Wait at least an hour before using the nuclei.

Note. Special attention should be paid to step 15: At the end of the centrifuge run there will be a red band of erythrocytes in the gradient; carefully remove the supernatant, avoiding contact between the pellet and this band (clean the tube walls with a cotton tip before resuspending). This helps in obtaining a purer preparation of nuclei (see Fig. 13.1).

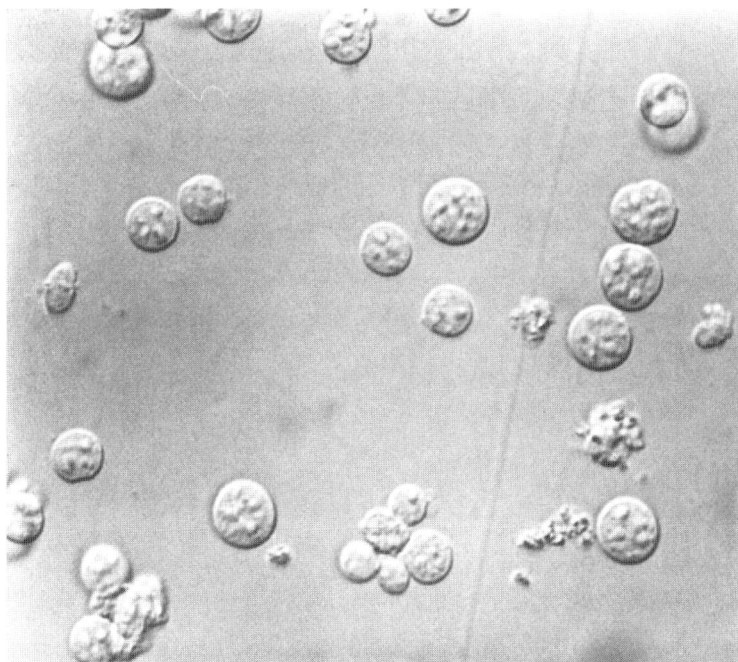

Fig. 13.1. Isolated liver nuclei obtained with the isolation procedure. The nuclei lie on the bottom of a polypropylene Petri dish and require no treatment to adhere to the surface. Mouse liver nuclei are about 10 μm in diameter

Isolation of Nuclei from Cultured Cells

To obtain a good yield, approximately 9–10 million cells should be collected. For example, three $75 \, cm^2$ flasks seeded with about $300\,000$ cells/ml can be used.

1. Prepare on ice:
   - A glass pestle (clearance $0.018 \, \mu m$)
   - Two 10 ml centrifuge glass tubes
   - Two 15 ml polypropylene tubes
   - A syringe with a long needle (e.g., made of glass from a Pasteur pipette), containing 3 ml of 2.3 M TKM solution

   Note. The next steps must be carried out at 4 °C.

2. Eliminate culture medium from one dish.

Note. If cells are in a flask, it is useful to make an incision on top of the flask with a welder: proceeding will then be simpler.

3. Scrape the bottom of the Petri dishes by either vigorously washing or by using a Teflon scraper in 2 ml of 0.25 M TKM solution.

4. Put the cells in the pestle using a Pasteur pipette, and rinse the dish with 1 ml of 0.25 M TKM to remove all the cells.

5. Repeat the last three steps for the other dishes.

6. Homogenate the cells with 100 up-and-down strokes by hand.

7. Spin the homogenate in a swing rotor for 5 min at 1000 rpm (200 g at r max = 16 cm)

8. Dispose of the supernatant.

9. Resuspend pellet in 3 ml of 0.25 M TKM.

10. Add 6 ml of 2.3 M TKM.

11. Mix with Vortex at 15–20 Hz.

12. Underlay with 1 ml 2.3 M TKM by syringe.

13. Spin for 48 min in a swing rotor at 4000 rpm (2900 g at r max = 16 cm)

14. Dispose of the supernatant.

15. Dilute pellet in 3 ml of High-K solution.

16. Spin for 5 min in a swing rotor at 1000 rpm (200 g at r max = 16 cm)

17. Repeat the last three steps twice (washing in High-K).

18. Wait at least an hour before using nuclei.

Micromanipulation of Mouse Zygote Pronuclei

Mouse zygote pronuclei are another useful model to study nuclear ionic permeability. They can be obtained from female mice (e.g., B6D2 F_1 mice) by superovulation with 5IU human choriogonadotropin (Sigma Chem. Co., St. Louis, U.S.A.) and individual mating with males.

Zygotes are collected about 12 h after mating. To obtain individual zygotes, fallopian tubes are surgically removed from the mouse and gently pierced with a fine stainless steel needle.

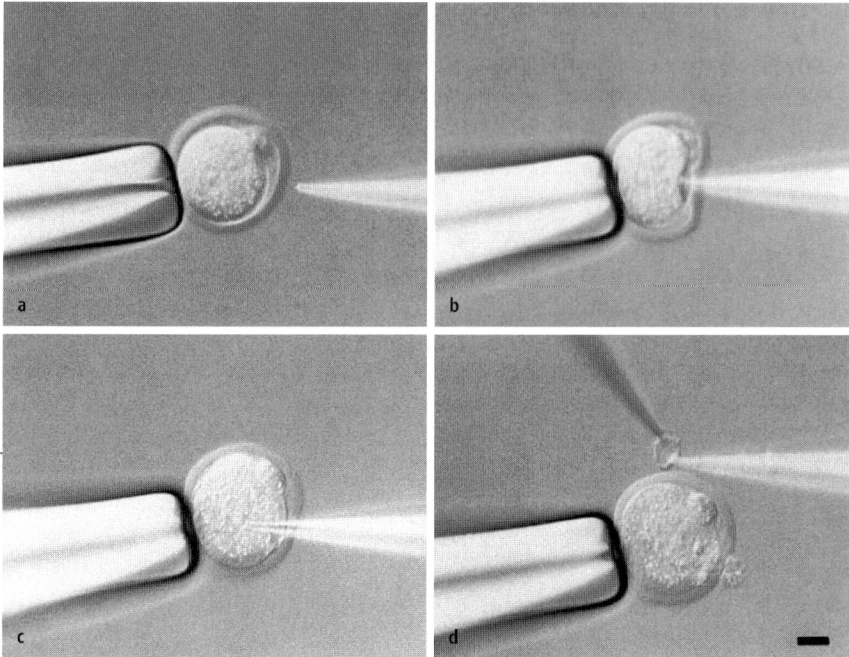

Fig. 13.2a–d. Micromanipulation of mouse zygote pronuclueus. A large-suction glass pipette holds the mouse zygote, while a smaller pipette approaches it (**a**). The small pipette enters the zygote (**b**), attaches to the pronucleus by gentle suction(**c**) and extracts it from the zygote (**d**)

Cumulus cells are removed from the zygotes by incubation in a 1 % hyaluronidase solution for several minutes.

After transferring the zygote to the bath solution, one of the pronuclei is extracted using micromanipulation (see Fig. 13.2).

Micromanipulating requires the installation of two pipettes: one large glass pipette fashioned from Pyrex glass (outer diameter 1 mm; inner diameter 0.65 mm) and one smaller glass pipette (outer diameter 75–100 µm). These pipettes are to be used in performing the following steps:

1. Connect both pipettes to micromanipulators and provide for suction systems.

2. Place the pipettes close to the microscope field (magnification 10x).

3. Hold the zygote by the zona with light suction via the large pipette.

4. Enter the zygote with the smaller suction pipette.

5. Attach the pipette to the pronucleus.

6. Extract and hold the pronucleus for patch-clamping.

Note. Steps 3–6 should take less than 10 min. There is no need to distinguish between sperm and oocyte pronuclei.

Xenopus Oocytes Preparation

Before starting oocyte preparation, the solutions required for both their isolation and maintenance must be prepared.

Collegenase
- Collagenase A, Boehringer-Mannheim, cat. number 103–586, from *Clostridium histoliticum*. Collagenase activity is crucial for avoiding damage to the oocytes. The enzyme should be diluted in sterile OR-2 solution (see below) at a concentration of 1 mg/ml (0.15 U/ml). It is convenient to prepare 50 ml of solution with 100 mg of collagenase. Store the solution at −20 °C in 10 ml aliquots in large (13 ml) polypropylene tubes.

Note. If you use another collagenase lot or type, try different concentrations between 0.5 and 2 mg/ml till you obtain good oocytes separation.
- OR-2 (calcium-free): 82.5 mM NaCl, 2 mM KCl, 1 mM MgCl2, 5 mM HEPES; pH 7.5 with NaOH.
- ND-96: 96 mM NaCl, 2 mM KCl, 1.8 mM CaCl2, 1 mM MgCl2, 5 mM HEPES, pH 7.5 with NaOH. Supplement ND-96 solution with 550 mg/liter sodium pyruvate and 50 µg/ml gentamicin (use 100x stock from Boehringer-Mannheim).

Note. It is critical to treat the oocytes with collagenase in the absence of calcium. Contaminant calcium in the solution is enough to activate the collagenase. Otherwise the calcium ions activate other proteases in the collagenase, which results in markedly decreased oocyte viability.

Acquisition of oocytes
Before starting this procedure, sterilize surgical instruments (bistoury, scissors, forceps), for example, in 70 % ethanol for at least 2 h.

1. Anesthesia: Anesthetize the frog using one of the following two anesthetics:
- Benzocaine (ethyl-*p*-aminobenzoate): prepared as a 100x stock in ethanol (3 % solution). Dilute this 1:100 (15 ml to 1.5 liters of water) for

anesthesia. Store the stock solution at 4 °C, and discard the anesthetic solution after use. To anesthetize the frog, immerse it in a container containing the 0.03 % Benzocaine, and carefully watch the effect of the anesthetic. (Benzocaine is very potent, and the frogs go from moving around to complete anesthesia very quickly.) The whole process takes 10–20 min. As soon as the frog is anesthetized, remove it immediately from the Benzocaine solution and immerse it in water to wash off residual anesthetic. (This is critical otherwise the Benzocaine will continue to be adsorbed through the skin).

– Tricaine (MS-222; is 3-amino benzoic acid ethyl ester). Dissolve in water at a concentration of 0.17 %, which is the appropriate concentration to use for anesthesia. The solution should be stored at 4 °C and can be reused many times. (This anesthetic is not as effective nor as consistent as Benzocaine and may require from 15 min to 1 h for complete anesthesia). After anesthesia the frog should be rinsed as described above for Benzocaine.

Note. For both anesthetics, the effect of anesthesia can be tested by turning the frog upside down. If she does not right herself, she is out. Do not leave the frog in either anesthetic for any length of time once anesthetized, otherwise she may not recover.

Note. Wear gloves while handling the frog in either anesthetic, since the solution is not healthy for humans either.

2. Place the frog upside down on a frozen surface (e.g., a basin with ice) to keep the frog anesthetized.

3. Pierce the skin in the abdomen with a bistoury (the skin is very tough and this requires some pressure).

4. Make a small incision with the scissors (about 5 mm in length) corresponding to the frog groin. Cut through the underlying fascia as well.

5. Remove the oocytes using forceps. Take as many as are needed; scissors are usually required to separate them while removing them. Place the oocytes in a 60 mm tissue culture dish containing OR-2.

6. Use one stitch to hold the two sides of the incision in the fascia and two stitches for the skin (tie with double square knot).

7. Place the frog in a container of glass-distilled water to allow her to recover. Full recovery takes a few hours, after which she can be returned to the vivarium.

Note. Since frogs must respire in air, they can drown when immersed in water. Therefore place the frog in a small amount of water after surgery, tilted so that the mouth is at the surface. Change the water if it becomes cloudy, and carefully monitor the frog's recovery.

Oocytes Preparation Protocol

1. Thaw one tube of collagenase at room temperature. This should be done before removing the oocytes from the frog. Divide the 10 ml into two equivalent aliquots in two polypropylene (13 ml) tubes. Store one tube at 4 °C and use the other for the first collagenase treatment.

2. Tease apart the oocytes using two pairs of fine forceps. It is not possible to separate the oocytes completely but it is necessary to at least partially separate them.

3. Remove as much of the OR-2 as possible, and pour the oocytes into the tube of collagenase.

4. Place the tube in a rack sideways on a rotator (to keep containers in motion) for 1–1.5 h at room temperature (around 20 °C).

5. Remove the collagenase from the tube with a Pasteur pipette, and replace it with the collagenase aliquot which was left at 4 °C.

6. Place the tube back on the rotator. The total time for collagenase treatment should be about 2.5–3 h. This depends on the lot of collagenase being used. Too short a time results in first stage oocytes hardly separable from the other types. Too long a time results in greatly decreased oocyte viability, such that the oocytes may not survive further manipulation. Oocytes should initially be monitored by microscopy to determine the optimal time for each lot of collagenase.

7. After the appropriate time, remove the collagenase by pipette. Thoroughly wash the oocytes with 5–10 ml OR-2, placing the tube on the rotator for few minutes; then pipet off the OR-2.

   Note. It is critical to remove all the collagenase and contaminating proteases by multiple washes because the final solution used for incubation contains calcium, which will activate the proteases and result in decreased viability. Four to five washes are generally adequate.

Note. Be careful not to pipette off the small first stage oocytes with each solution change. It is also not a good idea to ever pipette the oocytes, since this frequently results in squashing a large number of them.

8. After the OR-2 washes are complete, wash the oocytes in supplemented ND-96 for two to three washes. This is the final incubation solution.

9. After the ND-96 washes are complete, pour the oocytes into a 60 mm tissue culture dish with supplemented ND-96.

10. Select the oocytes to be used and transfer them into 35 mm tissue culture dishes using a glass pipette with a fire-polished glass tip.

13.2
Patch Clamp

Materials

Isolated nuclei exactly resemble cells, with a smooth surface and no **Equipment** branchings. Furthermore, it is possible to apply the same techniques used for ion channels in plasma membranes to the ionic permeability pathways of the nuclear envelope. With in situ nuclei there is the additional problem of reaching the nucleus beyond the cell membrane, but the required patch-clamp equipment is the same. The basical equipment needed for patch-clamping on isolated and in situ nuclei resemble the setup for single cell study (Sakmann and Neher 1995).

- Microscope. Due to the small size of the isolated nuclei, for optimal results at least 30x–40x magnification is required. An inverted phase-contrast microscope (e.g., Zeiss, Axiovert 35) is recommended as it provides a better view of the shape of the nucleus. In situ nuclei require a lower order of magnification (10x).

- Electrodes. Patch-electrodes are pulled from borosilicate glass capillaries, 1.75 mm in diameter, on a Sachs-Flaming puller (P-87, Sutter Instruments). They are coated with a resin (Sylgard, Dow Corning) that decreases tip capacitance and then fire-polished to an external final diameter of 1–2 μm; final electrode resistance is around 7–10 MΩ, as measured with High-K solution for both the bath and the patch pipette.

- Micromanipulators. Precise movements are essential for dealing with the small dimension of nuclei. Motor-driven manipulators may be preferred.

- Patch-Clamp Amplifier.

Single channel currents through the nuclear envelope can be in the range of those flowing through plasma membranes with a normal step protocol or even larger. This fact makes the normal patch-clamp amplifiers suitable for nuclear current recording. The nucleus-attached patch-clamp configuration is based on the same principle of the traditional cell-attached patch-clamp configuration described in several texts (e.g., Sakmann and Neher 1995).

Depending on nuclei type, configuration (isolated or in situ nuclei), and conditions, the nuclear envelope may or may not show a resting potential. When there is a resting potential, it is in the order of few millivolts (Loewenstein and Kanno 1963). Thus, steady state condition can be chosen, observing the spontaneous current flux. However, it is always useful to elicit nuclear currents via a voltage step protocol that reveals further features of the currents of interest. For doing this a stimulation apparatus (i.e., "the step command") that is built into the patch-clamp amplifier or, better, a programmable stimulator can be used.

Data Acquisition and Storage

Single channel recording always requires a high level of signal resolution. Electrophysiological experiments in nuclei present even larger problems than encountered with traditional application of the patch-clamp technique on cell membranes. The seal resistance is usually not higher than $20\,G\Omega$ and during in situ experiments there is an additional stray capacitance due to the pipette coming into contact with the cell membrane and the cytoplasm. Both these factors increase the noise level.

Nuclear single channel recording often shows current amplitudes that are bigger than those measured in conventional ionic channels.

A high-quality, electrophysiological data storage system, complete with a low-pass filter between 1 and 2 kHz, will be sufficient to obtain good quality recordings.

Procedure

On-Nucleus Patch-Clamp

Once isolated nuclei are obtained (see procedure above), they must be seeded in the experimental chamber and rinsed with High-K solution. The nuclei adhere to the bottom of the chamber, without any special treatment of the surface, in a few minutes. After this time they can be worked with and even used in a perfusion system without detaching from the chamber surface.

Under the microscope (magnification 32x) the nuclei appear similar to small dishes; sometime it is possible to distinguish them from the few erythrocytes remaining from the isolation procedure (Fig. 13.1).

The following steps are designed to study nuclear ionic permeability under the simplest conditions, with nuclei in a solution that mimics the ionic composition of the cytosol and with the electrode filled with the same solution. This protocol can be used as the basis for any further investigations, changing the external (bath) and electrode solutions.

1. Set the nucleus to be examined under the microscope light.

2. Fill the patch pipette with filtered High-K solution

3. Place the electrode into the holder.

4. Provide a little positive pressure into the electrode: this keeps the tip clean and facilitates seal formation

5. Focus the electrode tip next to the nucleus, in the High-K solution

6. A step voltage command (a few millivolts) will help to monitor the pipette conditions (accidentally broken tip, blocked tip, etc.)

7. Use the micromanipulator to gently place the electrode just onto the center of the nucleus.

8. Release the pressure and give gentle suction: a seal should be formed within 10–20 s.

Basic protocol

Recordings of current flux via voltage step protocols can now be carried out.

A perfusion system is necessary if the effects of changing the external solution, varying the solute concentrations and adding cytoplasmic modulation factors (ATP, calcium, etc.) are to be studied.

In Situ Nuclei

The presence of the cell membrane represents a clear obstacle in this configuration, as it has to be crossed with a glass electrode, avoiding any cytoplasmic loss (this is just the aim and advantage of this technique!) and forming a seal on the nuclear envelope. Accurate electrode (i.e., micromanipulator) movements and the presence of a system for controlling pipette pressure are critical for success in these kind of experiments.

The following protocol allows observation of current fluxes through the nuclear envelope under control conditions.

Solutions
– Ringer solution: 140 mM NaCl, 5 mM KCl, 0.5 mM $MgCl_2$, 1.5 mM $CaCl_2$, 5 mM glucose, 10 mM HEPES; pH 7.4 with NaOH
– Electrode solution: 120 mM KCl, 2 mM $MgCl_2$, 1.1 mM EGTA, 0.1 mM $CaCl_2$, 5 mM glucose, 10 mM HEPES, pH 7.4 with KOH.

In situ patch-clamping

1. Enzymatically isolated first stage oocytes are transferred in the Ringer solution

2. Using a micromanipulator, a glass holding pipette is placed near the oocyte, always focusing them both under the microscope light (magnification 10x).

3. A light negative pressure keeps the oocyte attached to the holding pipette; this allows easier operation with the patch electrode

4. The patch electrode, filled with the specified solution, is focused next to the oocyte border; maintaining a light positive pressure in the pipette to keep the tip clean.

5. A step voltage command (a few millivolts) helps in monitoring the pipette conditions (accidentally broken tip, blocked tip, etc.)

6. Using the micromanipulator, approach the plasma membrane until it is broken.

   Note. The oocyte membrane should adhere to the electrode without tearing apart; the presence of Sylgard on the tip helps considerably.

7. Once next to the nucleus surface, gently release a small amount of solution using the pipette: this partially cleans the surface.

8. Gently push the patch pipette against the nuclear surface: a seal should be formed in a few seconds.

Now the voltage step protocols can be delivered and ionic fluxes can be observed through the envelope.

Results

Examples

The patch-clamp technique allows the recording of ion channel activity on the nuclear envelope surface (Fig. 13.3). The nuclear ion channels are potassium selective on the basis of their response to ion species substitutions (Mazzanti et al. 1990).

Independent of nuclear preparation, the ion channels display a comparable behavior: outward currents shows an inactivation component, while inward currents do not change during the entire voltage challenge (data not shown). Even though different current amplitudes can be observed, ranging from tens of picoamperes to some nanoamperes, the ionic flow maintains similar kinetic characteristics. This observation suggests a homogeneous channel population.

Figure 13.4 shows current traces from an oocyte in the nucleus-attached configuration. In this case the current amplitudes recorded in the presence of millimolar ATP in the pipette are tenfold greater than those measured when ATP is absent, as shown by Mazzanti et al. (1994).

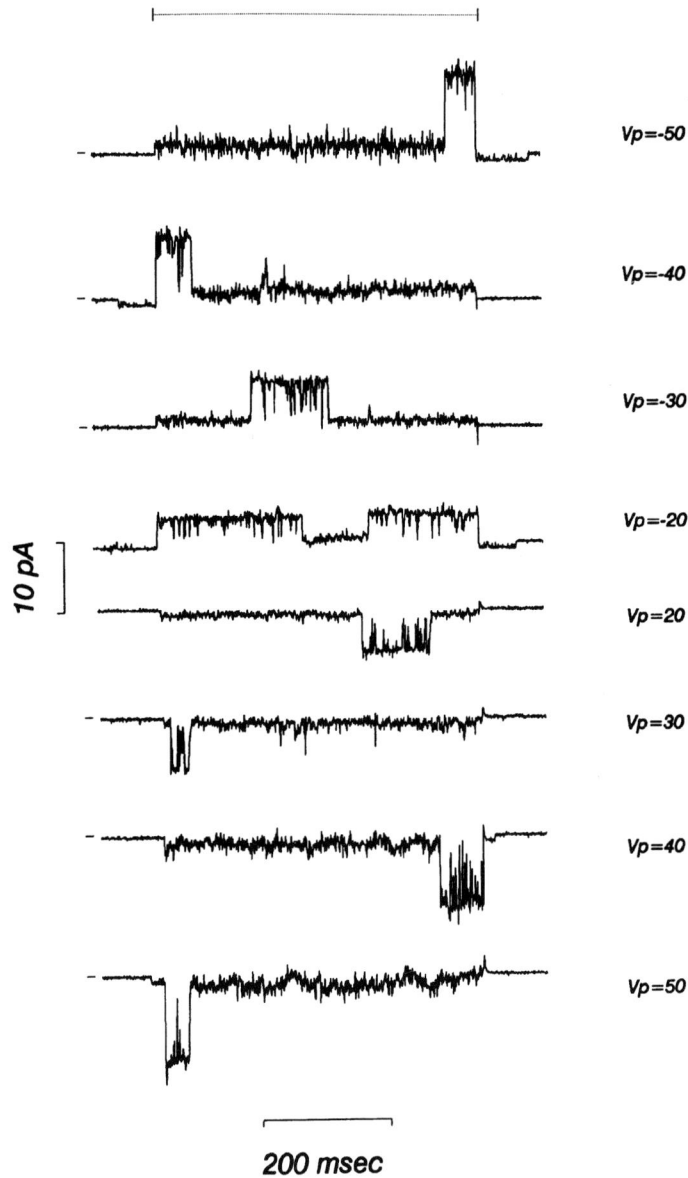

Fig. 13.3. Single channel recordings in the nucleus-attached configuration on isolated liver nuclei. Application of different voltage steps (from −50 mV to +50 mV, pipette voltage) elicits ion channel-like currents through the nuclear envelope with clear open-close transitions. In this example, the pipette and bath solutions were the same (High-K, see text). Other experiments, done under exactly the same conditions, showed currents with larger amplitudes but similar shapes

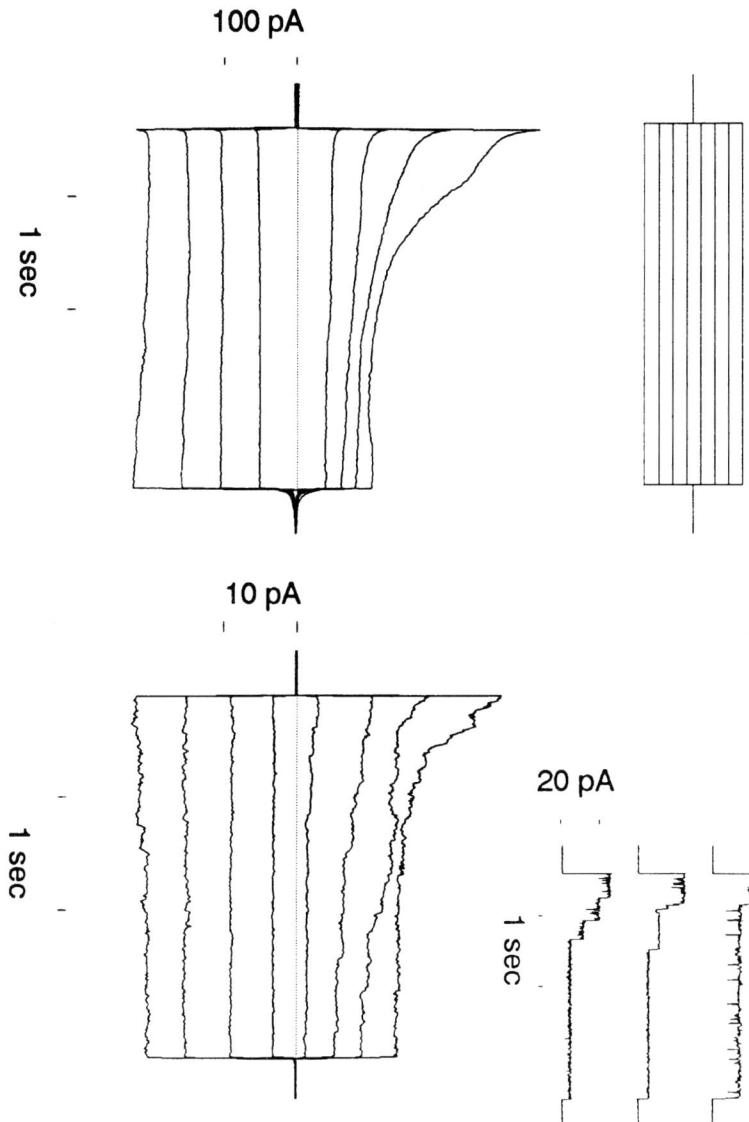

Fig. 13.4. Single channel recordings and average currents from in situ nuclei. A typical voltage step protocol, from $-25\,mV$ to $+25\,mV$ (*upper left*) evokes "New York Skyline-like" current traces: the *upper right panel* depicts three current traces at $Vp = -25\,mV$. These current recordings present multilevel transitions and often time-dependent kinetics. *Below* Averages of several recordings showing the difference in current amplitude whether ATP is (*lower left*) or is not present (*lower right*) in the pipette solution. Inactivation kinetics of the outward current remained in both situations

References

Bustamante JO (1992) Nuclear ion channels in cardiac myocytes. Pflueg. Arch. 421:473–485

Loewenstein WR, Kanno Y (1963) Some electrical properties of a nuclear membrane examined with a microelectrode. J.Gen. Physiol. 46:1123–1140

Kanno Y, Loewenstein WR (1963) A study of the nucleus and cell membranes of oocytes with an intracellular electrode. Exp. Cell Research 31:149–166

Ito S, Loewenstein WR (1965) Permeability of a nuclear membrane: changes during normal development and changes induced by growth hormone. Science 150:909–910

Matzke AJM (1990) Detection of a large cation-selective channel in nuclear envelope of avian erythrocytes. FEBS Lett. 271:161–164

Mazzanti M, De Felice LJ,Cohen J, Malter H (1990) Ionic channels in the nuclear envelope. Nature 343:764–767

Mazzanti M, Innocenti B, Rigatelli M (1994) ATP-dependent ionic permeability on nuclear envelope in in-situ nuclei of *Xenopus oocytes*. FASEB J. 8:231–236

Sakmann B, Neher E (1995) Single-channel recording (2nd edn), Plenum, New York

Monitoring Cytosolic Calcium and Membrane Potential from Cell-Attached Patch-Clamp Currents in Human T Lymphocytes

Jos A.H. Verheugen

Background

In the cell-attached configuration of the patch-clamp technique (Hamill et al. 1981) currents through ion channels present in the small area under the pipette are measured while the attached cell remains intact and capable of physiological functioning. Two aspects of the cell's physiology, the intracellular Ca^{2+} concentration ($[Ca^{2+}]_i$) and the membrane potential (V_m), can exert effects on ion channels in the cell membrane and thus modify the currents through the patch. This chapter describes a method to infer the state of $[Ca^{2+}]_i$ from the activity of Ca^{2+}-activated K^+ [K(Ca)] channels and V_m from the reversal potential of current through voltage-gated K^+ [K(V)] channels in cell-attached patches. At periods of elevated $[Ca^{2+}]_i$, V_m can also be estimated from the single channel amplitude of the K(Ca) channel, which allows one to check the reliability of the reversal potential measurements. In contrast to more conventional techniques for measuring $[Ca^{2+}]_i$ and V_m, e.g., intracellular fluorescent probes such as fura-2 and bis-oxonol, the use of K(Ca) and K(V) channel activity as a bioassay for $[Ca^{2+}]_i$ and V_m makes it possible to monitor both parameters simultaneously, and without disturbing the cytoplasmic environment. In addition, no equipment other than a patch-clamp setup is needed. The method has been validated and calibrated for activated human peripheral blood T lymphocytes (Verheugen et al. 1995), which express elevated levels of highly K^+ selective K(Ca) and K(V) channels (IK-type and $K_v1.3$, respectively; Table 14.1). However, in principle this method could be applied to any cell with Ca^{2+}-activated and voltage-gated channels.

Table 14.1. Summary of the characteristics of the voltage-gated $K_v1.3$ K(V) channel and the voltage-independent IK-type K(Ca) channel in human lymphocytes

Property	K(V)		K(Ca)	
	Low $[K^+]_{out}$	High $[K^+]_{out}$[a]	Low $[K^+]_{out}$	High $[K^+]_{out}$[a]
Single channel conductance	9–12 pS	35 pS	11 pS	30–50 pS
Activation midpoint	−40 mV	Identical	450 nM Ca^{2+}	Identical
Inactivation time constant[b]	160–300 ms	430 ms		
Recovery from inactivation[b]	30–60 s	5–10 s		
Steady state inactivation[b]	>−60 mV	Identical		
Ion selectivity	K^+ (1) >Rb^+ (0.8) >NH_4^+ (0.1) >Cs^+ (0.02) >Na^+ (0.01)		K^+ (1) >Rb^+ (0.96) >NH_4^+ (0.17) >Cs^+ (0.07)	
Blockers	Charybdotoxin, noxiustoxin, kaliotoxin (nM) 4AP (10^2 µM) TEA (mM), etc.		Charybdotoxin (nM)	

[a]Properties vary to some extent with the composition of the pipette and the bath solution, in particular with the extracellular K^+ concentration
[b]The K(V) channel inactivates upon opening with a time constant as indicated. Recovery from inactivation is slow and incomplete at depolarized potentials (steady state inactivation).

Outline

Short protocol

1. K^+ currents through K(Ca) and K(V) channels are measured in the cell-attached configuration of the patch-clamp technique. A high $[K^+]$ pipette solution is used, which gives a Nernst potential (see Chap. 6) for K^+ currents across the patch close to 0 mV.

2. After formation of a gigaseal between the pipette and the cell, the patch potential ($V_m + V_h$) is set at a hyperpolarized potential ($V_h = -60$ mV; i.e. a pipette potential of +60 mV with respect to "zero" bath potential).

3. At periods of elevated $[Ca^{2+}]_i$, activity from K(Ca) channels is observed as inward single channel currents (Fig. 14.1a). The open

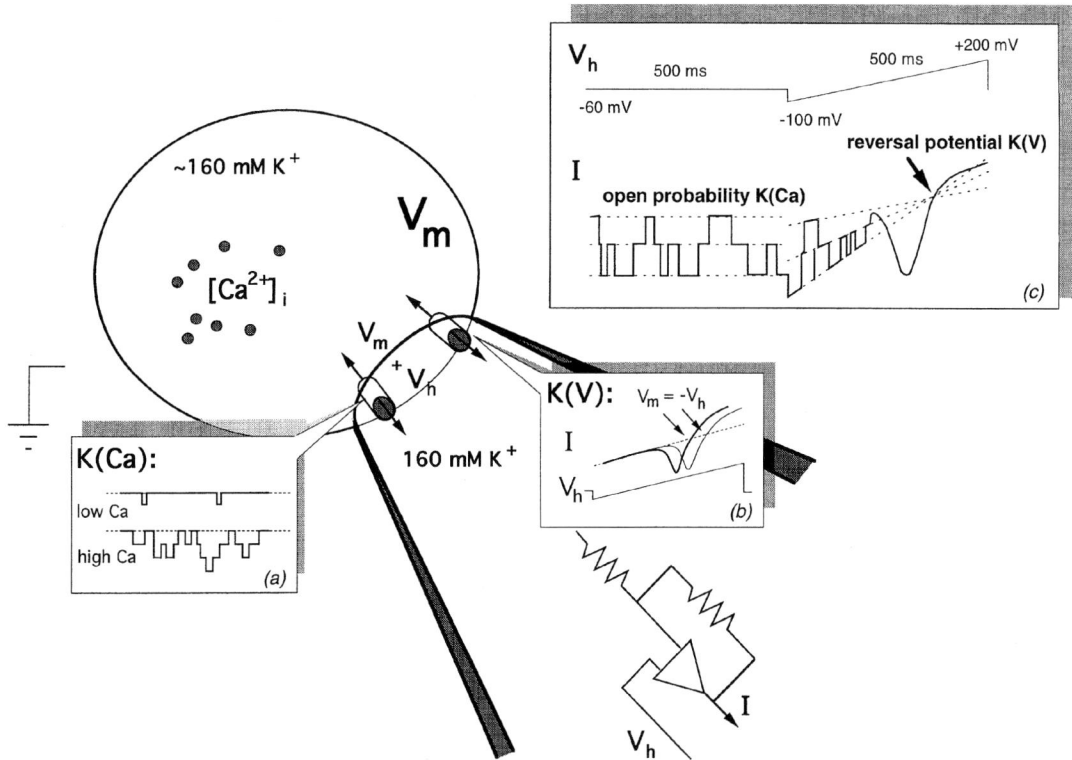

Fig. 14.1a–c. Use of the patch-clamp technique to infer $[Ca^{2+}]_i$ from the activity of Ca^{2+}-activated K(Ca) channels (**a**) and the membrane potential, V_m, from the reversal of K^+ current through voltage-gated K(V) channels (**b**) in cell-attached patches. Both parameters can be monitored simultaneously from the same cell by combined measurements of currents through the two kinds of K^+ channels (**c**).

probability of the K(Ca) channel gives a quantitative measure for $[Ca^{2+}]_i$.

4. K(V) currents are evoked by a depolarizing voltage ramp stimulation. The point at which the K(V) current reverses from inward to outward indicates the membrane potential of the cell (Fig. 1b).

5. In lymphocytes, combined $[Ca^{2+}]_i$ and V_m measurements can be obtained as often as one every 5 s by recording K(Ca) channel activity during a 0.5 s period at $V_h = -60$ mV and subsequent K(V) current recording during a 0.5 s voltage ramp from -100 to $+200$ mV (Fig. 14.1c), followed by a 4 s interval to allow recovery of the K(V) channel from inactivation caused by the depolarizing voltage ramp.

Materials

Cells   Lymphocytes are isolated from peripheral blood using a standard Ficoll-Paque (Pharmacia, Uppsala, Sweden) sedimentation technique. Isolated cells are cultured at a density of 0.5×10^6 cells/ml in RPMI supplemented with 10% heat-inactivated fetal calf serum, 1 mM glutamine, 5 mM pyruvate, 100 U/ml penicillin and 100 µg/ml streptomycin at 37 °C in a humidified atmosphere containing 5% CO_2. T lymphocytes are activated by a 24 h incubation with 10 µg/ml PHA-P (Difco Laboratories, Detroit, MI, USA or Wellcome, Dartford, UK), which is added to the culture medium from 10 mg/ml aliquots kept at −20 °C and thawed shortly before use. Medium is refreshed every 2–3 days. Activated T cells, which express higher levels of K(Ca) and K(V) channels compared to resting cells, can be used at least until 7–10 days after isolation.

Solutions   – Bath (low K^+) solution (in mM): 160 NaCl, 4.5 KCl, 2 CaCl$_2$, 1 MgCl$_2$, 5 HEPES, adjusted to pH 7.4 with NaOH and 330 mOsm with sucrose.
– Pipette (high K^+) solution (in mM): 160 KCl, 2 CaCl$_2$, 1 MgCl$_2$, 10 HEPES, adjusted to pH 7.4 with KOH and 310 mOsm with sucrose.

Electrophysiology

Equipment and software   Detailed description of the patch-clamp technique and single channel analysis can be found elsewhere (Hamill et al. 1981; Chap. 6). The stimulation equipment has to allow generation of voltage ramps. Analysis software should include routines for the calculation of single channel open time or the integral of single channel current traces, and for conversion of current recordings into amplitude histograms. Line fitting routines can be helpful. However, many details of the analysis of the cell-attached currents as described in this chapter are difficult to automate and involve manual examination of individual current records. Apart from the digitized data collection, recording of the analog current and voltage signal continuously on paper is recommended during the entire experiment as a rapid visual reference.

Patching   Pipettes (borosilicate glass, GC150, Clark) are pulled with a resistance of 2–6 MΩ, fire-polished and filled with the 160 mM K^+ solution described above. The junction potential between pipette and bath solution is compensated prior to patch formation (for possible errors arising from junction potentials see Barry and Lynch 1991). Membrane currents are recorded using low-pass filtering at 2 kHz. The prerequisite for accurate cell-attached K^+ channel measurements is a high-resistance seal, allowing low-noise single channel recording with a limited leak component during voltage ramp stimulation. In lymphocytes, cell-attached patches with resistances of 10–100 GΩ often remain stable for periods as long as 1 h. Deterioration of the seal is manifest as an increase of the leak current, which is especially apparent from the slope of the current during the voltage ramp. Abrupt disappearance of the K(V) component from the current evoked by voltage ramps usually indicates patch excision, whereas spontaneous break-in to whole-cell is seen as a transient increase in the holding current, arising from activation and subsequent inactivation of the K(V) current.

V_m is the cell's membrane potential; V_h the holding potential (in the cell-attached configuration the **pipette** potential, V_p, is $-V_h$). Note that in the cell-attached configuration the potential across the patch is determined by both V_m and V_h, while the potential across the remainder of the membrane area is V_m. Furthermore, in cell-attached recordings inward currents are currents which flow **out of** the pipette. The high K^+ pipette solution gives a Nernst potential for K^+ (E_K) across the patch close to 0 mV, but for the rest of the cell membrane E_K is about -89 mV in normal low K^+ bath solution.

Electrophysiological definitions

Procedure

Following gigaseal formation, V_h is set at -60 mV, i.e., the potential across the patch is 60 mV hyperpolarized relative to the cell's membrane potential, V_m. The hyperpolarized potential allows recovery from inactivation of K(V) channels which are partially inactivated at the resting cell membrane potential, but prevents spontaneous activity of this channel which would disturb K(Ca) channel recordings. In addition, it amplifies K(Ca) currents by increasing the driving force for K^+ influx across the patch, thereby improving the signal-to-noise ratio. With the above pipette solution, both the K(Ca) and the K(V) channel display a single channel conductance of $\sim$33 pS. The mean open time of the K(Ca) channel is $\sim$3 ms. Open events of single K(V) channels, which are often observed in the tail currents of depolarizing voltage stimulations, have a mean duration of $\sim$200 ms.

[Ca^{2+}]$_i$ Measurements from the K(Ca) Channel Activity

K(Ca) channel recording

The probability of K(Ca) channel opening increases with increasing intracellular Ca^{2+} concentrations. Thus, by following the K(Ca) channel open probability, P_o, in cell-attached patches the [Ca^{2+}]$_i$ of the attached cell can be monitored (Fig. 1a). IK-type K(Ca) activity of lymphocytes is recorded at $V_h = -60$ mV. At this potential, K(Ca) currents appear as inward single channel events of which the opening frequency is determined by [Ca^{2+}]$_i$, and the amplitude by the difference between E_K (~0 mV with high K$^+$ pipette solution) and the patch potential $V_h + V_m$. The K(Ca) activity should be sampled such that changes in P_o caused by fluctuations in [Ca^{2+}]$_i$ can be discriminated from stochastic variation in channel activity at a steady [Ca^{2+}]$_i$. The duration of individual current traces should be chosen such that they contain $10^2 - 10^3$ events at periods of high [Ca^{2+}]$_i$ and they should be sampled at a rate at least five times faster than the expected frequency of calcium fluctuations. In lymphocytes, current traces of 250–1000 ms, recorded every 2–20 s, provide the best compromise between a good time resolution for detection of changes in [Ca^{2+}]$_i$ and the amount of data needed for a reliable estimate for the K(Ca) channel open probability.

K(Ca) channel analysis

The P_o of the K(Ca) channel is calculated as the total open time of the channel divided by the observation time. The value for the open probability (formally $N \cdot P_o$, in which N is the number of channels in the patch), measured in this way, can become >1 when the patch contains more than one channel. The open time can be calculated from the integral of the K(Ca) current, which is the area under the closed level (dotted line Fig. 14.1a). The K(Ca) current integral represents the charge movement across the membrane through the K(Ca) channels. This parameter is influenced not only by the open time of the K(Ca) channels but also by the amplitude of the single channel currents, which varies with changes in V_m. Therefore, the integral has to be divided by the actual single K(Ca) channel amplitude at the time of the recording to obtain a correct measure for the channel open time. The single channel amplitude is most accurately determined when the current trace is converted into an amplitude histogram, as the difference between the closed level and the first open level. The open time can also be calculated directly from each amplitude histogram (open time = Σ[number of points × open level]/ total number of points) or from idealized current traces generated by the single channel analysis software, which both bypass the need to correct for V_m fluctuations.

In human T lymphocytes, the activity of the K(Ca) channel correlates with $[Ca^{2+}]_i$ according to the Michaelis-Menten equation

$$N.P_o \, /N.P_{o \, max} = 1/(1 + \{K_d/[Ca^{2+}]\}^n) \tag{1}$$

with a K_d of 450 nM and Hill coefficient, n, of 3 as determined by simultaneous measurements of cell-attached patch-clamp currents and fura-2 measurements of $[Ca^{2+}]_i$ of the same cell (Verheugen et al. 1995). The K_d is the $[Ca^{2+}]_i$ at which K(Ca) channel activity is half maximal, whereas the Hill coefficient or slope factor is generally considered to indicate the stoichiometry of molecular interaction, suggesting that approximately three Ca^{2+}ions have to bind to the K(Ca) protein to open the channel. The threshold for activation of the K(Ca) channel is between 100 and 200 nM. Maximum activity occurs at values close to 1 µM. Higher concentrations of $[Ca^{2+}]_i$ inhibit K(Ca) channel activity. The open probability of the K(Ca) channels can be used as a relative measure for $[Ca^{2+}]_i$. Alternatively, an absolute measure can be derived when $P_{o \, max}$ is known. $P_{o \, max}$ may be determined by gradually elevating $[Ca^{2+}]_i$ for instance by exposing the cells to increasing concentrations of 10–1000 nM thapsigargin.

Thapsigargin

Thapsigargin blocks the Ca^{2+} pump in the membrane of the endoplasmic reticulum, thereby unmasking a passive Ca^{2+} leak which empties these intracellular Ca^{2+} stores. In lymphocytes and many other cells emptying of the stores in turn activates a Ca^{2+} conductance in the cell membrane, the so-called Ca^{2+} release-activated Ca^{2+} current or I_{CRAC}, thus leading to a biphasic rise in $[Ca^{2+}]_i$. Thapsigargin (Calbiochem) is dissolved in DMSO and stored in 1 mM aliquots at $-20°$ C until use.

Example

During the sporadic K(Ca) channel activity in an unstimulated lymphocyte, a K(Ca) current integral of $0.09 \, 10^{-12}$ C is measured during a 500 ms recording period at $V_h = -60$ mV with a single K(Ca) channel amplitude amounting to 3.0 pA, which gives a total open time of $0.09 \, 10^{-12}$ C/ 3 pA = 30 ms (with a single channel conductance of 33 pS the single channel amplitude of 3.0 pA indicates a V_m of -31 mV because -3 pA/ 33 pS = -91 mV = $V_m + V_h$). During Ca^{2+} responses following T cell receptor stimulation, integrals of 2.0×10^{-12} C to $8.8 \, 10^{-12}$ C with a single channel amplitude of, respectively, 4.0 pA and 4.9 pA are measured, which yields open times of 500 ms (and a V_m of -61 mV) to 1800 ms (and a V_m of -89 mV). The $N \cdot P_o$, i.e., the open time divided by the 500 ms observation time, in these examples is 0.06, 1 and 3.6, respectively. The maximum K(Ca) channel activity, as determined by exposing the cells to an optimum concentration of thapsigargin, was found to be $12.3 \ 10^{-12}$ C/500 ms, with a single channel amplitude of 4.9 pS, giving a $N \cdot P_{o \ max}$ of 5. Using Eq. (1), these activities of the K(Ca) channel indicate values for $[Ca^{2+}]_i$ of 100, 280 and 620 nM, respectively. In the unstimulated cell, the absence of K(Ca) channel activity during most of the time indicates levels of $[Ca^{2+}]_i$ below the threshold for K(Ca) channel activation. The hyperpolarization of V_m at periods of K(Ca) channel activity, as is manifest from the increase in single channel amplitude, agrees with a shift towards E_K (which is approximately -89 mV for the cell in normal bath solution) as a consequence of an increased K^+ conductance of the cell membrane.

Optimum number of K(Ca) channels

The sensitivity to detect small changes in $[Ca^{2+}]_i$ is increased if the patch contains a large number of K(Ca) channels. However, the single channel amplitude and closed level are difficult to determine when the K(Ca) channel activity is too strong. The number of K(Ca) channels in the patch can be modulated to some extent by changing the size of the pipette tip (which is inversely related to the pipette resistance). In human lymphocytes, additional control over the number of K(Ca) channels can be achieved by using cells at different stages of activation, which influences the level of K(Ca) channel expression. In lymphocytes, maximum levels of K^+ channel expression occur 3–5 days following PHA activation, when channel numbers >10 per patch are not unusual. Since almost invariably a small transient increase of $[Ca^{2+}]_i$ occurs following contact between the cell and the patch pipette, a rough indication of the number of channels in the patch is obtained from K(Ca) channel activity occurring immediately after seal formation. If seal formation is followed by a short period of modest K(Ca) channel activity ($N \cdot P_o$ of ~0.5–2) one can assume that the patch contains a suitable number of channels for $[Ca^{2+}]_i$ measurements. If initial K(Ca) channel activity is absent or too extensive, one might prefer to try another cell.

$[Ca^{2+}]_i$ Measurements from Ca^{2+}-Activated Channels in Other Cells and Limitations of the Method

In principle, $[Ca^{2+}]_i$ can be monitored from the activity of any Ca^{2+}-activated channel in cell-attached patches, provided that its Ca^{2+} sensitivity has been characterized. Voltage-sensitive Ca^{2+}-dependent channels, however, can only be used as calcium monitor if changes in V_m are taken into account. The Ca^{2+} dependence of Ca^{2+}-activated channels can be assessed by exposing excised inside-out patches that contain one or more channels to solutions buffered at various Ca^{2+} concentrations. Alternatively, if the experimental conditions allow, the Ca^{2+} sensitivity can be evaluated from simultaneous measurements of $[Ca^{2+}]_i$ with fluorescent Ca^{2+} probes and cell-attached patch-clamp currents of the same cell (Verheugen et al. 1995). The latter approach has the advantage that the Ca^{2+} sensitivity of the channel is determined under conditions identical to those during recordings of the channel to monitor $[Ca^{2+}]_i$ as described in this chapter, thus avoiding errors resulting from possible changes in channel properties upon patch excision. To assess the Ca^{2+} sensitivity in this manner $[Ca^{2+}]_i$ is manipulated during the combined fura and cell-attached recordings either pharmacologically, e.g., by ionomycin or thapsigargin, or physiologically by stimulating receptors which generate a Ca^{2+} signal, e.g., the T cell receptor in lymphocytes. The K(Ca) channel activity is then plotted against the corresponding $[Ca^{2+}]_i$. The relation between $[Ca^{2+}]_i$ and the P_o of the channel is described by a dose-response relation of the form shown in Eq. (1), by adjusting the values for the K_d and Hill coefficient to obtain the best fit of the experimental data.

Especially for Ca^{2+}-activated channels with a steep Ca^{2+} dependence (i.e., a high Hill coefficient), reliable estimates for $[Ca^{2+}]_i$ from the channel activity are limited to concentrations in a narrow range around the K_d, which are usually physiologically the most relevant concentrations. The determination of $[Ca^{2+}]_i$ from the channel activity is further limited by the threshold for activation, while very high Ca^{2+} concentrations may inhibit the channel, as is the case in human lymphocytes.

V_m Measurements from the K(V) Current Reversal Potential

The method to infer the state of the cell's membrane potential, V_m, from the current through K(V) channels in cell-attached patches is illustrated in Fig. 14.1b. The potential across the patch is determined by the applied holding potential, V_h, controlled by the patch-clamp amplifier, and V_m, which can fluctuate according to the activity of the cell. Since the Nernst potential for K^+ across the patch is ~ 0 mV with high $[K^+]$ in the pipette, the V_h at which the direction of K^+ currents reverses ($V_h + V_m = 0$ mV) indicates the V_m at that moment.

When a voltage ramp from -100 to $+200$ mV (0.6 mV/ms) is applied, a K(V) current is evoked (superimposed on a linear leak current), which is initially inward beyond the threshold for activation. After reaching a

Determination of K(V) current reversal

peak, the current declines and reverses when the pipette potential cancels V_m. To determine the reversal potential, a correction for the linear leak current has to be made by extrapolation of the slope of the closed level at potentials below the threshold for K(V) channel activation (Fig. 14.1b, dotted line). At periods of K(Ca) channel activity, the determination of the leak current may be hampered by the absence of a closed level (Fig. 14.1c). However, in ramp recordings with many K(Ca) channel openings the point of K^+ current reversal can also be found by determining the intersection between the current trace and the extrapolated linear fits through the distinct open levels of the K(Ca) channel (Fig. 14.1c).

Because voltage-dependent inactivation of the K(V) channel occurs at depolarized potentials, an interval of at least 5 s at $V_h = -60\,mV$ between subsequent ramp stimulations has to be observed to allow recovery from this inactivation.

Validation of the V_m Measurements from the K(V) Current Reversal Potential

As was discussed above, the activity of the K(Ca) channel current not only indicates $[Ca^{2+}]_i$, but the value of V_m can be inferred from its single

Fig. 14.2A, B. A K(Ca) and K(V) activity in a cell-attached patch of an activated human T lymphocyte, recorded at rest, in the presence of 1 µM ionomycin, in high K^+ bath solution and after returning to low K^+ solution. Currents were recorded with high K^+ solution in the pipette at intervals of 20 s at $V_h = -60\,mV$. **B** Time course of the changes in reversal potential of the K(V) current, and in the open probability ($N \cdot P_o$) and single channel amplitude of K(Ca) channels in the same cell as in **A**. Initially, the K(V) current reversal potential fluctuated around +60 mV (+60±6 mV) indicating membrane potential fluctuations around −60 mV in the unstimulated cell. K(Ca) activity was essentially absent. After addition of 1 µM ionomycin (t = 500 s), a dramatic increase of $N \cdot P_o$ of the K(Ca) channel activity occurred, concomitant with a shift of the K(V) reversal potential to +89 ± 3 mV, pointing to a rise of $[Ca^{2+}]_i$ and a hyperpolarization of V_m to the K^+ Nerst potential, E_K, as a result of the increased K^+ conductance of the cell membrane. Wash out of ionomycin (t = 800 s) reversed these effects. Changing the bath solution to high K^+ solution (t = 1400 s) caused the K(V) reversal to shift to lower values, indicating a membrane depolarization, and partial inhibition of the K(V) current. After returning to the normal external solution from high K^+, a transient burst of K(Ca) activity occurred, resulting in an overshoot of V_m beyond the resting value. The changes in V_m are also reflected in the single channel amplitude of K(Ca) channels, which activate during periods of elevated $[Ca^{2+}]_i$ (*lower panel*). (Reproduced from Verheugen and Vijverberg, Cell Calcium, 1995, 17:287–300 with the permission of Churchill Livingston Medical Journals)

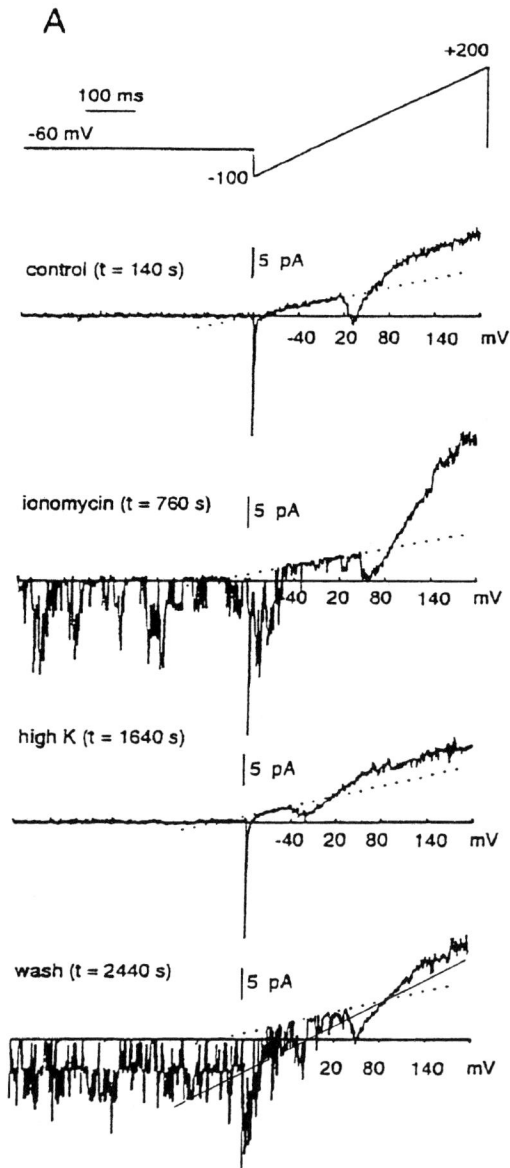

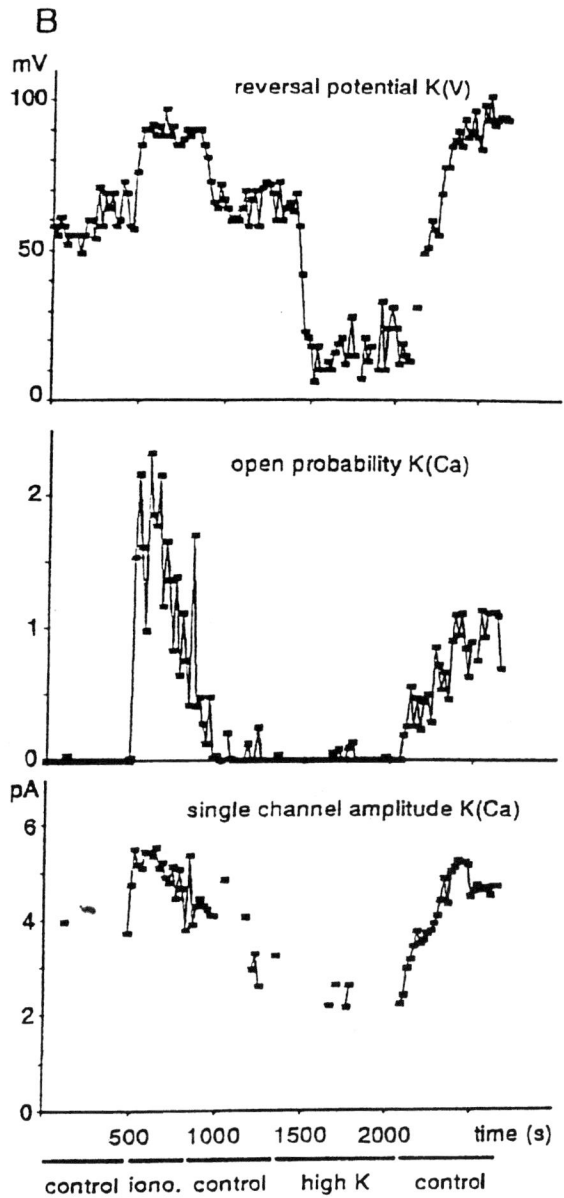

channel amplitude. While these V_m measurements are restricted to periods of K(Ca) channel activity (see Fig. 14.2B, lower panel) and thus are not as generally applicable as using the K(V) current reversal potential, current records which contain activity from both channels can be used to verify the reliability of the estimates from the K(V) current reversal potential. Recordings from which the single channel conductance of the K(Ca) channel can be determined, i.e., when K(Ca) channel openings occur during the ramp (Figs.14.1C, 14.2A, lower trace), are particularly useful for this purpose because from these current traces two complete and independent quantitative V_m estimates can be obtained.

Example   The amplitude of the single K(Ca) channel current depends on the membrane potential according to Ohm's law $U = I * R$, where U is the the potential difference between the Nernst potential for K^+ and the patch potential, I is the single channel current and R is the reciprocal of the single channel conductance (1/S). Thus:

$$I = S * (E_K - (V_m + V_h)) \tag{2}$$

The single channel conductance, S, can be calculated from the change in K(Ca) channel amplitude during the ramp stimulation as $S = \Delta I/\Delta V$.

   In the lower current trace of Fig. 2a, the slope through the first open level of the K(Ca) channel during the initial phase of the ramp stimulation, corrected for the leak current (straight line – dotted line), is 3.4 pA per 100 mV yielding a single channel conductance of 34 pS. The single channel amplitude at $V_h = -60$ mV amounts to 5.0 pA. This indicates a difference between E_K and the patch potential, $V_m + V_h$, of 5.0 pA/34 pS = 147 mV. With $E_K \approx 0$ mV and $V_h = -60$ mV, this yields a V_m of -87 mV. Reversal of K(V) current occurred at 88 mV, suggesting a membrane potential of -88 mV. The close agreement between the two estimations confirms the validity of using K(V) current reversal as monitor of V_m in this cell.

V_m Measurements from Voltage-Gated Channels in Other Cells and Limitations of the Method

The method as described in this chapter could be applied to any cell which expresses voltage-gated channels. However, channels that are selective for an ionic species which is present at a high intracellular concentration, i.e. K^+ and Cl^-, are preferred since in that case even fairly large unspecified concentration fluctuations of the ion or deviations from the estimated intracellular concentration will affect the Nernst potential across the patch only to a limited extent. The kinetics and voltage dependence of steady state inactivation and use-dependent inactivation of the channel should be assessed to determine the required minimum interval and holding potential between subsequent stimulations to allow recovery from these two kinds of inactivation.

At strongly depolarized membrane potentials, inhibition of the voltage-gated channel may occur as a result of steady state inactivation (see Fig. 14.2A, current recorded with high K^+ bath solution). This can easily be overcome by switching V_h in between stimulations to more negative values than -60 mV. Another problem which may be encountered when applying the present method is inhibition of the voltage-gated channel by intracellular factors, such as cAMP, protein kinase C (PKC) or high concentrations of Ca^{2+}, as has been reported for the K(V) channel in lymphocytes. However, in lymphocytes significant effects of Ca^{2+} are usually seen only at levels beyond that attained during most physiological Ca^{2+} elevations, whereas the effects of cAMP and PKC are in general quite modest.

The accuracy of the determination of the reversal potential improves with the amplitude of the voltage-gated current. However, a large inward current evoked during the voltage ramp stimulation may compromise V_m measurements by causing an artefactual depolarization. The extent of the disturbance of V_m by cell-attached currents depends on the input resistance of the attached cell (see Barry and Lynch 1991). In small cells with high input resistances, such as resting lymphocytes, the method has to be applied with caution. In activated lymphocytes, a K(V) current at the inward peak between 3 and 12 pA yields good results. In large cells with lower input resistances, larger currents will still have limited effects on the membrane potential. One can have some control over the current amplitude by adjusting the size of the pipette tip, thus modulating the number of channels in the patch. Effects of the voltage-gated current on V_m will be apparent from deviations between estimations based on single channel amplitude and current reversal potentials as described above. Another sign that currents through the patch are affecting V_m is the distortion of the normal rectangular shape of single channel openings into events with an exponential decay of the current amplitude (Barry and Lynch 1991).

Optimum number of K(V) channels

Results

By combining periodic measurement of K(Ca) channel P_o at $V_h = -60$ mV with the determination of the K(V) current reversal potential during subsequent voltage ramp stimulations, the $[Ca^{2+}]_i$ and V_m of the same cell can be followed in the course of time. This is illustrated in Fig. 14.2, where the interaction between both parameters in a human T lymphocyte is shown upon addition of ionomycin, which elevates $[Ca^{2+}]_i$, and after changing the bath solution to a high K^+ solution, which depolarizes V_m. In this example current traces were recorded during 500 ms periods at $V_h = -60$ mV followed by a 500 ms voltage ramp from -100 to $+200$ mV.

References

Barry PH, Lynch JW (1991) Liquid junction potentials and small cell effects in patch-clamp analysis. J Membrane Biol 121:101–117

Hamill OP, Marty A, Neher E, Sakmann B, Sigworth FJ (1981) Improved patch-clamp techniques for high-resolution current recording from cells and cell-free membrane patches. Pflügers Archiv 391:85–100

Verheugen JAH, Vijverberg HPM, Oortgiesen M, Cahalan MD (1995) Voltage-gated and Ca^{2+}-activated K^+ channels in intact human T lymphocytes: noninvasive measurements of membrane currents, membrane potential, and intracellular calcium. J Gen Physiol 105:765–794

Verheugen JAH, Vijverberg HPM (1995) Intracellular Ca^{2+} oscillations and membrane potential fluctuations in intact human T lymphocytes: role of K^+ channels in Ca^{2+} signaling. Cell Calcium 17:287–300

Ion Channel and Membrane Potential Measurements Using the Patch-Clamp Technique

II-3 Whole Cell Measurements

Membrane Potential and Action Potential Measurements in Whole Cell and Perforated Patch Configurations

Zheng Wang, Dirk L. Ypey, and Rutgeris J. Van den Berg

Background

The cell membrane is a permeability barrier separating the intracellular, cytoplasmic, space from the extracellular environment. In neurons, as in any other cell type, the intracellular ionic composition differs from that of the extracellular fluid. Of the four major ions in neurons, Na^+ and Cl^- are more concentrated outside the cell, and K^+ and organic anions are more concentrated inside the cell. The organic anions are negatively charged amino acids and proteins, which are nonpermeant. Due to the selective permeability of the membrane for different ions, an electrical potential difference across the plasma membrane is present under resting conditions. This potential difference is, therefore, called the resting membrane potential (RMP). The RMP is important in regulating neuronal excitability (Hille 1992).

Direct measurement of the neuronal membrane potential can be achieved by using diverse techniques, e.g., the microelectrode and patch-clamp techniques (Hille 1992; Allen and Burnstock 1987). The patch-clamp technique is based on electrical isolation of a small cell membrane patch by using glass micropipettes (patch pipettes), which permit the recording of ionic currents flowing through single ion channels. One of the advantages of this technique is that different measurement configurations (cell-attached patch, whole cell, inside-out and outside-out) can be obtained for membrane potential or current measurement. In the current-clamp whole cell mode, resting membrane potentials or action potentials can be recorded by holding the current at zero or by applying depolarizing current steps. In voltage-clamp whole cell mode, the ionic currents through the whole cell membrane can be measured at the clamped membrane potential (Hille 1992).

In this chapter, we will describe the patch-clamp method we used for measuring membrane potentials and action potentials of rat dorsal root ganglion (DRG) neurons. Firstly, the preparation of DRG neurons will be described. Then, the conventional and perforated patch, whole cell experimental (Hamill et al. 1981) procedures for membrane potential measurement will be described in a step-by-step fashion.

Resting Membrane Potential

To understand the origin of the RMP (see also Chap. 6) we firstly suppose that the neuronal membrane is only permeable to K^+ ions. Since the K^+ concentration is larger inside than outside the cell, K^+ will tend to diffuse out of the cell. However, the progressive build-up of negative charges inside the cell (the remaining anions) counteracts the further movement of K^+ out of the cell due to electrostatic retraction. Thus, the diffusional force causing the charge separation results in an electrical potential difference. The difference in electrical potential increases as the diffusion of K^+ out of the cell continues, until it reaches a value that has an effect on K^+ equal and opposite to the effect of the concentration gradient. This value is called the K^+ equilibrium or Nernst potential, E_K, and can be calculated according to the Nernst equation:

$$E_K = \frac{RT}{zF} \ln \frac{[K^+]_o}{[K^+]_i} \tag{1}$$

where R is the gas constant, T the absolute temperature, z the valence of K^+, F the Faraday constant and $[K^+]_0$ and $[K^+]_i$ the concentrations of K^+ at the outside and inside of the cell, respectively. Under these conditions, RMP is determined solely by the K^+ equilibrium potential (about -80 mV for physiological K^+ concentrations). Usually, the number of displaced charges to establish the Nernst potential is so small that the existing intra- and extracellular concentrations can be considered as unchanged (Junge 1992; Hille 1992).

Since the neuronal membrane is permeable to Na^+ and Cl^- too, both ions will have an influence on RMP. For Na^+, the concentration gradient and electrical gradient act on it to propel it into the cell. For a cell with a resting membrane potential of -80 mV, the equilibrium potential for Na^+, E_{Na}, calculated from the Nernst equation may be, for example, about $+55$ mV. Under these conditions RMP is influenced by both E_K and E_{Na} and develops to a value at which outward movement of K^+ balances the inward movement of Na^+. However, since the resting neuronal membrane is highly permeable to K^+ but only slightly permeable to Na^+, the membrane potential actually moves somewhat away from E_K but never comes close to E_{Na}. For the neuron to have a steady resting membrane potential the net influx and efflux of charges are balanced by a Na^+/K^+ pump actively transporting Na^+ ions from the cell into the extracellular fluids and K^+ ions in the opposite direction. The Na^+/K^+ pump is not neutral but electrogenic, because it pumps three Na^+ out for two K^+ ions into the cell. Thus, it produces a net flux of charge across the membrane. In some neurons, like DRG neurons, the potential generated by the electrogenic pump hardly contributes to the resting membrane potential (Junge 1992; Gorman and Marmor 1970; Wang et al. 1994). The Cl^- ions are only subject to passive forces (electrical and chemical gradients). Thus, the concentration gradient of Cl^- across the membrane usually is in equilibrium with the existing RMP.

The Goldman-Hodgkin-Katz Equation

Taking all the ion concentration gradients and permeabilities of the membrane to these ions together, while ignoring the contribution of the Na^+/K^+ pump, the neuronal RMP can be described by the Goldman-Hodgkin-Katz equation (Hodgkin and Huxley 1952):

$$RMP = \frac{RT}{F} \ln \frac{P_K[K^+]_o + P_{Na}[Na^+]_o + P_{Cl}[Cl^-]_i}{P_K[K^+]_i + P_{Na}[Na^+]_i + P_{Cl}[Cl^-]_o} \tag{2}$$

where P_K, P_{Na} and P_{Cl} are the membrane permeabilities for the ions K^+, Na^+ and Cl^-, respectively, and $[I]_i$ and $[I]_o$ the intra- and extracellular concentrations of the indicated ions. This equation applies only to the already existing RMP. If RMP would change for some reason, the P values may also change since they are voltage-dependent. If the membrane potential becomes more negative or positive than RMP, these changes are termed hyperpolarization or depolarization, respectively.

A depolarization of the membrane by current pulses to a certain voltage level (threshold V_{th}) causes a rapid opening (activation) of Na^+ channels, resulting in an inward Na^+ current. This current causes further depolarization, which leads to the generation of the action potential. After the peak of the action potential, the membrane potential rapidly repolarizes back to the resting level. This is caused by the inactivation of Na^+ and the activation of K^+ channels. After this, some hyperpolarization follows due to prolonged opening of K^+ channels. Action potentials are important for intercellular communication (Hille 1992).

15.1
Preparation of Dorsal Root Ganglion Neurons

Materials

The dissection and dissociation procedures require the following:

- 1 day old postnatal Wistar rats.
- Laminar flow hood (DLF/BSS6, Clear Air Tech. Woerden, NL).
- A dissection microscope (objective magnification 5–20x).
- One 37 °C incubator with humidified 5 % CO_2 atmosphere (Forma-science Brouwer Scientific, de Meern, NL).
- One sterile surgical knife.
- One pair of sterile scissors.
- Two fine pairs of sterile tweezers.
- Sterile plastic Petri dishes with a diameter of 35 mm (Greiner, Alphen a/d Rijn, NL).
- Coverslips (diameter of 24 mm, Boon, Meppel. NL) coated with poly-D-lysine hydrobromide (MW 70 000–150 000; Sigma, St Louis, USA).
- 50 ml 50 % alcohol.
- 100 ml sterile physiological salt solution.
- 10 ml culture medium composed of 9 ml F14 culture medium (Imperial Lab, UK) enriched with 10 % horse serum (Gibco, NL; Wood et al. 1988).

Procedure

The whole procedure is performed under sterile conditions in a laminar flow hood and includes the following steps:

Dissection of the Dorsal Root Ganglia

1. Rinse the newborn rats in alcohol for 1 min to ensure sterile conditions during the dissection.

2. Kill the rats by decapitation using a surgical knife. Rinse the bodies of rats in sterile physiological salt solution to wash away the blood. The rat is placed on its venter.

3. Remove the back skin using a pair of scissors until the spinal column becomes visible (Fig. 15.1A). The spinal canal is opened by making

four cuts: one dorsal-longitudinal through the vertebral arch, one ventral-longitudinal through the vertebral body, allowing 3–4 mm on each side of the canal, and two transversal ones, one made rostrally just above the first rib and one caudally to eliminate the tail. The spinal cord becomes visible and can be taken off. Now, the dorsal root ganglia can be seen in the intervertebral foramina under the dissection microscope (Fig. 15.1B).

4. DRGs are collected with fine tweezers from different levels of the spinal cord in a Petri dish with 2 ml culture medium.

   Note. This procedure was approved by the Animal Ethics Committee at the Leiden Medical Faculty (approval number: FFF-10), as required by Dutch law. Please check the regulations in your institute before performing the dissection

Dissociation of Neuronal Somata from the Ganglia

1. Three or four ganglia are transferred to a drop of 0.5 ml of culture medium on a poly-D-lysine coated coverslip in a Petri dish.

2. Fine pairs of tweezers are used to disrupt the capsules of the ganglia and to dissociate the cells by triturating the ganglia into minute fragments. The minute fragments containing a few to about a hundred cells are used together with the dissociated single cells for culturing.

Culturing the DRG Dissociate

1. The dissociated neurons and non-neuron cells (Schwann cells) are allowed to attach to the coverslips in an incubator for a period of 4 h.

2. 2 ml of culture medium is added to the coverslip and the cell dissociate is cultured in a 37 °C incubator with humidified 5 % CO_2 atmosphere. The culture medium is replenished every day.

   Note. Freshly dissociated neurons have a spherical shape, sometimes with a stump of the disrupted axon. These cells have a bright appearance under the phase contrast microscope in the patch-clamp setup (Fig. 15.1C). The neurons sprout neurites within a few hours and grow neurites of 50–500 μm length in 24 h (Fig. 15.1D). Thus, if one wants to measure the soma properties before the outgrowth of significant pro-

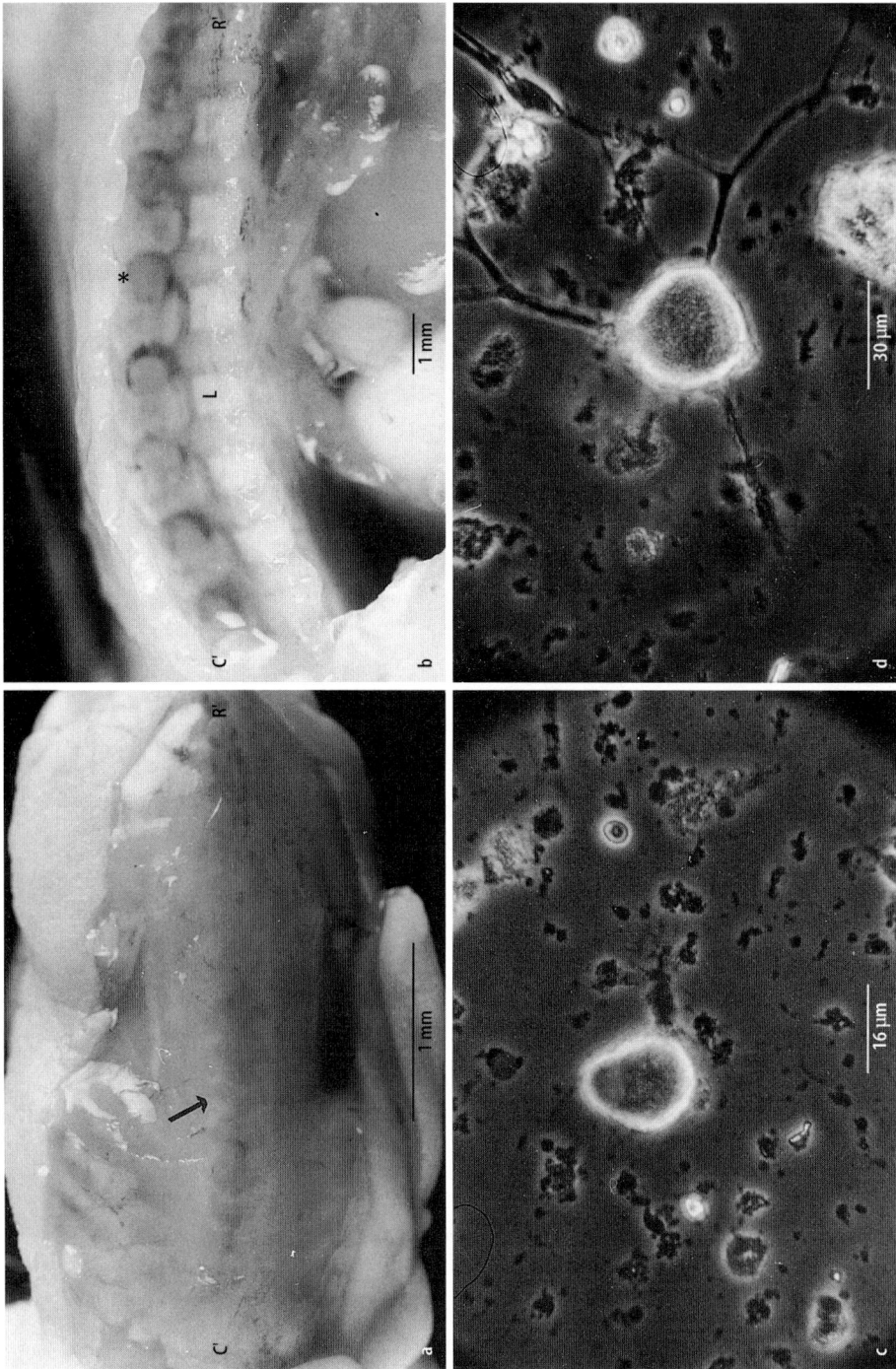

Fig. 15.1A–D. Dissection of dorsal root ganglia from a 1-day old neonatal rat. **A** Dorsal view of the vertebral column (*arrow*) after the removal of back skin. *R'* is the rostral end and *C'* is the caudal end. **B** Dissected lumber part of the vertebral column left side. View from medial to lateral after removal of the cord. Parts of the lumber can be identified: *L* is the vertebral corpus. *DRG* is indicated as an *asterisk*. *R'* is the rostral end and *C'* is the caudal end. **C** A spherical neuron (average diameter of 20 μm) with its nucleus centrally positioned, after a culture period of 5 h. The beginning of process formation can be seen. **D** Multipolar DRG neurons (diameter of 30 μm) after a culture period of 24 h. The newly formed processes can be seen

cesses, as in the present experiments, one should measure within a few hours of isolation. If one wishes to measure the properties of the growth cones of the neurites, one may choose older cultures.

15.2
Conventional Whole Cell Patch-Clamp Experiments

Materials and Procedure

Preparation of the Neuron Culture for Recording

Coverslips with cultured DRG neurons are transferred to an open bottom chamber (Ince et al. 1983) which can be attached to the stage of an inverted phase contrast microscope (Zeiss Invertoscope D). The cells are covered with a 400 μm layer of extracellular-like solution (ECS; Table 15.1). To study the origin of RMPs, membrane potentials are measured with intracellular and extracellular solutions of different ion compositions. For example, pipettes are filled with low Cl^- ICS and the cells can be bathed in high K^+ ECS or in Na^+-free ECS (Table 15.1). With these solutions the contributions of Cl^-, K^+ and Na^+ ions to the setting of RMP can be studied. To explore the roles of Na^+ and K^+ channels during

Table 15.1. Extracellular and intracellular solutions used for investigating the role of specific ion conductances in resting membrane potential and action potential properties of DRG neurons[a]

	ECS	K^+ ECS	4-AP ECS	TEA ECS	Na^+-free ECS	ICS	Low Cl^- ICS	ICS nys
NaCl	140	5	140	130	–	10	10	–
Choline Cl	–	–	–	–	140	–	–	–
KCl	5	140	5	5	5	140	–	–
CaCl$_2$	1	1	1	1	1	1	1	1
Glucose	6	6	6	6	6	–	–	–
MgCl$_2$	1	1	1	1	1	2	2	2
HEPES	10	10	10	10	10	10	10	10
4-AP	–	–	3	–	–	–	–	–
TEA	–	–	–	10	–	–	–	–
EGTA	–	–	–	–	–	10	10	10
KCH$_3$SO$_4$	–	–	–	–	–	–	140	140
Choline Cl	–	–	–	–	–	–	–	10
pH	7.2	7.2	7.2	7.2	7.2	7.2	7.2	7.2

[a] Concentrations given in mM.

action potential generation, we study action potential properties in Na^+-free ECS, 4-aminopyridine ECS (4-AP ECS) and tetraethylammonium-ECS (TEA-ECS, Table 15.1). Note that all the solutions must be filtered through a nuclepore filter (0.22 μm, Molsheim, France) before use. This procedure improves sealing properties.

Preparation of Patch Pipettes

Glass capillaries of about 1.5 mm outside diameter with a small glass filament in the lumen (drawn from Clark GC 150-TF 15 borosilicate glass capillaries, UK) are used to make patch pipettes. By heating the capillary in the middle, two pipettes can be pulled by a vertical or horizontal two-stage electrode puller. The pipette tips are heat-polished under the microscope to clean and round off the tips for better sealing properties. The external tip diameter of the heat-polished pipettes should be approximately 3 μm. Such pipettes are filled with an intracellular-like salt solution (ICS; Table 15.1) by using a thin plastic needle with a tip diameter of around 0.5 mm. The pipette is placed in a pipette holder, which is connected to a micromanipulator.

Note. EGTA containing solutions can acidify in a steel needle due to binding of metal ions from the steel.

Patch-Clamp Setup

For membrane potential measurements, the conventional whole cell configuration of the patch-clamp technique is used with an EPC-7 amplifier (List Electronic, Darmstadt/Eberstadt, Germany). The amplifier is connected to the pipette solution by an Ag/AgCl pellet electrode. Another Ag/AgCl electrode makes contact between the bath and the ground of the amplifier. These Ag/AgCl pellets convert currents carried by ions to currents carried by electrons. The recordings are stored on computer disk after analog to digital conversion using a TL-100 interface (Axon Instruments, Burlingame, CA). For stimulus protocols and data acquisition we use the software package "pClamp" (version 6.01, Axon Instruments, Burlingame, CA).

15.3
Conventional Whole Cell Membrane Potential Measurements

Procedure

Electrode Offset Compensation

The pipette is placed in the bath solution and positioned above the cells using a micromanipulator. A slight positive pressure is applied inside the pipette so that the stream of fluid out of the tip keeps the tip clean from attaching particles in the bath solution. Prior to patching the cell, the pipette potential due to electrode offset voltages and to diffusion potentials (junction potentials) is compensated by using the amplifier V_p offset knob in voltage-clamp mode.

The pipette potential may vary between 1 and 10 mV and adds to the measured membrane potential. Thus, it is important to take it into consideration. Under our experimental conditions with or without slight overpressure in the pipette (ICS), the junction potential is probably small (~ -3 mV, Neher 1992) and can be neglected. With low Cl$^-$ ICS in the pipette, the junction potential will be increased (~ -6 mV, Neher 1992). Because the precise value is not known and the correction made in ECS is in fact not correct for the whole cell configuration, the measured membrane potential values are not corrected for this value, but nullifying the initial -6 mV offset implies that the actual resting membrane potential may have been ~ 6 mV more negative.

Junction potential

Continuous voltage pulses of 1 mV lasting 50 ms are applied through the pipette by a command voltage generator to measure pipette resistance R_{pip} in voltage-clamp mode. In the whole cell configuration the input resistance of the amplifier R_{amp} is in series with R_{pip}. Usually R_{amp} is low compared to R_{pip}, so that the voltage across the former can be neglected. Thus, the value of R_{pip} can be calculated by dividing the voltage stimulus by the steady state current response. The value of R_{pip} ranges from 1 to 4 MΩ.

Measuring pipette resistance

Sealing and Measuring Seal Resistance

The pipette is slowly brought down to touch the membrane of a neuron without damaging or penetrating it, while applying 1 mV voltage pulses in voltage-clamp mode. The contact between the pipette and the cell membrane is noticed as a reduction of the current response on the oscilloscope. The positive pressure is then immediately turned off and a slight underpressure is applied to the pipette, usually by mouth suction. A further reduction of the current response is then observed due to a better sealing between the pipette tip and the membrane. This seal can be represented by a resistance, R_{seal} (Fig. 15.2A). When R_{seal} is increased in the range of several to tens of giga-ohm, the cell-attached patch (CAP) configuration is achieved. In this configuration, the resistances of the patch, R_{patch}, and the seal, R_{seal}, are in parallel (Fig. 2A). When R_{patch} is very high as compared to R_{seal} in the CAP, current flow through the former can be neglected. Under these conditions R_{seal} can be measured in voltage-clamp mode by for example, applying hyperpolarizing pulses (V_s) of 20 mV lasting 100 ms. The value of R_{seal} is then calculated by dividing the applied ΔV_s by the stationary ΔI_s. Under our experimental conditions, the estimated seal resistance is around 10 GΩ.

Note. These estimates are always **under**estimates since the exact value of R_{patch} is unknown.

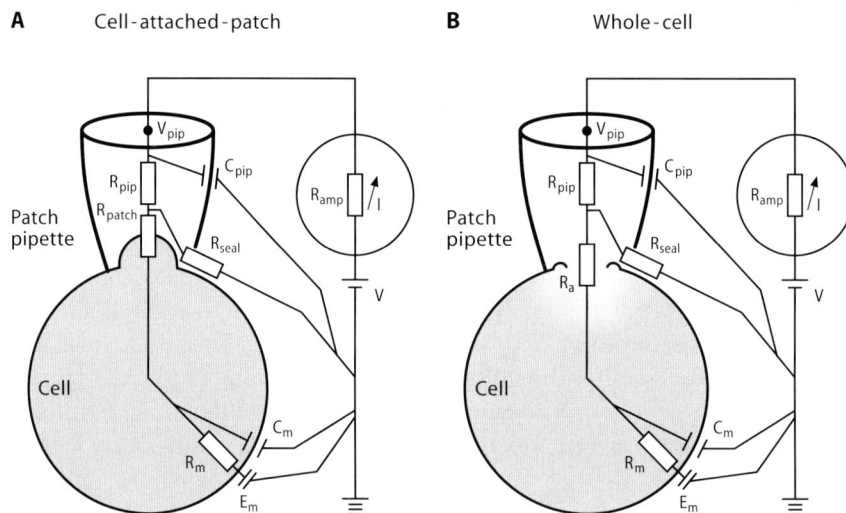

Fig. 15.2A, B. Electrical equivalent circuits of the cell-attached patch (**A**) and whole cell (**B**) configurations. V is the potential applied through the amplifier, E_m is the intrinsic membrane potential of the neuron, V_{pip} pipette potential, R_{amp} the output resistance of the amplifier, R_{pip} the pipette resistance, R_{patch} the resistance of the patch, R_{seal} the seal resistance, R_a the access resistance, R_m the membrane resistance, C_{pip} the pipette capacitance and C_m the membrane capacitance

C-Fast (Pipette Capacitance) Transient Cancellation in Voltage-Clamp Mode

Using voltage pulses of 1 mV lasting 50 ms the pipette capacitance charging currents are monitored. The voltage pulses and current responses are visualized on an oscilloscope. Since the pipette has a capacitance C_{pip} in parallel to the resistance $R_{pip} + R_{seal}$, but in series with the R_{amp} (Fig. 2A), a charging current response is observed as a fast spike that relaxes exponentionally from a peak value to a steady state level. Using the fast capacitance transient cancellation knob (C-FAST) of the EPC-7 amplifier, the spike due to C_{pip} is removed. This procedure "cleans" the current response record and prevents the amplifier from going into saturation. The time constant of this fast transient is $\tau_f = R_{amp} \times C_{pip}$, which is in the order of 1 µs.

Note. The transient is in fact too fast to be faithfully recorded by the EPC-7 because of its maximal low-pass cutoff frequency of ≤ 50 kHz.

Making a Whole Cell and Measuring Series Resistance and Membrane Capacitance

After the establishment of a seal, a short suction pulse (10–50 kPa) is applied inside the patch pipette to rupture the patch membrane, while the high-resistance seal between the membrane and pipette remains intact. This configuration is termed normal or conventional whole cell configuration (Fig. 15.2B). In the whole cell configuration, the total resistance through the pipette tip opening to the whole cell membrane, the series resistance R_s and the cell capacitance C_m can be measured in voltage-clamp mode (VC). R_s is the sum of R_{pip} and the intracellular access resistance, R_a, from the pipette tip to the plasma membrane (Fig. 2B). The neurons are clamped at the holding voltage of -70 mV. During a hyperpolarizing voltage pulse (V), for example to -73 mV, the whole cell membrane current I_m is measured in a frequency band of 50 kHz and 25–36 current traces are averaged to obtain an average membrane current record, $<I_m(t)>$ (Fig. 3A). This record is fitted according to:

$$\langle I_m(t) \rangle = I_c(0)e^{-\tau} + \Delta I_L \tag{3}$$

where $I_c(0)$ is the instantaneous capacitive current (dis)charging the membrane capacity, τ its time constant and ΔI_L the change in the leakage current in response to the voltage step ΔV. The magnitude of $I_c(0)$

equals to $\Delta V / R_s$. Since τ is approximately given by the product of R_s and the membrane capacitance C_m, both parameters (R_s and C_m) can be obtained from the capacitive current transient (if $R_{seal} \gg R_s$ and $R_m \gg R_s$). In our studies we find values for R_s and C_m of around 7 MΩ and 23 pF, respectively. R_s values are usually two to three times R_{pip}.

Series Resistance Compensation (C-Slow)

The values of R_s and C_m can also be measured from the G-SERIES and C-SLOW knobs of the EPC-7 amplifier, when they are set in the right combination to cancel the transient signal upon a small hyperpolarizing voltage ($\Delta V = -3$ mV) stimulation. Finally, the R_s COMP knob can be used to reduce the distorting effect of the slow capacitive transient on fast membrane current changes. This procedure speeds the current response by electronically reducing R_s, thereby limiting errors in fast current measurements. Thus, it is important to keep R_s as small as possible. Under our experimental conditions, R_s can usually be compensated for 50 %–80 % (Fig. 15.3B), resulting in a critically damped response with a (compensated) time constant smaller than 100 µs. Thus, a relatively fast control of the membrane potential is obtained when doing voltage-clamp experiments.

Note. During current-clamp experiments R_s compensation in our amplifier (EPC-7) is disabled. Therefore, in experiments in which we measured membrane resistance and action potential parameters we selected R_s values $< 0.05 R_m$.

Measuring the Resting Membrane Potential

By switching the EPC-7 amplifier to the current-clamp command mode (CC+COM) and holding the clamped current at zero, the recorded voltages are the measured resting membrane potentials (RMP$_m$) of the DRG neurons. The value of RMP$_m$ was found to be -57 ± 8 mV (mean $\pm$SD, $n = 68$, Wang et al. 1994).

Fig. 15.3A, B. Series resistance measurement, compensation and cancellation under voltage-clamp conditions. **A** Series resistance was measured during a hyperpolarizing voltage pulse of 3 mV from a holding potential of −70 mV. Thirty six current traces were averaged to obtain $<I_m(t)>$. The continuous line is a curve fitted to the data points according to Eq. (3), where τ was 150 μs. The value of R_s calculated according to $\Delta V/\Delta I$ was 4.8 MΩ. **B** The remaining capacitive current after optimal transient cancellation and R_s compensation. The 60% reduction of R_s could be read from the compensation dial on the amplifier. The record shows a smaller and faster remaining capacitive transient

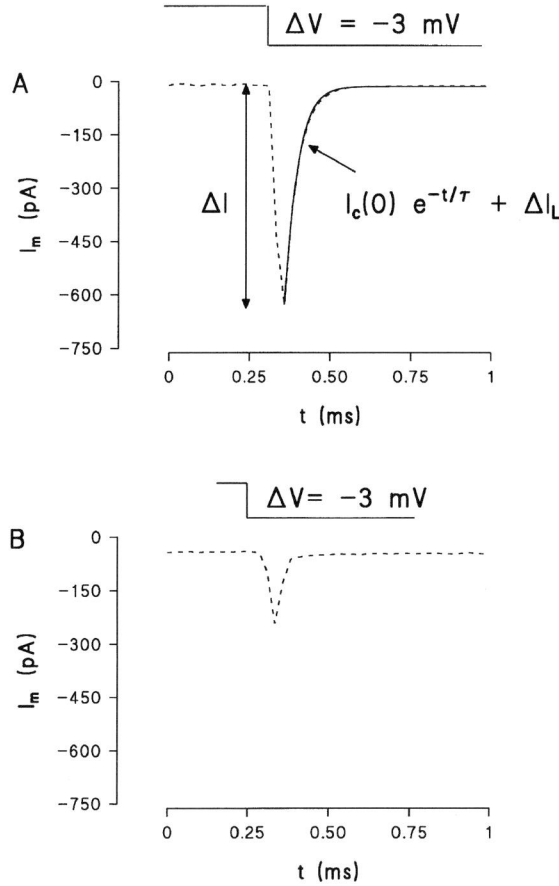

Measuring the Membrane Resistance

Close to RMP, a hyperpolarizing current (ΔI_m) stimulus of 10 pA lasting 200 ms (Fig. 15.3A) was used to measure the membrane resistance (R_m). The resistance R^* calculated from the voltage change (ΔV_m) to a current stimulus ($\Delta V_m/\Delta I_m$) is, in fact, the parallel resistance of R_{seal} and R_m. The value of R_m can be derived from the parallel resistance equation:

$$R^* = \frac{R_m R_{seal}}{(R_m + R_{seal})} \tag{4}$$

Under our experimental conditions, R_m was found to be 700 ±200 MΩ ($n = 35$, Wang et al. 1994; cf. Fig. 15.4A).

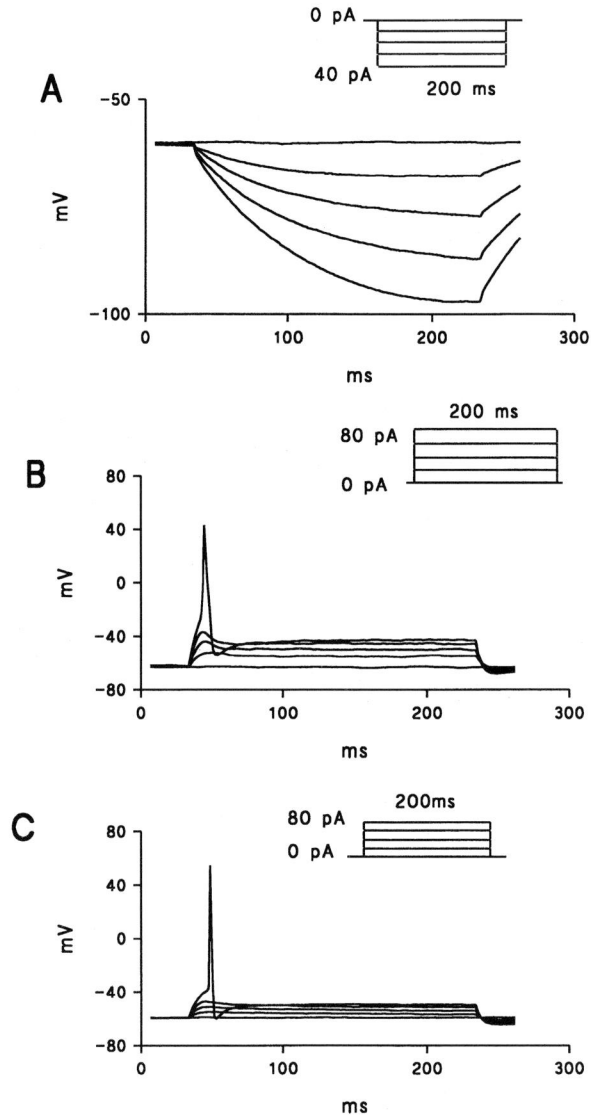

Fig. 15.4A–C. The voltage responses to the application of hyperpolarizing (**A**) and depolarizing current (**B, C**) injections. **A** and **B** are obtained in the conventional whole cell configuration, while **C** is obtained in the perforated patch configuration. The R_m value was 600 MΩ. In **B**, the action potential threshold was −35 mV, overshoot 48 mV and half-width 4 ms. In **C**, the action potential threshold was −38 mV, overshoot 55 mV and half-width 2 ms. In this case, the action potential in the perforated patch configuration is faster than that in the conventional whole cell configuration. However, the mean values of the half-width in both configurations are not significantly different from each other

Calculation of the True Resting Membrane Potential

As the DRG neurons have a high membrane resistance R_m, the possibility of a shunt due to a relatively low parallel seal resistance R_{seal} should be recognized (cf. Fig. 15.2B). This shunt will cause an underestimation of RMP. Since R_m is approximately linear around RMP_m (deviations $\leq 10\,mV$), the true (i.e., unattenuated) resting membrane potential (RMP_t) can be obtained from:

$$RMP_m = \frac{RMP_t \cdot R_{seal}}{(R_m + R_{seal})} \tag{5}$$

After correction according to Eq. (5) and using mean values for R_m and R_{seal}, the unattenuated RMP_t is estimated as $-64\,mV$ (Wang et al. 1994).

Solution Changes To Study the Ionic Origin of the Resting Membrane Potential

Upon replacing control ECS (5 mM KCl) with K^+ ECS (140 mM KCl), the membrane potential can be reversibly depolarized towards 0 mV (Fig. 15.5A). Upon a change of the control ECS to Na^+-free ECS a mem-

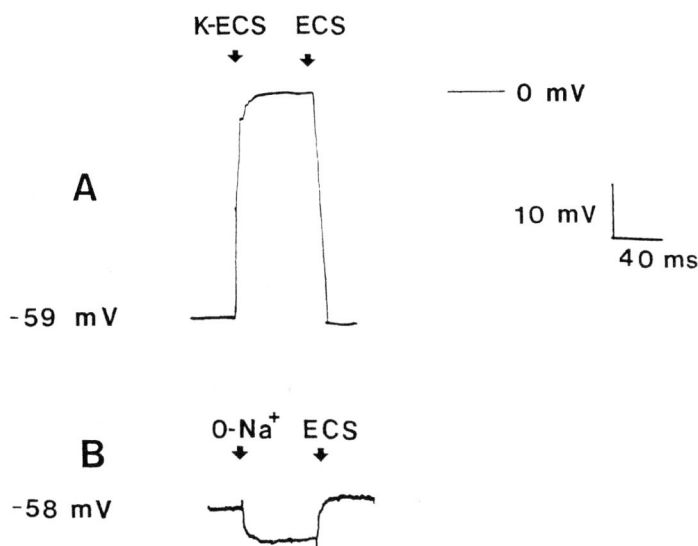

Fig. 15.5A, B. Resting membrane potential changes upon application of K^+-ECS (**A**) and Na^+-free ECS (**B**). In K^+-ECS the membrane potential was reversibly depolarized to 0 mV, while Na^+-free ECS reversibly hyperpolarized the membrane by 10 mV

brane potential hyperpolarization of 10 mV is observed (Fig. 15.5B). These observations illustrate that the RMP of DRG neurons is mainly determined by K^+ and to a minor degree by Na^+. In low Cl^- ICS the value of RMP corrected for junction potentials is not significantly different from that in control indicating that Cl^- ions do not contribute significantly to the RMP level. From the Goldman-Hodgkin-Katz equation (Eq. 2), and neglecting Cl^- permeability, we estimate the permeability ratio P_K/P_{Na} as 1:0.004 (Wang et al. 1994).

Measuring Action Potential Properties

To evoke action potentials and to determine the firing threshold, depolarizing current pulses of 200 ms with an amplitude increasing from subthreshold to suprathreshold levels are used (Fig. 15.4B). The membrane voltage, at which the action potential appeared for the first time, is taken as the firing threshold (V_{thr}) of the neurons (Wang et al. 1994). The value of V_{thr} is found to be -25 ± 8 mV ($n = 35$) in DRG neurons. The peak value of an action potential is the overshoot over the zero potential. In rat DRG neurons, overshoot ranges from 20 to 45 mV with a mean value of 31 ± 13 mV ($n = 35$). The action potential duration at half height (half-width) is 6 ± 4 ms ($n = 35$, Wang et al. 1994).

Pharmacology of the Action Potential

Upon changing the control ECS to Na^+-free ECS, the action potentials usually disappear (Fig. 15.6A, B). This indicates that Na^+ channels play an important role in the generation of action potentials. Extracellular application of the fast transient K^+ channel blocker 4-AP (4-AP ECS, Table 15.1) induces repetitive firing and broadening of the action potential (Fig. 15.5C). In 4-AP ECS the action potential threshold is -19 ± 1 mV, half-width 17 ± 5 ms and overshoot 40 ± 11 mV ($n = 3$, Wang et al. 1994). Thus, K^+ channels are able to modulate firing properties of rat DRG neurons. Upon application of the delayed outward rectifying K^+ channel blocker TEA (TEA-ECS, Table 15.1), the action potential half-width is largely increased (Fig. 15.6D) from 4 ± 2 ms to 43 ± 5 ms ($n = 3$, Wang et al. 1994) indicating that delayed outward rectifying K^+ channels contribute to the repolarization phase of the action potentials in rat DRG neurons.

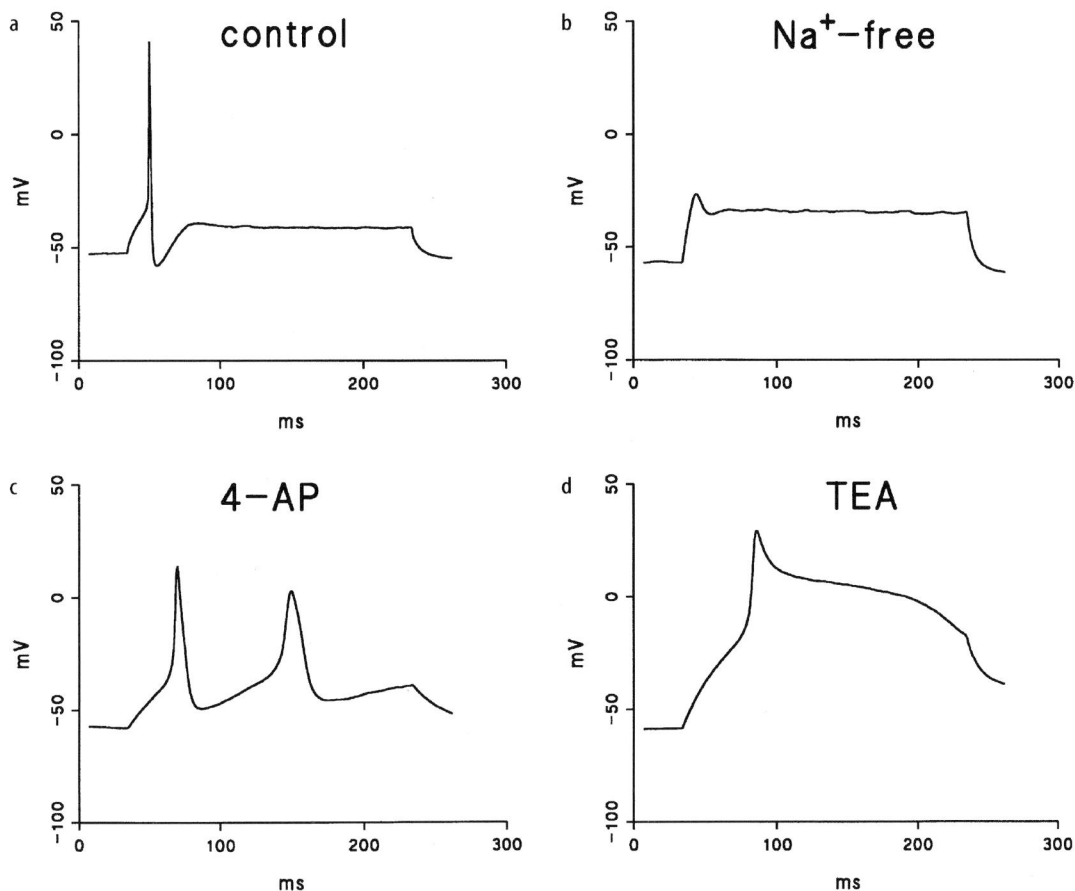

Fig. 15.6A–D. Effects of Na$^+$-free ECS and K$^+$-channel blockers on the action potential of rat dorsal root ganglion neurons. **A** and **B** are from the same cell. The controls for **C** and **D** are similar to those in **A**. Since Na$^+$-free ECS hyperpolarizes the membrane and K$^+$ channel blockers depolarize the membrane potential, depolarizing or hyperpolarizing currents are injected to bring the membrane potential close to the original resting membrane potential, in order to prevent changes in the action potential due to altered resting membrane potentials. Action potentials are recorded upon depolarizing current pulses stimuli of 80 pA and lasting 200 ms

A. Action potential measured in control ECS

B. Na$^+$-free ECS completely blocks the generation of action potentials under our experimental conditions and hyperpolarizes the resting membrane potential by 10 mV (not shown).

C. Repetitive firing during depolarizing current injection is induced by 4-AP ECS. The resting membrane potential was depolarized by 18 mV (not shown).

D. TEA-ECS greatly broadens the action potential and depolarizes the resting membrane potential by about 40 mV (not shown).

15.4
Perforated-Patch Whole Cell Technique in Membrane Potential Measurements

With the conventional whole cell method of opening the CAP the cell will be perfused with the pipette fluid. The exchange between pipette solution and the cell interior takes about 1–5 s under our experimental conditions. As a consequence, the intracellular molecules are diluted by the pipette solution and the intracellular ion concentrations and Ca^{2+} buffering are altered (Kettenmann and Grantyn 1992). Thus, membrane potential measurements may be hampered by artefacts that are due to unwanted cell dialysis. This problem has led to the invention of the perforated patch method by Horn and Marty (1988) (see also Chap. 16). To perforate the patch membrane, pore-forming antibiotics were added to the pipette solution. Antibiotics used include nystatin and other polyene antibiotics, such as amphotericin B (see Rae et al. 1991), which form ion channels in lipid membranes. These channels are permeable for monovalent cations and to a lesser extent for monovalent anions, while Ca^{2+}, Mg^{2+}, multivalent anions and larger molecules and organelles are retained. Thus, this method prevents loss of cytoplasmic constituents to the pipette. Furthermore, the nystatin-formed channels display little lateral diffusion away across the sealing glass rim of the pipette (Horn 1991).

Use of Nystatin   To prepare a nystatin stock solution, 100 mg of pluronic F127 (Molecular Probes Inc., Eugene, Oregon, USA) is firstly dissolved in 100 μl DMSO (Sigma, St Louis, USA). Then, 5 mg of nystatin is added to this solution and dissolved by vortexing. Finally, 20 μl nystatin stock solution is added to 4 ml intracellular solution (ICS-nys; Table 15.1). Thus, the final nystatin concentration should exceed 250 μg/ml. Care should be taken while filling the pipette with the nystatin containing intracellular solution. Air bubbles should be avoided since they may disturb drug activity and reduce pore formation (Kettenmann and Grantyn 1992). After making a seal, small hyperpolarizing pulses of −20 mV are applied every second. The gradual formation of the whole cell configuration during patch permeabilization is evident from a decrease in the measured input resistance and from the appearance of the resting membrane potential. The duration of patch permeabilization usually is about 5–10 min. After establishment of a whole cell, RMP and V_{thr} can be measured in the same way as in the conventional whole-cell configuration. Values found by us are -57 ± 1 mV and -26 ± 1 mV ($n = 10$; Wang et al. 1994), respectively. The overshoot and half-width of action potentials amounted to 29 ± 2 mV and 6 ± 1 ms, respectively (Fig. 15.4C). These values are not significantly different from those obtained in the conventional whole cell recordings. Thus, the ion channels underlying these phenomenal are not significantly altered by intracellular perfusion. However, the variability of the RMP and V_{thr} values is much smaller than that during the normal whole cell measurement, suggesting that intracellular factor(s) could have an influence.

Note. The value of R_a in the perforated patch configuration is around 20 MΩ, which is larger than that in the conventional whole cell configuration.

Comments

The present chapter shows that patch-clamp measurements in rat DRG neurons can provide information about the resting membrane potential and action potential properties of these cells. However, there are several sources of error (discussed above) which may disturb the membrane potential measurement. The use of an electrical circuit representing the measurement conditions is of great use in understanding and dealing with these problems.

References

Allen TGJ, Burnstock G (1987) Intracellular studies of the electrophysiological properties of cultured intracardiac neurons of the guinea-pig. J Physiol (Lond) 388: 349–366

Gorman ALF, Marmor MF (1970) Contributions of the sodium pump and ionic gradients to the membrane potential of a molluscan neuron. J Physiol (Lond) 310: 897–917

Hamill OP, Marty A, Neher E, Sakmann B, Sigworth FJ (1981) Improved patch clamp techniques for high resolution current recording from cells and cell free membrane patches. Pflügers Arch. (Europ. J. Physiol.) 391: 85–100

Hille B (1992) Ionic channels of excitable membranes. Sinauer Associates, Sunderland

Hodgkin AL, Huxley AF (1952) A quantitative description of membrane current and its application to conduction and excitation in nerve. J Physiol (Lond) 117: 500–544

Horn R and Marty A (1988) Muscarinic activation of ionic currents measured by a new whole cell recording method. J Gen Physiol 92: 145–159

Horn R (1991) Diffusion of nystatin in plasma membrane is inhibited by a glass-membrane seal. Biophys J 60: 329–333

Ince C, Ypey DL, Diesselhoff-den Dulk MMC, Visser JAM, De Vos A, Van Furth R (1983) Micro CO_2 incubator for use on a microscope. J Immunol Meth 60: 269–275

Junge D (1992) Nerve and muscle excitation. 3rd edn. Sinauer Associates, Sunderland

Kettenmann H, Grantyn R (1992) Practical electrophysiological methods: a guide for in vitro studies in vertebrate neurobiology. Wiley Liss, New York, 274–278

Neher E (1992) Correction for liquid junction potentials in patch clamp experiments, In: B Rudy , LE Leverson (ed) Ion channels. Academic, San Diego, 123–131

Rae J, Cooper K, Gates P, Watsky LM (1991) Low access resistance perforated patch recording using amphotericin. B. J Neurosci Meth 37: 15–26

Wang Z, Van den Berg RJ, Ypey DL (1994) Resting membrane potentials and excitability at different regions of rat dorsal root ganglion neurons in culture. Neurosci 60: 245–254

Wood JN, Winter J, James IF, Rong HP, Teats J, Bevan S (1988) Capsaicin-induced ion fluxes in dorsal root ganglion cells in culture. J Neurosci 8: 3208–3220

Chapter 16

Perforated Patch-Clamp Technique in Heart Cells

E. Etienne Verheijck

Background

The perforated patch-clamp technique (Horn and Marty 1988; Korn and Horn 1989) was developed as an alternative for the whole cell patch-clamp technique to prevent changes in membrane currents due to diffusional exchange between the cytoplasm and the content of the pipette, a phenomenon called "washout". Washout can be caused by the loss or alteration of enzymes, second messengers and phosphorylation states critical for the functioning of ion channels. These changes can occur within minutes after obtaining the whole cell configuration and thus prevent long-lasting (up to 90 min) stable recordings.

Perforated patch recording uses pore-forming antibiotics such as nystatin or amphotericin B to gain electrical access to the cell. The antibiotics make selective pores in the membrane of a cell-attached patch and thereby limit the washout of intracellular components. An important feature of nystatin is that it does not cross the cell membrane (Korn et al. 1991). Lateral diffusion of nystatin through the cell membrane is inhibited by the pipette glass-membrane seal (Horn 1991). Thus, antibiotic pores will only be formed in the membrane facing the lumen of the pipette. It is likely, but not yet known, that amphotericin B shows a similar behavior. Both antibiotics form pores in the cell membrane that are permeable to monovalent cations and anions but exclude multivalent ions such as Ca^{2+} and Mg^{2+} and substances larger than 0.4 nm such as glucose and sucrose. The permeability for monovalent cations is about nine times that of monovalent anions. Additionally, the conductance of amphotericin B-created pores is approximately twice as large as that of nystatin-created pores (Holz and Finkelstein 1970; Horn and Marty 1988).

The nystatin and amphotericin B perforated patch-clamp technique has successfully been used in a large variety of cell types, i.e., lacrimal gland cells (Horn and Marty 1988), pituitary cells (Horn and Marty

1989), brown fat cells (Lucerno and Pappone 1990), and cardiac cells (Kurachi et al. 1989; Verheijck et al. 1995). It has become clear that for antibiotic pore formation the cell membrane should contain cholesterol, which thus excludes plant cells. As an alternative the bacterial porin haemolysin from *Escheria coli* has successfully been used in protoplasts to gain electrical access to the cell (Schroeder 1988).

In our laboratory the perforated patch-clamp technique worked out well for heart cells isolated from ventricle, atrium and sino-atrial node. Amphotericin B steadily resulted in a lower access resistance than obtained with nystatin, most likely due to its higher single pore conductance; it is also easier to use. Therefore, this chapter will limit its scope to the use of amphotericin B for perforated patch recordings in heart cells.

Materials

Solutions
- Amphotericin B Stock Solution
  Prepare a series of microcentrifuge tubes containing 6 mg amphotericin B (Sigma Chemical Co., St. Louis, USA). These tubes can be stored in the freezer ($-20\,°C$) without loss of activity. Shortly before each experiment a stock solution is made in which the 6 mg batch of amphotericin B is diluted in 100 µl DMSO. The stock solution needs to be shielded from light because amphotericin B is photolabile.

- Pipette Solution
  As the antibiotic-created pores in the cell membrane are not equally permeable for all ions (see above), a careful consideration of the pipette filling solution is needed. It is recommended that the pipette filling solution is similar to the internal solution of the cell, especially for the monovalent ions. In our perforated patch whole cell experiments the electrode solution contained (in mmol/l) K^+-gluconate 120, KCl 20, HEPES 5, $MgCl_2$ 5, $CaCl_2$ 0.6, EGTA 5 (pH adjusted with KOH to 7.1).

  Note. Although Ca^{2+} and Mg^{2+} are impermeable through the antibiotic-created pores, addition of these ions is recommended as they seem to enhance membrane stability.

Patch-clamp pipettes
Electrodes can be made from different types of borosilicate glass. We prefer to use thin walled borosilicate glass with a filament in its lumen. This type of glass allows variation of tip diameters and can be filled without the occurrence of air bubbles.

Note. The series resistance, which comprises the resistance of the pipette itself, and, in series, the resistance of the patch, should be kept as low as possible. The patch resistance depends on the total number of pores formed by the antibiotic and the single pore conductance. As the total number of pores depends on the surface area of the patch it is advisable to use low resistance (i.e., large diameter) pipettes ($1-3\,M\Omega$). A disadvantage in using low pipette resistances is that the patch becomes more fragile.

Procedure

Cell Isolation

For the isolation of sino-atrial nodal cells the reader is referred to Verheijck et al. (1995), for the isolation of ventricular cells to Veldkamp et al. (1994), and for the isolation of atrial cells to Tytgat et al. (1994).

Perforated Patch Procedure

A step by step procedure for obtaining reliable, perforated patch whole cell recordings from heart cells is presented below. With this procedure different types of heart cells, isolated from rabbit, guinea pig, sheep and humans, can be patched. It should be kept in mind that, when cells are poorly patched by the conventional whole cell method, stable seal formation will also be difficult in the perforated patch configuration. When the quality of the isolated cells is good, the success rate can be about 60 %.

Preparation of a pipette

Different types of glass capillaries can be used. Pipettes can be fabricated by a puller in one or more steps. For heart cells we normally use a vertical one-stage patch electrode puller. After pulling, the pipette is gently fire-polished such that the resistance is between 1 and $3\,M\Omega$ when filled with pipette solution and placed in the recording solution.

Pipette filling

1. Shortly before use, 6 mg of amphotericin B is diluted in $100\,\mu l$ DMSO. This will take some time, but with gentle pipetting with an $100\,\mu l$ micropipette, a yellow translucent solution can be prepared.

2. $10\,\mu l$ of this solution is added to 3 ml filtered electrode filling solution. Because of the photoliability of the solution, the syringe containing the final pipette solution is wrapped in aluminium foil. It is also possi-

ble to filter the final pipette solution with a 0.22 or 0.45 μm filter to remove small nondiluted amphotericin B particles that can block the pipette.

3. The pipette tip is immersed for about 1 s in normal electrode solution; it is then backfilled with the amphotericin B containing solution.

Tip. Do not immerse borosilicate glass pipette tips much longer than 1 s in the normal electrode solution. Longer immersion times will drastically enhance diffusion time of the antibiotic to reach a tip concentration comparable to the pipette filling solution. It will take some time to adjust the immersion time of the pipette tip in normal pipette solution. The ideal time will also depend on the time needed to approach a cell and to form a giga-ohm seal.

Note. It is also possible to fill the pipette with only the amphotericin B containing solution. The advantages are that this procedure will speed up seal formation and does not depend on the dilution of the antibiotic. The disadvantages are: (1) when the antibiotic leaks out of the pipette upon approaching the cell, antibiotic pores will be formed in the entire cell, many times resulting in cell death; (2) pore formation is very fast which will mask the formation of a giga-ohm seal. This can result in the application of too much negative pressure and a whole cell configuration will be formed, resulting in cell death.

Note. The stock solution will start to lose its perforating activity after about 3 h. Because perforated patch whole cell recordings can last up to 2 h, the stock solution and the amphotericin B containing pipette filling solution should be replenished every hour.

Giga-ohm seal formation

After filling the pipette with the amphotericin B containing solution it should be used immediately. Upon approaching the cell, **no** positive pressure should be applied to the pipette as is normally done when using the whole cell method.

Tip. Since no positive pressure can be applied to the pipette when approaching the cell, the solution with which the cells are superfused should not contain particles that can adhere to or block the pipette tip. Therefore, cells should be superfused for at least 10 min before an approach with the micropipette is started.

After the pipette is placed on the surface of the cell, a giga-ohm seal is formed in the usual way by applying negative pressure (Fig. 16.1a). After obtaining a high-resistance seal, the incorporation of the antibiotic in

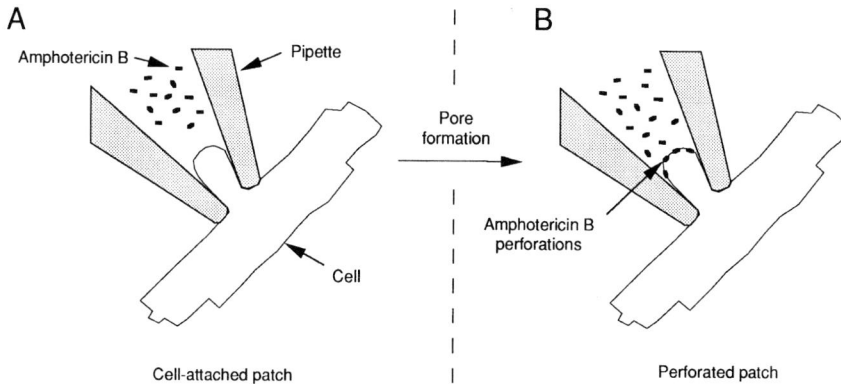

Fig. 16.1A, B. The whole cell perforated patch-clamp technique. **A** The formation of a giga-ohm seal. **B** The incorporation of an antibiotic in the cell membrane by which electrical access to the cell is obtained

the cell membrane (Fig. 16.1b) will result in the formation of electrical access to the cell interior. The formation of electrical access can be monitored in either the voltage- or current-clamp mode. Both procedures will be described.

Monitoring the Formation of Electrical Access in the Voltage-Clamp Mode

After obtaining a high-resistance seal, the pipette potential is set at $-80\,\text{mV}$, and voltage pulses ($-10\,\text{mV}$ amplitude, 20 ms duration) are periodically (2–5 Hz) applied to monitor access resistance. When the antibiotic diffuses through the pipette tip and pores are formed in the membrane, series resistance will decrease and a slow capacitative current will appear, which indicates the formation of electrical access to the cell interior. This process normally occurs within 30 s, but, depending on the diffusion process, can take about 3 min. Longer times needed for the initial electrical access to the cell normally result in longer times (20 min and more) to reach a steady state level of the series resistance. Generally it will take between 10 and 15 min to reach a steady state series resistance of 5–10 MΩ, which subsequently can electrically be compensated to a large extent.

Formation of electrical access

Tip. The gradual decrease in series resistance should be monitored in every experiment until a steady state level is reached. When this level is reached, series resistance will become very stable over a period of at least 90 min. Nevertheless a periodic check is advisable.

Monitoring the Formation of Electrical Access in the Current-Clamp Mode

Formation of electrical access

In the current clamp mode the formation of a high resistance seal is reflected in a large decrease in voltage change upon a current-clamp pulse (-10 pA amplitude, 100 ms duration, 1–2 Hz). During the incorporation of antibiotic pores in the membrane, this voltage change becomes gradually smaller, and the recorded potential will gradually become more negative. The initial jump in voltage reflects the series resistance, which should be monitored until a steady state level is reached and which can subsequently be canceled. Times needed to reach a stable and low series resistance is the same as mentioned in the previous section.

Note. Rupture of the perforated patch will result in a conventional whole cell situation and subsequent rapid cell death.

Note. When properly performed, the success rate is as high as or even higher than the conventional whole cell recording technique. This is most likely due to the fact that no suction (negative pressure) is needed to disrupt the membrane patch to form a whole cell configuration.

References and Further Reading

Holz R, Finkelstein A (1970) The water and non electrolyte permeability induced in thin lipid membranes by the polyene antibiotics nystatin and amphotericin B. J Gen Phys 56:125–145

Horn R, Marty A (1988) Muscarinic activation of ionic currents measured by a new whole cell recording method. J Gen Physiol 92:145–159

Horn R (1991) Diffusion of nystatin in plasma membrane is inhibited by a glass-membrane seal. Biophys J 60:329–333

Kurachi Y, Asano Y, Takikawa R, Sugimoto T (1989) Cardiac Ca current does not run down and is very sensitive to isoprenaline in the nystatin-method of whole cell recording. Archiv Pharm 340:219–222

Korn SJ. Horn R (1989) Influence of sodium-calcium exchange on calcium current rundown and the duration of calcium-dependent chloride currents in pituitary cells, studied with whole cell and perforated patch recording. J Gen Physiol 94:789–812

Korn SJ. Marty A, Conner JA and Horn R (1991) Perforated patch recording. In: Conn PM (ed) Electrophysiology and microinjection. Methods in neuroscience. vol. 4. Academic, San Diego, pp 364–373

Lucero MT, Pappone PA (1990) Membrane responses to norepinephrine in cultured brown fat cells. J Gen Physiol 95:523–544

Rae J, Cooper K, Gates P, Watsky M (1991) Low access resistance perforated patch recordings using amphotericin B. J Neuroscience Methods 37:15–26

Schroeder JI (1988) K^+ transport properties of K^+ channels in plasma membrane of *Vivia faba* guard cells. J Gen Physiol 92:667–683

Tytgat J (1994) How to isolate cardiac myocytes. Cardiovasc Res 28:280–283

Veldkamp MW, van Ginneken ACG, Bouman LN (1993) Single delayed rectifier channels in the membrane of rabbit ventricular myocytes. Circ Res 72:865–878

Verheijck EE, van Ginneken ACG, Boerier J, Bouman LN (1995) Effects of delayed rectifier current blockade by E-4031 on impulse generation in single sinoatrial nodal myocytes of the rabbit. Circ Res 76:607–617

Zhou Z (1991) Pacemaker current, i_f recorded from latent atrial pacemaker cells using two different whole-cell recording methods. Circ (Suppl) 84, 4: II-178

Measurement and Analysis of Different Aspects of Potassium Currents in Human Lymphocytes

Rezso Gáspár Jr., Zoltan Varga, György Panyi,
Zoltan Krasznai, Carlo Pieri, and Sandor Damjanovich

Background

Lymphocytes possess a wide variety of receptors and ion channels which play an important role in the perception and processing of information during the course of the immune response. Many distinct types of K^+, Cl^- and Ca^{2+} channels, as well as a fast Na^+ channel, have been characterized in human lymphocytes (Lewis and Cahalan 1990; Gallin 1991; Lee et al. 1992). The existence of a voltage activated K^+ channel in human peripheral blood T lymphocytes was first demonstrated in 1984 (Matteson and Deutsch 1984; DeCoursey et al. 1984). This channel is referred to as n-type K^+ channel, the ion channel underlying almost exclusively whole cell K^+ currents in a wide variety of lymphocytes including quiescent and mitogen activated human T and B lymphocytes (Deutsch et al. 1986; Sutro et al. 1989).

Several lines of evidence suggest the role of K^+ channels and membrane potential in lymphocyte activation.

Potassium Channels

- Voltage-gated potassium channels are required for human T lymphocyte activation (Chandy et al. 1984).

- Mitogen-induced hyperpolarization is accompanied by an increase in the K^+ permeability of the cell membrane (Grinstein and Dixon 1989).

- The K^+ channel density (both K_n and K_{Ca}) is higher in the cell membrane of mitogen-activated lymphocytes (Deutsch et al. 1986; Sutro et al. 1989; Grissmer et al. 1993)

- K^+ channel blockers quinine, tetraethylammonium (TEA), charybdotoxin (CTX), noxiustoxin (NxTx) and margatoxin (MgTx) inhibit K^+

current and mitogen stimulated proliferation of T cells with similar rank order and potency (DeCoursey et al. 1984; Chandy et al. 1984; Price et al. 1989; Drakopoulou et al 1995). The blockage of K^+ conductance is sufficient to inhibit T cell activation (Lin et al. 1993).

Membrane Potential

- Activation causes hyperpolarization in human T and B cells, but also may have a depolarizing or even no effect on mixed human peripheral blood lymphocyte (HPBL) populations (Grinstein and Dixon 1989).

- A substantial depolarization (from $-70\,mV$ to $-15\,mV$) inhibits PHA-induced T cell activation (Gelfand et al. 1987).

The most plausible theory for the role of K^+ channels in the proliferation of lymphocytes is that K^+ channels are responsible for the maintenance of the right membrane potential for initiation and propagation of the mitogenic signal. K^+ channel blockers depolarize T cells (Grinstein and Smith 1990; Leonard et al. 1992) and depolarization inhibits Ca^{2+} signaling (Oettgen et al. 1985; Lin et al. 1993, Damjanovich et al. 1994), thereby K^+ channels may indirectly facilitate Ca^{2+} influx through mitogen-stimulated Ca^{2+} channels by maintaining a permissive negative membrane potential (Cahalan et al. 1985).

Recently, experimental evidence has shown that expression of calcium- and voltage-activated potassium channels varies when the cells are activated by different signals that lead to cell proliferation. Stimulation via Ag receptors increases the expression of both types of potassium channels, whereas a bacterial mitogen, LPS, only enhances the expression of the voltage gated one (Pariseti et al. 1993). There are also suggestions that multimeric complexes of the plasma membrane protein CD20, expressed exclusively by B lymphocytes, may form a Ca^{2+} conducting ion channel (Bubien et al. 1993). The binding of a CD20-specific monoclonal antibody(mAb) to $CD20^+$ lymphoblastoid cells also enhanced the transmembrane Ca^{2+} conductance and CD20 is structurally similar to several ion channels.

Considering these observations, it is relevant to study the effect of new drugs and mAb binding on the K^+ channel activity of human lymphocytes and lymphoblastoid cell lines (Gáspár et al. 1994; Gáspár et al. 1995).

In the rest of this chapter the basic tools for studying such effects on the whole cell K^+ currents of human lymphocytes are outlined. Examples of experimental results are outlined together with the relevant whole cell patch-clamp techniques and the methods of analysis of the primary results. The type of analysis described in this chapter for the potassium currents of human lymphocytes may be applied with due caution to a wide variety of other channels in different other cell types and may be used to complement other successful measuring techniques, e.g., the fluorescence based ones, for studying the membrane potential and ion channel activities in the plasma membrane of live cells.

Materials

Cell Isolation and Chemicals

The measurements and analysis described in this chapter have been performed exclusively on HPBLs donated by healthy volunteers. HPBLs can be isolated from heparinized blood using the standard Ficoll-Hypaque sedimentation technique (Gupta et al. 1978). The cells can be used on the day of isolation and stored until the measurement at room temperature in a PBS solution (150 mM NaCl, 3.3 mM KCl, 8.6 mM Na_2HPO_4, 1.69 mM KH_2PO_4, pH 7.4) supplemented with 5 mM glucose. We purchased all chemicals from Sigma (St. Louis, MO, USA.). Cyclosporin A was the kind gift of Dr. A. Aszalós, FDA, Washington, USA, but may be purchased, e.g., from Sandoz Pharma Ltd.(Basel, Switzerland).

Equipment

We recorded membrane currents with Axopatch-200 and 200A patch-clamp amplifiers in conjunction with Axon Instruments TL-1-125, or Digidata 1200 computer interfaces with varying sampling rates. Low-pass filtering was usually applied at half of the sampling frequency.

Note. In some cases the applied sampling rate was much lower than that required by theory, because of the lowest possible filter setting available on the patch-clamp amplifiers. However, in these cases the proper recording of frequency responses of the membrane currents or membrane potential changes was not critical from the point of view of the results.

It is generally a good idea to do the first piloting experiments with a relatively high sampling rate and decrease it later, when the properties of the currents and the exact goal and type of analysis one wants to perform on the primary data become clear. Patch-clamp membrane potential measurements were performed in the whole cell configuration, in $I = 0$ current-clamp mode, using a home-made EPC-5-like patch-clamp amplifier with a specially adjusted feedback resistor to avoid abusing membrane potential oscillations and to provide stable measuring conditions at the very high input impedance of the lymphocytes. For a detailed description of the operational modes of patch-clamp amplifiers and connecting equipments, we refer the reader to the introductory chapters and the instruction manuals of the instruments.

Patch electrodes of $3-4$ MΩ resistance, as measured in 4.5 mM K$^+$-ECS (see below) are an ideal size for patch-clamping human lymphocytes. They are fabricated from GC 150F-15 borosilicate glass capillaries (Clark Electromedical Instruments, Reading, England). Any reliable two-stage puller should be suitable for fabricating patch pipettes. The pipettes were heat-polished with a glass-coated heated platinum wire. For single channel measurements the pipettes were coated with Sylgard close to their ends.

Electrophysiology

Throughout the patch-clamp measurements described in this chapter, lymphocytes are bathed in extracellular-like solutions (ECS), patch pipettes are filled with intracellular-like solutions (ICS). The ion compositions of the solutions are given in Table 17.1.

At the beginning of the measurement move the lymphocytes stored in PBS into a 35 mm diameter plastic Petri dish containing about 1 ml ECS and leave them alone for approx. 5–10 min to adhere to the bottom of the dish.

Note. We found the HPBLs adhere better to plastic Petri dishes than to glass and the resulting immobilization of the cells made their patch-clamping easier.

Note. After this waiting period damaged lymphocytes show a marked visual change and become black. Cells in good and stable condition remain bright yellow.

Select a viable cell for patch-clamping under the inverted field phase contrast microscope of the setup using a 40X objective.

After approaching the cell under investigation with the patch pipette, a gigaseal in the order of $20-100$ GΩ can be formed by applying

Table 17.1. Solutions

Solutions	Ionic composition (mM)							
	K^+	Na^+	Ca^{2+}	Mg^{2+}	Cl^-	F^-	EGTA	HEPES
4.5 mM K^+-ECS	4.5	160	2	1	170.5	0	0	5
160 mM K^+-ECS	160	0	2	1	166	0	0	10
KF-ICS	162	0	1	2	6	140	11	10
KCl-ICS	162	0	1	2	146	0	11	10

The pH was 7.4 for the extracellular solutions (ECS) and 7.2 for the intracellular (ICS) ones. The measured osmolarity of the solutions was 290–330 mOsm.

50–70 H_2O cm negative pressure to the interior of the pipette via a U-shaped glass tube filled with water.

Note. The success rate of obtaining high quality gigaseals varies. We observed a strong dependence of the success rate upon the ICS used and the individual who donated the lymphocytes for the experiments. Seal formation proved much more likely with the KF-ICS than with the KCl one and an earlier loss of the stable patch-clamp measuring configuration could be expected with the KCl-ICS. Otherwise experiments with the KCl internal solution yielded results identical to those obtained with KF.

After the formation of the gigaseal, it usually proves to be a good strategy to wait approx. 1 min before entering whole cell. Whole cell measuring configuration is established by the application of a series of approx. 250–350 H_2O cm negative pressure pulses with the help of a syringe. Transition into whole cell is checked by the usual voltage-clamp protocol, however, instead of waiting for the channels to recover completely from inactivation a quick repetition rate (5 s) can be used. Start control experiments at least 12 min after entering whole cell. With the KF internal solution, K^+ channel properties of T lymphocytes can remain stable for up to several hours.

We do not use series resistance compensation during the experiments. After the initial setting, there is usually no need to alter the capacitive transient compensation, which means that after the initial equilibration period no tendentious changes in R_s happen during the experiments. As far as time control comparisons are concerned, peak K^+ current amplitudes at a given depolarizing potential usually remain within ±5 % of the control value up till several hours. In most cases the experiment ends when an abrupt loss of the measuring configuration occurs. In most of the experiments leak current is negligible compared to the K^+ currents, therefore during analysis of the results leak subtraction is not usually required.

Results shown are from patch-clamp experiments carried out at room temperature (25 °C).

Note. Room temperature measurements are preferred in most of the cases because of the simplicity of the experimental setup; however, one must be aware of the strong temperature dependence of the kinetic parameters of human lymphocyte K^+ channels (Lee and Deutsch 1990).

We use the pClamp6 program package (Axon Instruments Inc., Foster City, CA, USA) for data acquisition and analysis.

In this chapter we describe different aspects of the potassium channels of human lymphocytes to be investigated with the whole cell patch-clamp technique. Each aspect is illustrated with a figure and a description of the purpose of the applied protocol and the subsequent analysis of the results.

Procedure

Single Channels in the Whole Cell Configuration

Due to the high input resistance of lymphocytes and the almost complete inactivation of the K^+ conductance at depolarizing potentials it is possible to record **single channel events in the whole cell configuration** of lymphocytes. Figure 17.1 demonstrates the result of such an experiment. Before starting the measurement the cell was held at 0 mV for 5 s, a time sufficient for the inactivation of most of the K^+ channels. The lymphocyte was bathed in 4.5 mM K^+-ECS, the pipette was filled with KF-ICS. Before fire polishing and filling, the pipette had been coated with Sylgard to reduce pipette capacitance and noise. The recording of the single channel currents was done with an Axopatch 200A patch-clamp amplifier in the capacitive feedback mode to minimize instrument noise during recording. The trace was recorded practically unfiltered, with a sampling interval of 500 μs, via a Digidata 1200 computer interface using the Fetchex routine of the pClamp6 program package. The holding potential was kept at 0 mV. For this presentation the trace was digitally filtered with a Gaussian type filter with cutoff frequency of 500 Hz.

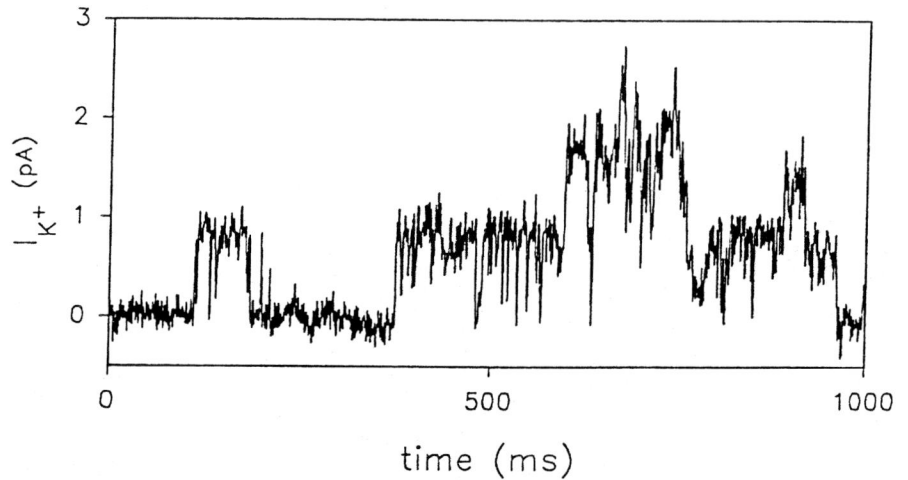

Fig. 17.1. Single channel recording in a human lymphocyte in whole cell configuration

Whole Cell Currents

Depolarization-activated current

Figure 17.2a illustrates that under typical voltage-clamp conditions, with 4.5 mM K^+-ECS in the bath and KF-ICS in the pipette, HPBLs display strong outward whole cell currents upon step depolarizations and the current activates more rapidly with increasing depolarization. After reaching a peak, the current declines due to inactivation toward a small steady state level. The current traces in the figure were evoked by voltage steps lasting for 1000 ms, applied every 50 s from a holding potential −80 mV in 10 mV increments from −50 to +60 mV (Fig. 17.2a insert). The records were taken with a sampling interval of 2000 µs. Ion channels contributing to the transient outward current have been identified as voltage-gated potassium channels on the basis of their K^+ selectivity, having a linear single channel current-voltage relationship with a slope conductance of approx. 10 pS at pipette potentials between −30 to +40 mV, and properly responding to quinine (100 µM) and 4-aminopyridine (5 mM) (Matteson and Deutsch 1984; DeCoursey et al. 1984; Lee et al. 1992).

Note. Be careful when applying agents such as quinine and 4-aminopyridine to the extracellular medium. Some of them may need apolar solvents to be dissolved, others may change the pH of the extracellular medium dramatically if applied without caution.

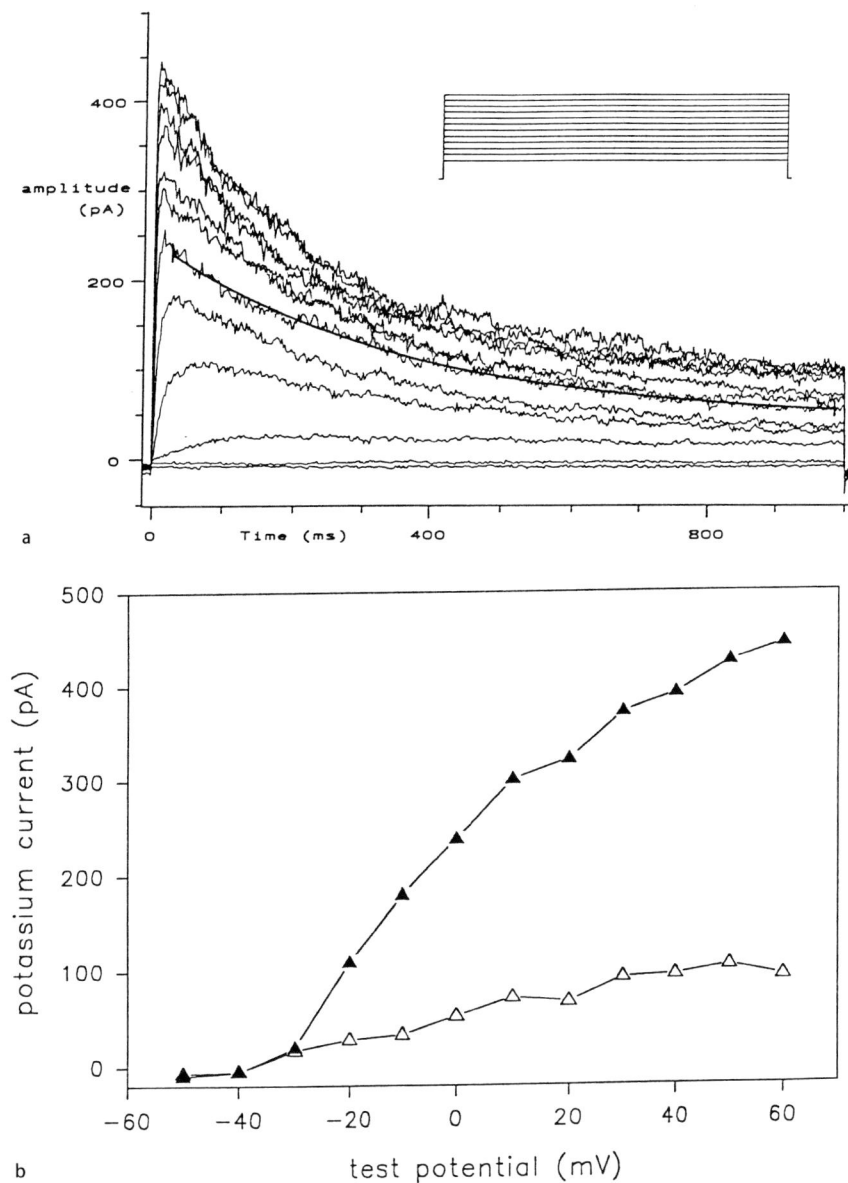

Fig. 17.2a, b. **a** Whole cell K$^+$ currents in a human lymphocyte evoked by a typical voltage-clamp protocol with 4.5 mM K$^+$ ECS in the bath and KF-ICS in the patch pipette. **b** Peak (*filled triangles*) and steady state (*open triangles*) current-voltage relationship (I–V) curves

As general advice the final concentration of apolar solvents should be kept as low as possible and their effect on the channel activity should be checked in control experiments. Following the application of an agent, it is also advisable to check the pH of the extracellular medium in a control experiment and readjust it if necessary. We found that in most of the cases the buffering capacity of the ECS used proved to be enough to overcome the latter problem.

Inactivation

During maintained depolarization the inactivation process of whole cell potassium currents is well characterized by a single exponential function in the form of:

$$I_K = A1.\exp\{-(t-\varkappa)/\tau_1\}+C \tag{1}$$

where τ_1 is the time constant of inactivation of the K^+ current, $A1$ is the current value at time $\varkappa$, the beginning of the fitted interval, and C is the noninactivating current component. Figure 17.2a displays a fitted curve corresponding to the above formula through data points recorded at $V_{test} = 0\,mV$, with a single exponential time constant of $\tau_1 = 355.85\,ms$. The fitted values of the other parameters were: $A1 = 185.47\,pA$; $\varkappa = 26.01\,ms$; $C = 41.62\,pA$.

Note. We obtained the fitted values by using the curve fitting routine of pClamp6 here and throughout the rest of the chapter.

Current-voltage relationship

The current-voltage (I/V) relationships in Fig. 17.2b illustrate the peak (filled triangles) and noninactivating (open triangles) components of the whole cell K^+ currents in human lymphocytes. In HPBLs the amplitude of both components, as well as the inactivation rate, vary from donor to donor ($n \geq 200$). During construction of the curves leak current has not been subtracted.

Activation

The whole cell voltage-clamp current traces of Fig. 17.3a were evoked by voltage steps lasting for 100 ms and recorded with a 10x higher sampling rate than those of Fig. 17.2a. The activation of lymphocyte whole cell K^+ currents proceeds with a sigmoidal time course. The rising phase of K^+ currents is fairly well described by a Hodgkin-Huxley type n^4 model in the form of:

$$I_K = A1*(1-\exp\{-t/\tau_1\})^4+C \tag{2}$$

where τ_n is the time constant of activation of the K^+ current, $A1$ is the maximal I_K and C is a small constant current component at the begin-

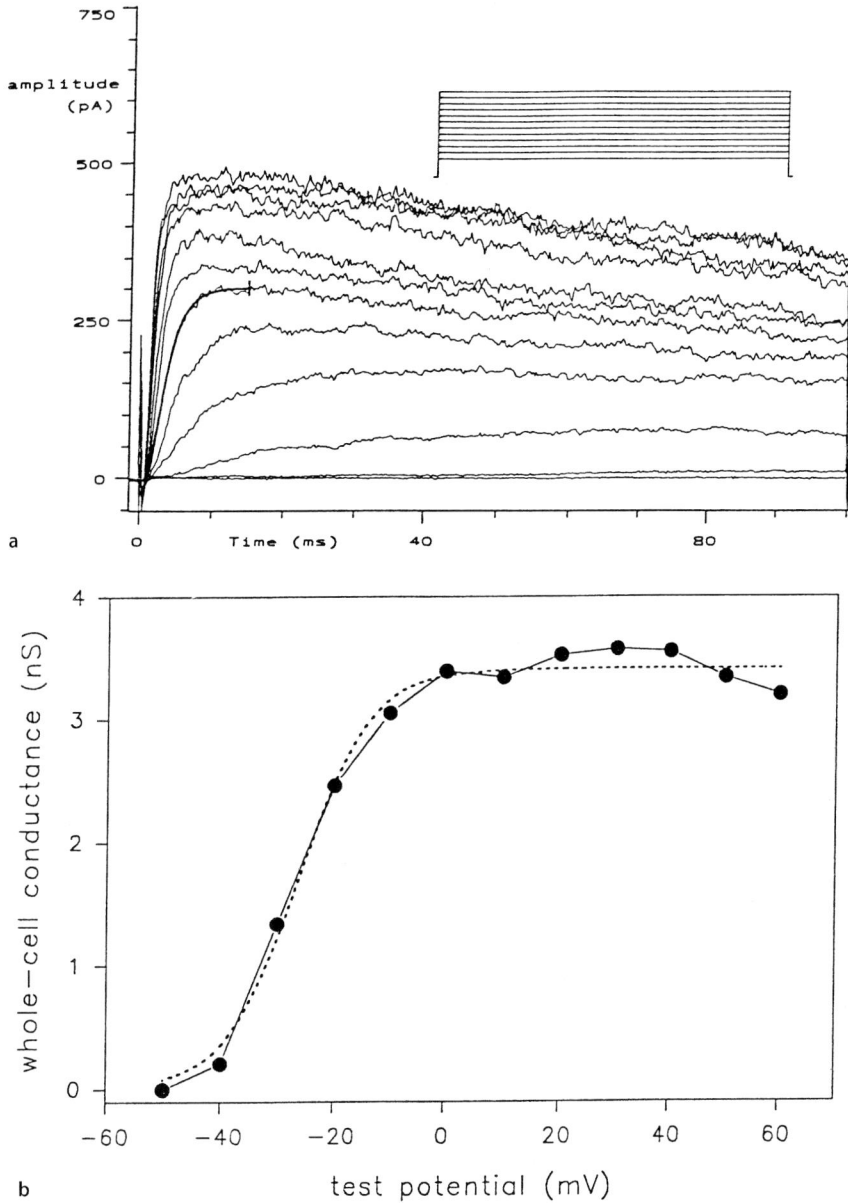

Fig. 17.3a, b. a Activation of whole cell K$^+$ currents in a human lymphocyte. **b** Peak chord conductance-voltage relationship curve (*filled circles*)

ning of the curve. Figure 17.3a displays a fitted curve corresponding to this formula recorded at $V_{test} = 0$ mV and characterized by a fitted exponential time constant of $\tau_n = 1.86$ ms. The fitted values of the other parameters were: $A1 = 305.10$ pA; $C = -4.75$ pA.

Note. Recording conditions throughout the rest of the experiments in this chapter were similar to those of Fig. 17.2, unless otherwise stated.

The underlying process governing the activation of the whole cell K^+ currents in Fig. 17.3a is the voltage-dependent activation of the whole cell K^+ conductance $G_K(V)$, based on the functioning of many (a few hundred per cell) n-type K^+ channels characterized by an approx. 10 pS single channel conductance.

Voltage dependence of activation

Voltage dependence of the activation of whole cell K^+ conductance is best characterized by a peak chord conductance-voltage relationship curve. An example is given for such a curve in Fig. 17.3b, where the chord conductance value (filled circles) at a given test potential was calculated on the basis of the measured peak current value (filled triangles) of the appropriate current trace of Fig. 17.3a. An −89 mV reversal potential value, estimated by linear interpolation from the peak current-voltage relationship curve of Fig. 17.2b, was used. The sigmoidal behavior of $G_K(V)$ is conveniently described by a first order Boltzmann function in the following form:

$$G_K(V) = G_{K,max}/(1+\exp(V-V_n)/K_n) \tag{3}$$

where $G_{K,max}$ is the maximum conductance, V_n is the voltage at the midpoint of the curve and K_n characterizes the steepness of voltage dependence. The dashed curve in Fig. 17.3b illustrates the Boltzmann distribution that best fits the experimental data with the following parameters: $G_{K,max} = 3.42 \pm 0.05$ nS; $V_n = -26.26 \pm 0.83$ mV; $K_n = -6.40 \pm 0.72$ mV. Recording conditions were similar to those of Fig. 17.2, unless otherwise stated. Due to the negligible leak current no leak correction was applied during analysis of the data.

Voltage dependence of inactivation

It is a property of the voltage-gated K^+ channels studied that during a longer time period they inactivate to a different degree at different depolarizing potentials. Figure 17.4 demonstrates the voltage dependence of steady state inactivation of whole cell K^+ currents in human lymphocytes. In this case steady state inactivation of the K^+ channels was studied by utilizing depolarizing test pulses several minutes long at those potentials which elicit K^+ currents in lymphocytes and subsequently testing

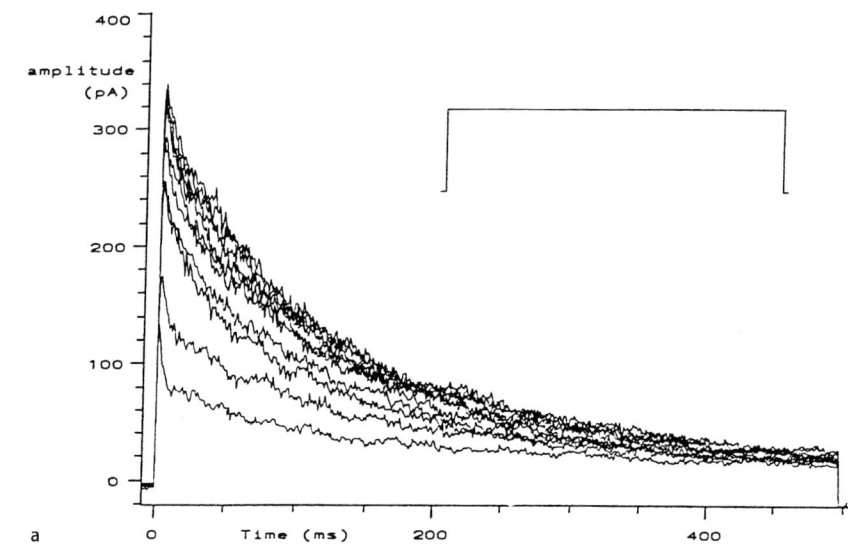

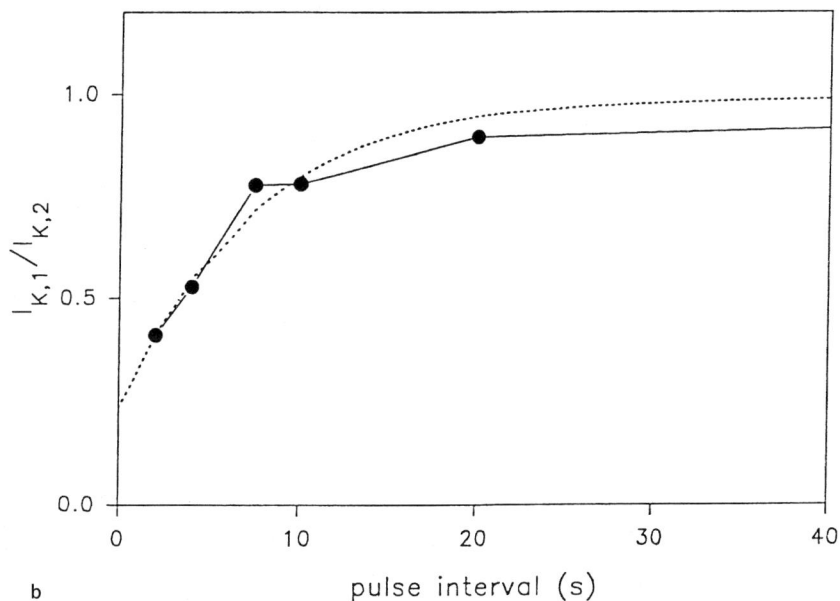

Fig. 17.4 a, b. a Recovery from inactivation of human lymphocyte K^+ channels. **b** Dependence of normalized peak (*filled circles*) currents on the time interval separating the pairs of test pulses

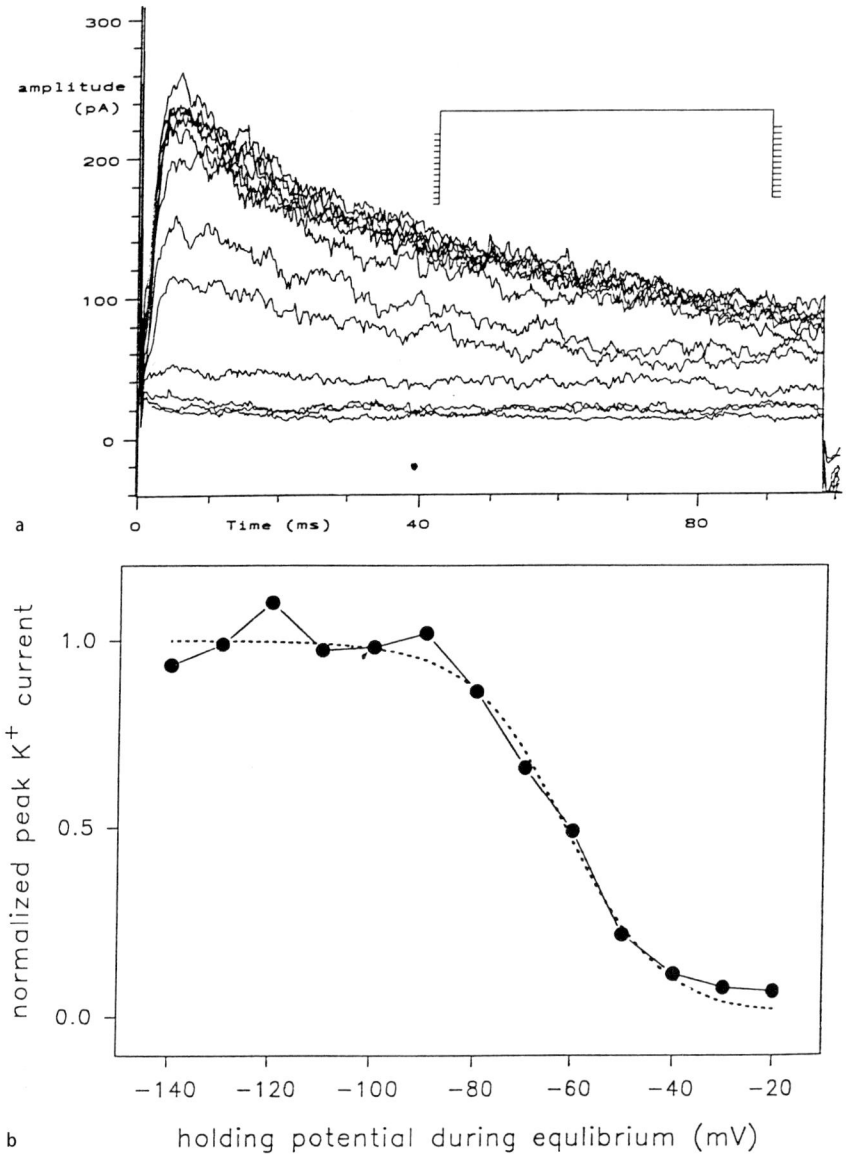

Fig. 17.5 a, b. **a** Voltage dependence of steady state inactivation of whole cell K⁺ currents in a human lymphocyte. **b** Voltage dependence of normalized peak (*filled circles*) currents

the noninactivating fraction of the whole cell current by a short pulse to a highly depolarizing potential. The current traces of Fig. 17.5a were evoked by +20 mV test pulses lasting for 100 ms, following the equilibration of the K$^+$ channels at different holding potentials for 3 min. The records were taken with a sampling interval of 200 μs). Figure 17.5b demonstrates that the normalized peak currents (filled circles) -derived from the current traces of Fig. 17.4a – can be fitted reasonably well to a first order Boltzmann function in the form of:

$$I_K(V)/I_{K,max} = 1/(1+\exp(V-V_j)/K_j) \qquad (4)$$

where $I_{K,max}$ is the maximum current, V_j characterizes the midpoint of the curve and K_j is the slope factor and V is the holding potential preceding the +20 mV test pulse. Parameters characterizing the best fit are: $I_{K,max} = 323.78 \pm 21.05$ pA; $V_j = -61.48 \pm 1.23$ mV; $K_j = 9.92 \pm 1.10$ mV. Data points were normalized to the fitted $I_{K,max}$. Recording conditions were similar to those of Fig. 17.2, unless otherwise stated. No leak correction has been applied during the analysis of the data.

Following their inactivation voltage-gated K$^+$ channels need a certain amount of recovery time to be activated again by depolarizing voltage pulses. Pairs of identical depolarizing pulses applied from a holding potential and separated by varying time intervals can be used to test the time course of recovery from inactivation of the K$^+$ channels. Current traces of Fig. 17.4 were evoked by pairs of identical test pulses 500 ms to +40 mV from a holding potential of −100 mV, separated by various intervals. The successive pairs of pulses were separated by 3 min equilibrating intervals. Current traces were recorded during both pulses of the pairs with a sampling interval of 1000 μs. The recovery process can be described by a single (or more complicated) exponential recovery law in the form of:

Recovery from inactivation

$$I_{K,2}/I_{K,1}(t) = 1 - A.\exp(-t/\tau_1) \qquad (5)$$

as can be seen in Fig. 17.4b. In the above formula $I_{K,1}$ and $I_{K,2}$ are the current amplitudes after the first and second test pulses, A describes the starting amplitude of the curve at time 0, and τ_1 is the time constant characterizing the recovery process. The fitted parameters are: $A = 0.76 \pm 0.07$; $\tau_1 = 7.59 \pm 1.16$ s. Data points were normalized to the mean $I_{K,1}$ value ($n = 6$). Recording conditions were similar to those of Fig. 17.2, unless otherwise stated. No leak correction was applied during analysis of the data.

Note. It is very important to note that one should allow at least $5 \times \tau_1$, corresponding to a 99 % recovery, between successive voltage-clamp experiments in order to avoid experimental error due to incomplete recovery of the channels from their inactive state.

Tail currents

Voltage-gated K^+ channels in the plasma membrane of lymphocytes can be "operated" in a special way by the voltage-clamp protocol of Fig. 17.6, where the current traces were evoked by a pulse sequence containing a 10 ms prepulse to $+60$ mV to maximally activate the K^+ conductance, followed by a 200 ms test pulse. The amplitude of the test pulse was changed between 0 and -110 mV in 10 mV steps during the successive pulse sequences following each other by 50 s intervals. Sampling intervals were 40 µs and 800 µs during the recording of the prepulse and test pulses, respectively. This "tail current experiment" is designed to maximally activate the ion channels with a short depolarizing pulse and, immediately after this pulse, drive the permeant ions through the open channels at another test potential, before the channels could react to the applied voltage changes.

Instantaneous I–V relationship

From the tail current experiment an instantaneous current-voltage relationship curve can be constructed and the reversal potential of the whole cell current may be determined (see Fig. 17.6b). Voltage dependence of instantaneous currents measured at 2 ms after the beginning of the test pulse is displayed in the figure. The currents reverse at -73.5 mV. No correction was made for an approx. -5 mV junction potential between the KF and 4.5 mM K^+-ECS. The reversal potential of human lymphocytes, measured in 4.5 mM K^+-ECS with KF-ICS in the pipette and derived from a tail current experiment, usually falls between -70 and -80 mV.

Deactivation time course

The time course of deactivation of the K^+ channels after a sudden change in the test potential can be obtained by studying the shape of the tail currents. The time constant of deactivation of the K^+ current can be determined by fitting the formula

$$I_K = A.\exp\{-(t-\varkappa)/\tau_d\}+C \tag{6}$$

to the measured data points, where τ_d is the time constant of deactivation and the other parameters have their usual meanings. (See as an example the fitted curve through data points recorded at $V_{test} = -30$ mV, with a single exponential tail current time constant of $\tau_d = 65.38$ ms.) The fitted values of the other parameters were: $A = 136.26$ pA; $\varkappa = 13.96$ ms; $C = 50.41$ pA.

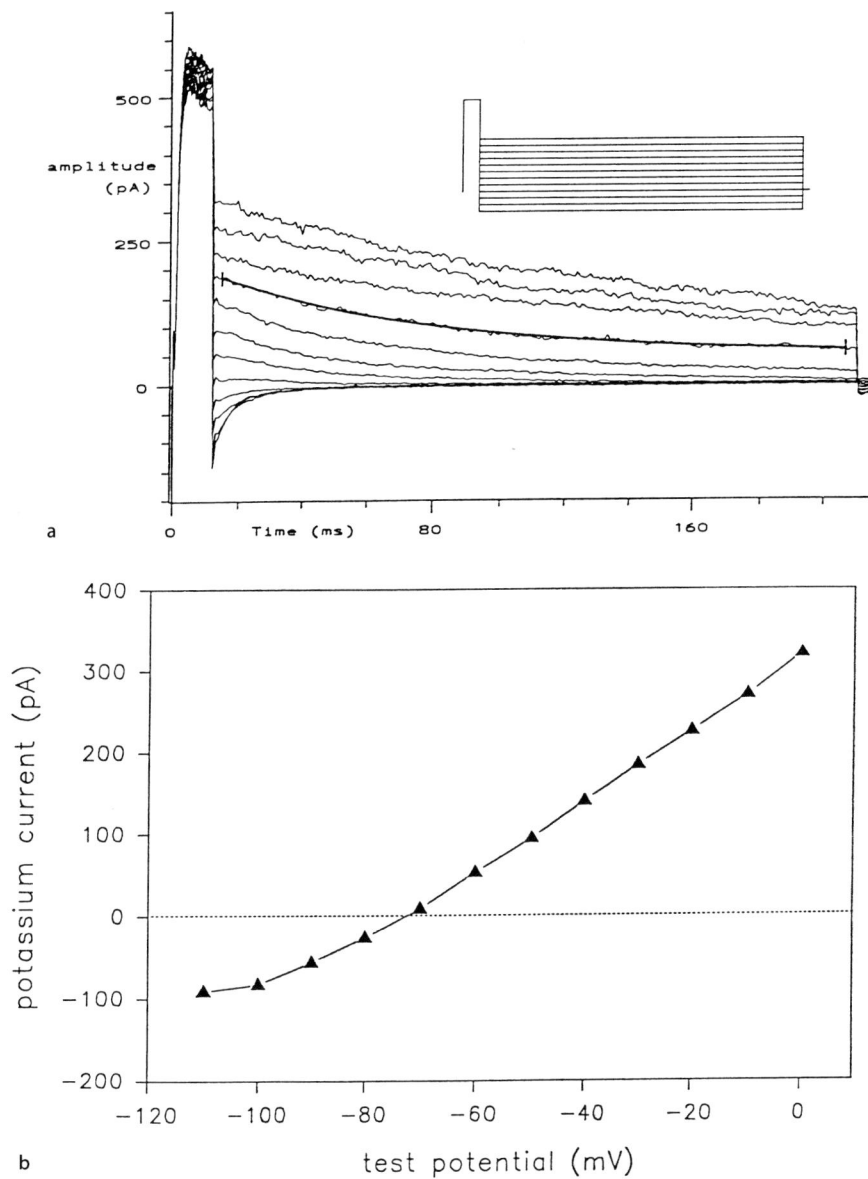

Fig. 17.6 a, b. a Current traces evoked by a typical tail current experiment. **b** Instantaneous current-voltage relationship in a human lymphocyte in 4.5 mM K$^+$ ECS

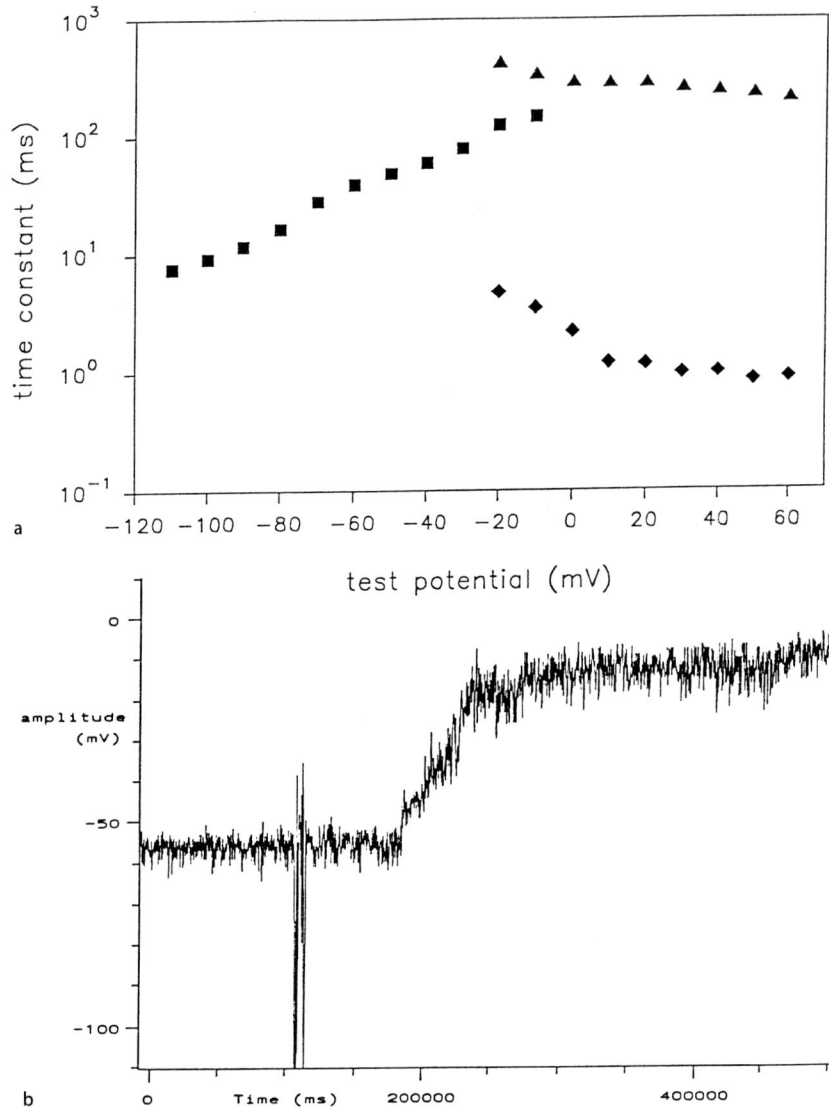

Fig. 17.7 a, b. a Summary of the voltage dependence of time constants characterizing K^+ current kinetics in human lymphocytes. **b** Patch-clamp membrane potential measurement on a human lymphocyte under $I = 0$ current-clamp condition in the whole cell configuration

A summary of the voltage dependence of time constants characterizing K^+ current kinetics in human lymphocytes is given in Fig. 17.7a. Time constants of activation τ_n (filled diamonds), inactivation τ_j (filled triangles) and deactivation τ_n (filled squares) are displayed as determined from the unprocessed data of Figs. 17.3, 17.2 and 17.6, respectively.

Effects of Changing the Ionic Composition

The alteration of the ionic composition of the extracellular environment, and also that of the intracellular one, causes dramatic changes in the shape and voltage dependence of the recorded current traces. Figure 17.8 demonstrates what happens if one increases the extracellular K^+ concentration to 160 mM (160 mM K^+-ECS) and records the whole cell currents of lymphocytes with equal K^+ concentration on both sides of the plasma membrane (KF-ICS in the pipette). Current traces were evoked by voltage steps lasting for 1000 ms, applied every 50 s from a holding potential −80 mV in 10 mV increments from −60 to +40 mV and recorded with a sampling interval of 2000 µs. In this case the reversal potential of the whole cell current is +2.7 mV, close to the expected 0 mV, and the direction of the whole cell current becomes inward at depolarizing potentials more negative than the reversal potential and outward above 0 mV (see Fig. 17.8b). The observed change in the reversal potential – from −89 mV to 0 mV – can be calculated from the Nernst equation. This fact strongly supports the idea that K^+ ions are responsible almost exclusively for charge conduction through ion channels in the plasma membrane of the lymphocytes under these conditions.

Effects of K^+

Membrane Potential Measurement on Human Lymphocytes in I = 0 Current-Clamp Mode

The membrane potential of resting HPBLs is maintained by functioning n-type K^+ channels. At the resting potential (V_{rest}) only one or two n-type K^+ channels are open out of the total number of 300–400 and the random opening of a few channels causes a considerable fluctuation in V_{rest} (Cahalan et al. 1985). The actual value of V_{rest} of unstimulated mammalian lymphocytes has been found to vary between −50 and −70 mV, close to the activation range of n-type K^+ channels, as determined by fluorescent methods (Gallin 1991; Damjanovich et al. 1994) and patch-clamp

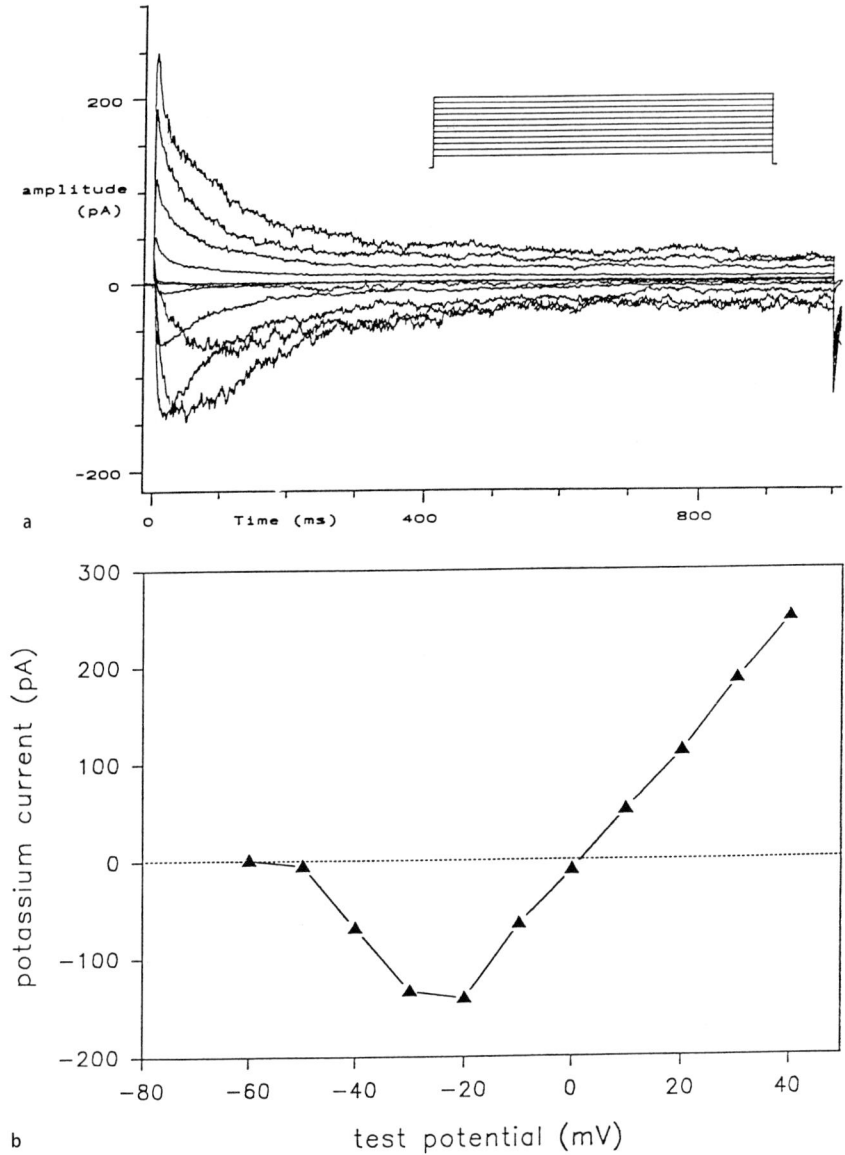

Fig. 17.8 a, b. a Human lymphocyte whole cell K⁺ currents recorded with equal K⁺ concentrations in the bath (160 mM K⁺-ECS) and the patch pipette (KF-ICS). **b** Peak (*filled triangles*) current-voltage relationship curve derived from the current traces of **a**

experiments in the whole cell (Gáspár et al. 1992) and cell-attached patch (CAP) configurations (Pahapill and Schlichter 1992).

Figure 17.7b demonstrates the result of a patch-clamp membrane potential measurement on a human lymphocyte under $I = 0$ current-clamp condition recorded with 4.5 mM K^+ ECS in the bath and KF-ICS in the pipette in the whole cell configuration.

The resting potential of the cell corresponds to the average value of the fluctuating membrane potential values at the initial part of the recording. The small fluctuations in the recording correspond to the openings and closings of individual K^+ channels, substantially influencing the membrane potential of the cell. The sampling interval was 300 ms, the pipette was filled with KCl-ICS, otherwise the experimental conditions were similar to those of Fig. 17.2. The large fluctuations in V_m around 110 s indicate the addition of cyclosporin A to the sample, which resulted in a marked depolarization of the cell. According to our experience immunosuppressor cyclosporin A (10 µg/ml) effectively reduces the whole cell n-type K^+ current in HPBLs by approximately 40 % without any change in the reversal potential of the current. The voltage dependence of steady state inactivation of the K^+ channels is not altered by this drug although the rate of K^+ current inactivation is increased. Our findings reveal that the inhibition of voltage-gated K^+ currents of HPBLs by the above immunosuppressor is responsible for the depolarized membrane potential observed in the presence of the drug.

Resting membrane potential

References

Bubien JK, Zhou LJ, Bell PD, Frizzell RA, Tedder TF (1993) Transfection of the CD20 cell surface molecule into epitopic cell types generates a Ca^{2+} conductance found constitutively in B lymphocytes. J Cell Biol 121:1121–1132

Cahalan MD, Chandy GK, DeCoursey TE, Gupta S (1985) A voltage-gated potassium channel in human T lymphocytes. J Physiol 358:197–237

Chandy GK, DeCoursey TE, Cahalan MD, McLaughlin C, Gupta S (1984) Voltage-gated potassium channels are required for human T lymphocyte activation. J Exp Med 160:369–385

Damjanovich S, Szöllosi J, Trón L, Edidin M (1994) Mobility and proximity in biological membranes. CRC, Boca Raton

DeCoursey TE, Chandy KG, Gupta S, Cahalan MD (1984) Voltage-gated K^+ channels in human T lymphocytes: a role in mitogenesis? Nature 307:465–468

Deutsch C, Krause D, Lee SC (1986) Voltage-gated potassium conductance in human T lymphocytes stimulated with phorbol ester. J Physiol 372:405–423

Drakopoulou E, Cotton J, Virelizier H, Bernardi E, Schoofs AR, Partiseti M, Choquet D, Gurrola G, Possani LD, Vita C (1995). Chemical synthesis, structural and functional characterization of noxiustoxin, a powerful blocker of lymphocyte voltage dependent K$^+$ channels Biochem Biophys Res Commun 213: 901–907

Gallin E K (1991) Ion channels in leukocytes. Physiol Rev 71:775–811

Gáspár R. Jr, Krasznai Z, Márián T, Trón L, Recchioni R, Fallasca M, Moroni F, Pieri C, Damjanovich S. (1992) Bretylium induced voltage gated sodium current in human lymphocytes. Biochim Biophys Acta 1137:143–147

Gáspár R Jr, Panyi Gy, Ypey DL, Krasznai Z, PieriC, Damjanovich S (1994) Effects of bretylium tosylate on voltage gated potassium channels in human T lymphocytes. Mol Pharmacol 46:762–766

Gáspár R Jr, Bene L, Damjanovich S, Muñoz-Garay C, Calderon-Aranda ME, Possani LD (1995) β-scorpion toxin 2 from *Centruroides noxius* blocks voltage-gated K$^+$ channels in human lymphocytes. Biochem Biophys Res Commun 213: 419–423

Gelfand EW, Cheung RK, Mills GB (1987) Role of membrane potential in the response of human T lymphocytes to phytohemagglutinin. J Immunol 138: 1115–1120

Grinstein S, Dixon JS (1989) Ion transport, membrane potential and cytoplasmic pH in lymphocytes: changes during activation. Physiol Rev 69:417–481

Grinstein S, Smith JD (1990) Calcium-independent cell volume regulation in human T lymphocytes. Inhibition by charybdotoxin. J Gen Physiol 95:97–120

Grissmer S, Nguyen AN, Cahalan MD (1993) Calcium-activated potassium channels in resting and activated human T lymphocytes. J Gen Physiol 102: 601–630

Gupta S, Good RA (1978) Subpopulations of human T lymphocytes. III. Distribution and quantitation in peripheral blood, cord blood, tonsils, bone marrow, thymus, lymph nodes, and spleen. Cell Immunol 36:263–270

Lee SC, Deutsch C (1990) Temperature dependence of K$^+$ channel properties in human T lymphocytes. Biophys J 57:49–62

Lee SC, Levy DI, Deutsch C J (1992) Diverse K$^+$ channels in primary human T lymphocytes. J Gen Physiol 99:771–793

Leonard R, Garcia M L, Slaughter R S, Reuben J P (1992) Selective blockers of voltage-gated K channels depolarize human T lymphocytes: Mechanism of the antiproliferative effect of charybdotoxin. Proc Natl Acad Sci USA 89:10094–10098

Lewis RS, Cahalan M D (1990) Ion channels and signal transduction in lymphocytes. Annu Rev Physiol 52:415–430

Lin CS, Boltz RC, Blake JT, Nguyen M, Talento A, Fischer PA, Speinger MS, Sigal NH, Slaughter RS, Garcia ML, Kaczorowski GJ, Koo GC (1993) Voltage-gated potassium channels regulate calcium dependent pathways involved in human T lymphocyte activation. J Exp Med 177:637–645

Matteson DR, Deutsch C (1984) K$^+$ channels in T lymphocytes: a patch clamp study using monoclonal antibody adhesion. Nature 307:468–471

Oettgen H C, Terhorst C, Cantley LC, Rosoff PM (1985) Stimulation of the T3-T cell receptor complex induces a membrane potential sensitive calcium influx. Cell 40:583–590

Pahapill PA, Schlichter LC (1992) Modulation of potassium channels in intact human T lymphocytes. J Physiol 445: 407–430

Pariseti M, Korn H, Choquet D (1993) Pattern of potassium channel expression in proliferating B lymphocytes depends upon the mode of activation. J Immunol 151:2462–2470

Price M, Lee SC, Deutsch C (1989) Charybdotoxin inhibits proliferation and interleukin 2 production in human peripheral blood lymphocytes. Proc Natl Acad Sci USA
86:10171–10175
Sutro JB, Vayuvegula B, Gupta S, Cahalan MD (1989) Voltage-sensitive ion channels
in human B lymphocytes. Adv Exp Med Biol 254:113–122

Chapter 18

Ionic Conductances in Chicken Osteoclasts

Zoltan Krasznai, Adam F. Weidema, Rezsö Gáspár, György Panyi, and Dirk L. Ypey

Background

Osteoclasts

Osteoclasts are bone resorbing cells which play an important role in bone formation. In adult life, osteoclasts are essential for bone remodeling (upon loading and following fractures) and calcium homeostasis. Upon stimulation, the osteoclast seals itself to the bone surface whilst forming a resorption lacuna between the cell membrane and the bone matrix. The osteoclast membrane facing this lacuna has a ruffled border through which both lysosomal enzymes and protons are actively secreted (Blair et al. 1989). This results in a dissolution of organic compounds and of calcium phosphate from the bone matrix. The resorption of bone minerals by osteoclasts involves a variety of ion transport processes. An electrogenic proton (H^+)-ATPase (Bekker and Gay 1990; Delaisse and Vaes 1992) may play a key role in the resorption process. Electrogenic extrusion of the positively charged protons hyperpolarizes the resting membrane potential (E_m). This hyperpolarization would oppose further proton pumping out of the cell. It may be compensated for by fluxes of other ions through ion channels that prevent excessive hyperpolarization and permit a continued efflux of protons. Thus,the osteoclast could control the efficiency of proton pumping by changing the resting membrane potential or by activating ionic channels.

In recent years, different K^+ conductances have been demonstrated in osteoclasts from different species, e.g., in chicken, rat, rabbit and mouse (Ravesloot et al. 1989a; Ravesloot et al. 1989b; Sims et al. 1991; Ypey et al. 1992; Weidema et al. 1993). The inward rectifying K^+ conductances are present in osteoclasts of all species and are more or less comparable. For the outward rectifying K^+ conductances the situation is more complex. Outward currents were only described in chicken osteoclasts but recently a transient outward rectifying K^+ conductance has been found in mammalian osteoclasts as well (Arkett et al. 1994). The expression of this out-ward conductance was influenced by the substrate on which the osteo-clasts were cultured.

The osteoclast is a hematopoietic cell whose lineage is still under de-bate. Osteoclasts may be closely related to macrophages. Alternatively they may be derived from a specific precursor whose commitment takes place very early in myeloid differentiation. As a consequence, osteoclasts may express some ion channels that are common to other hematopoietic cells as well. However, some of the channels involved in the process of bone resorption may be exclusively expressed in osteoclasts.

Ionic Conductances

The following conductances are known to occur in chicken osteoclasts:

- Inward rectifying potassium conductance (G_{Ki})

- Transient outward rectifying potassium conductance (G_{Kto})

- Sustained outward rectifying potassium conductance (G_{Ko}), which has been shown to be a calcium-activated potassium conductance (G_{KCa})

- Fast sodium conductance (G_{Nafast})

- Stretch-activated conductance ($G_{stretch}$)

- ATP-activated conductance (G_{ATP})

(Ravesloot et al. 1989a; Ravesloot et al. 1989b; Ypey et al. 1992; Weidema et al. 1993; Wiltink et al. 1995; Gáspár et al. 1995)

18.1 Isolation of Chicken Osteoclasts

Materials

- 10–15 chicken embryos 18–19 days old
- 90 μm sieve
- 10 cm Petri dish
- Two siliconized test tubes
- Four siliconized Pasteur pipettes
- Scissors, forceps, scalpels

- PBS (140 mM NaCl, 5 mM KCl, 1 mM $CaCl_2$, 10 mM glucose and 10 mM HEPES, titrated to pH 7.4 at 20 °C with NaOH)
- Iscove's modified DMEM (Gibco). Add 1.766 g Iscove's, 0.476 g HEPES, and 0.004 g $NaHCO_3$ to 100 ml distilled water and adjust the pH with NaOH to 6.9.
- α-MEM (Gibco) pH 7.2
- Gentamicin/glutamine stock solution: 42.5 mg gentamicin +292 mg glutamine in 10 ml PBS, pH 7.2
- Fetal calf serum (FCS) (Sigma)

For complete medium add 10 ml FCS and 2 ml gentamicin/glutamine stock solution to 88 ml of Iscove's modified Dulbecco's medium.

Note. For care and use of laboratory animals apply for permission and carefully follow the guide of the Committee of Scientific Ethics for the Protection of Animals.

Procedure

Isolation   Fill the 10 cm Petri dish with 20 ml PBS. Decapitate the embryos with a sharp pair of scissors, remove the tibiae and femora and place them in the PBS. Squeeze the bone marrow with a silicon stop of a sterile test tube, and remove large particles from the cell suspension by sieving it through the 90 μm sieve. Put the cell suspension in 10 ml siliconized test tubes and leave for 30 min. Meanwhile prepare the culture media: supplement both the Iscoves and α-MEM with 10 % FCS, 85 μg/ml gentamicin and 584 μg/ml glutamine.

Remove the top 75 % of the cell suspension. Centrifuge the remaining cells for 5 min at 80 g (approx. 600 rev/min with 20 cm rotor radius) then remove the supernatant. Resuspend the pellet in the supplemented Iscove (0.8 ml/embryo). Put the cell suspension on coverslips (e.g. diameter 24 mm) and let the cells adhere for 1 h at 37 °C, in 5 % CO_2. Remove nonadherent cells by gently washing with PBS and culture the cells in complete Iscove's or α-MEM for up to 3 days. Six hours after isolation the cells are flattened enough to be identified as osteoclasts. Osteoclasts can be selected by their multinuclear appearance ($n > 3$), and positive neutral red staining, which has no effect on the conductances of the cells. Prior to the experiments place the cells in extracellular-like solution (ECS, see Table 18.1) in the experiment dish. A Teflon culture dish developed for high magnification microscopy (Ince et al. 1985) allows high resolution phase contrast microscopy on an inverted microscope.

Table 18.1. Composition of extracellular (ECS) and intracellular (ICS) solutions

	NaCl	KCl	CaCl$_2$	MgCl$_2$	HEPES	EGTA	glu-cose	Na-gluco-nate	K-gluco-nate	NaF
S-ECS	145	5	1	1	10	0	5	0	0	0
Na-glu ECS	30	0	1	1	10	0	5	120	0	0
NaF-EC S	0	0	1	1	10	0	5	0	0	145
Na-ICS	140	0	4	1	10	10	0	0	0	0
K-glu ICS	0	30	1	1	10	0	5	0	120	0
7-ICS	5	140	4	1	10	10	0	0	0	0
Ba-TEA ECS[a]	127	5	1	1	10	0	5	0	0	0
15-Na-Kglu-ICS	15	15	1	1	10	0	5	0	120	0

ECS, extracellular solutions, pH 7.4; ICS, intracellular solutions, pH 7.2. The K-glu-ICS and 15-Na-Klgu-ICS solution do not contain EGTA, therefore it can be used only in nystatin-whole cell configuration. PBS contains 140 mM NaCl, 5 mM KCl, 1 mM CaCl$_2$, 10 mM glucose and 10 mM HEPES, titrated to pH 7.4 at 20 °C with NaOH.
Note. Keep the solutions in the refrigerator, filter and add glucose only before use.
[a]The Ba-TEA ECS solution contains an additional 5 mM of barium and 10 mM TEA.

18.2
Electrophysiology

Materials

Patch-clamp experiments require delicate pipette fabrication. Patch pipettes can be pulled from borosilicate glass (GC150TF-15, Clark, UK) or from hematocrit glass capillaries (Kimax-51, Kimble products, USA). A minimum two-stage pulling process should be used to produce pipettes with an opening diameter of 0.5–1 µm. Store the manufactured pipettes under cover and heat polish just before use. To improve signal to noise ratio coat the pipette tip with the special hydrophobic substance Sylgard 184 (Dow Corning Corporation, Midland, MI USA). By weight, add one part of Sylgard curing agent to ten parts of the Sylgard 184 Encapsulating Resin, coat the tip of the pipette and apply a short heat pulse to it (approx. 400 °C for 1–2 s to polymerize the product). The pipettes should be filled up from the back with filtered ICS solution before use. Good pipettes have a resistance of 2–5 MΩ as measured in ECS. The electrodes in the bathing solution and in the interior of the patch pipette are Ag/AgCl wire or pellet electrodes. The software pro-

Patch pipettes

gram pClamp (version 5.5.1 or 6.0, Axon Instruments, CA, USA) can be used for applying voltage protocols and recording the currents. There are several different types of patch-clamp amplifiers commercially available.

The cell under investigation is observed under a high magnification phase-contrast microscope. The patch pipette is attached directly to the pipette holder, through which external pressure can be applied to the pipette interior. The pipette holder is connected mechanically as well as electrically to the preamplifier of the current voltage converter. The pipette is positioned throughout the experiment via movement of the preamplifier-pipette holder complex, which is governed by precision micromanipulators.

The I/V converter is a very sensitive device capable of measuring currents in the range of a few picoamperes. Preset voltage levels can be applied to the pipette through the I/V converter. The voltage is always measured with respect to the reference electrode placed in the bathing solution. The resistance of the pipette and the seal formed between the pipette and the membrane surface can be monitored by applying small amplitude (1 mV) voltage pulses to the pipette and observing the current flowing through the pipette. The stimulus and the resulting current pulses are displayed on the screen of an oscilloscope and the computer monitor. The current and voltage changes during the experiment are continuously recorded on line by a computer.

Procedures

Protocol example

1. Set up patch clamp instrument, test working conditions with electric equivalent circuit.

2. Prepare chicken osteoclasts for patch clamp experiment following the description.

3. Prepare patch pipettes, bathing and pipette solutions according to the previous description.

4. Put a coverslip with adhered osteoclasts on it onto a Teflon Petri dish developed for high magnification microscopy. Pour 1 ml of filtered ECS into the Petri dish. Observe cells under high magnification and select a cell for the experiment.

5. Heat polish one pipette and fill it with filtered intracellular-like solutions (ICS) (see Table 18.1).

6. Apply positive pressure in the pipette and cross fluid surface of the bath with pipette tip and find the pipette position.

7. Measure the pipette resistance by applying a test pulse of 1 mV.

8. Electronically cancel the offset of the pipette potential.

9. Approach selected cell with care; the pipette is still under positive pressure. Touch cell with pipette, remove positive pressure. Observe change in amplitude of current due to 1 mV stimulus. Apply negative pressure to pipette interior according to instructions (approx. 10 cm water) and observe formation of gigaseal. The gigaseal usually develops within 1–3 min. If this does not occur, try applying a negative holding potential (-70 mV) on the cell, which usually facilitates development of the seal. Use cell only if a tight seal with a resistance greater than 5 GΩ is formed. Electronically cancel the fast capacitive transient which appeared after gigasealing.

10. Observe single channel currents flowing through membrane patch in this cell attached patch (CAP) configuration.

11. Apply short pulses of suction to disrupt the membrane patch. During the process observe the change in the capacitive transient. At sudden change of the transient, discontinue the suction. You have now obtained the (conventional) whole cell measurement configuration. Electronically cancel the slow capacitive transient which appeared after disrupting the patch.

The conductances present in the cell membrane can be measured by applying nystatin in the pipette solution (nystatin-whole cell configuration). Nystatin, a polyene antibiotic, is a channel forming ionophore. These channels have little selectivity, conduct monovalent anions and cations and also conduct neutral molecules such as water. The channels formed by nystatin behave as if they are 40–80 nm in diameter and are not permeable for bivalent cations such as Ca^{2+} or Mg^{2+}. Using nystatin in the pipette solution follow the description:

Dissolve 100 mg of pluronic acid (F-127) in 1 ml of DMSO. Add 50 mg nystatin into this stock solution. The stock solution should be stored in the dark at $-12\,°C$. Before use make a 200x dilution of this stock solution in pipette solution. Dissolution of nystatin can be promoted by sonication. Incorporation of nystatin in the cell membrane within the patch pipette starts during seal formation and reaches steady state within 2–10 min. Before measuring membrane conductances the development

Nystatin perforated patch

of the nystatin incorporation should be checked. In the nystatin-whole cell configuration stable membrane currents can be measured for up to 1–3 h, which is about twice as long as in the normal whole cell configuration. A series resistance of 15–30 MΩ in nystatin whole cell configuration is considered to be acceptable.

12. Set current-clamp condition and read value of membrane potential.

13. Use optimal series resistance compensation to correct for the voltage drop during voltage clamping.

14. Set voltage clamp condition and observe whole cell potassium current.

15. During long lasting experiments measure membrane potential regularly.

16. Analyze the results by making an I–V (current-voltage) curve. Correct for specific leakage.

Applications

Technical Approaches to Studying Specific Ion Channels in Chicken Osteoclasts

- For measuring the three potassium conductances present in chicken osteoclasts use S-ECS and 7-ICS solutions.

- While applying test pulses from −70 mV holding potential, all three potassium conductances appear. By keeping the cell at 0 mV holding potential, the G_{Kto} inactivates and the G_{Ko} can be measured (Fig. 18.1).

- Verapamil (50 µM) and barium (1 mM) are specific blockers of G_{Kto} and G_{Ki}, respectively. Other channel blockers [tetra-ethyl-ammonium (TEA), 4-amino-pyridine (4-AP)] may have complex effects on the potassium conductances of the chicken osteoclasts. They cause simultaneous blocking of different currents (i.e.,TEA blocks both G_{Kto} and G_{Ko}) or, while blocking one of the conductances, a stimulatory effect on another conductance can be observed (4-AP blocks only G_{Kto} and stimulates G_{Ko}).

- Studying G_{Kto} activation kinetics, use short (100 ms) test pulses.

- Studying G_{Kto} inactivation kinetics, use long (5 s) test pulses separated by long (20 s) recovery intervals.

- The fast sodium current is not present in all of the cells. When studying this current fill the pipette with 15Na-Glu-ICS and apply 5 mM

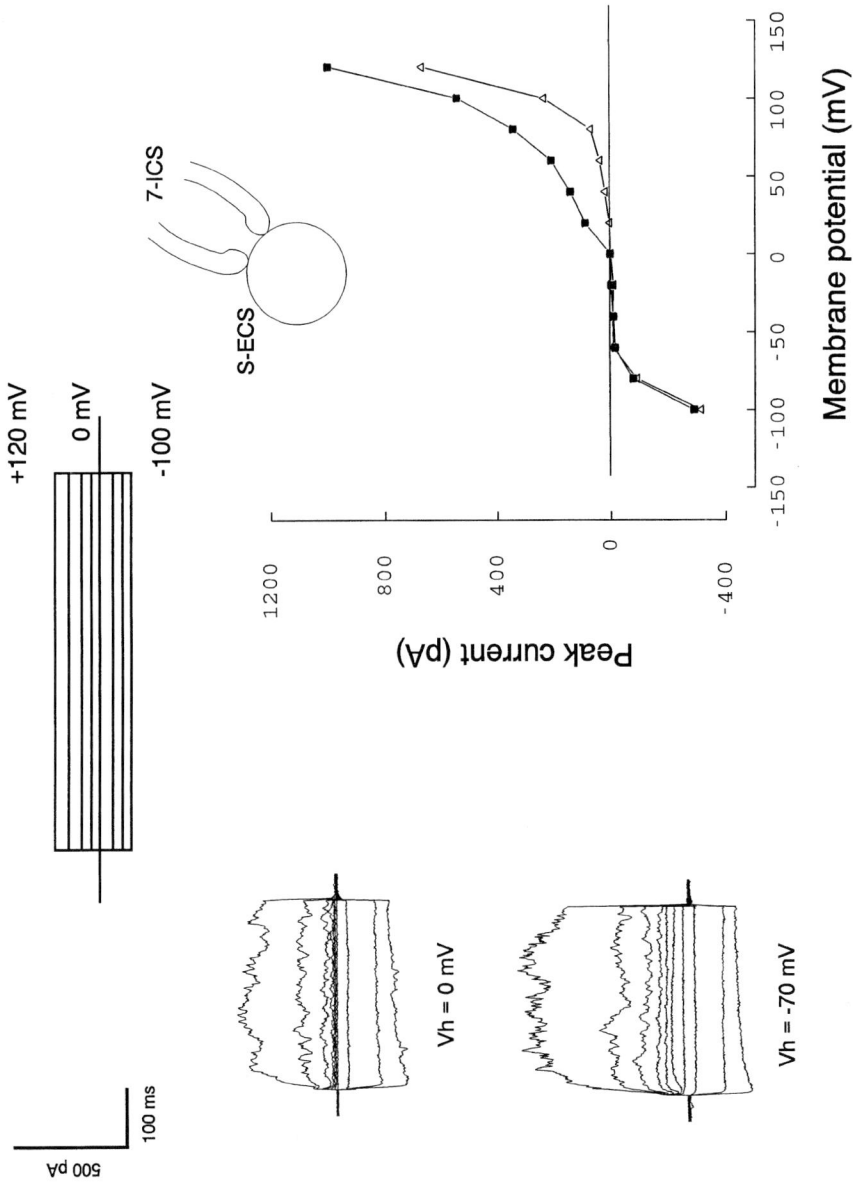

Fig. 18.1. Characteristic potassium currents present in embryonic chicken osteoclasts. Membrane potential was held at 0 mV and −70 mV, respectively and was stepped 1 s in duration to potentials from −100 to +120 mV by 20 mV increments. The pipette was filled with 7-ICS solution and the external medium contained S-ECS

barium and 10 mM TEA in the external solution to block potassium conductances both in the inward and outward directions. Use a very fast protocol and make a perfect transient cancellation. The G_{Nafast} inactivates within 5 ms. The current is tetrodotoxin-(TTX) sensitive (see Fig. 18.2). (Be careful, TTX is highly poisonous!)

- Adjust the osmolality of the extracellular and intracellular solutions to the same value with glucose! Different osmolality may result in initiating osmo-induced current!

References

Arkett SA, Dixon J, Yang JN, Sakai DD, Minkin C, Sims SM (1994) Mammalian osteoclasts express a transient potassium channel with properties of Kv1.3. Receptors and Channels 2:281–293

Bekke PJ, Gay CV (1990) Biochemical characterization of an electrogenic vacuolar proton pump in purified chicken osteoclast plasma membrane vesicles. J Bone Mineral Res 5: 569–579

Blair HC, Teitelbaum SL, Ghiselli R, Gluck S (1989) Osteoclastic bone resorption by a polarized vacuolar proton pump. Science 245:855–857

Delaisse JM, Vaes G (1992) Mechanism of mineral solubilization and matrix degradation in osteoclastic bone resorption. In: Rifkin BR, Gay CV (eds) Biology and physiology of osteoclasts. CRC , Boca Raton, pp 289–314

Gáspár R, Weidema AF, Krasznai Z, Nijweide PJ, Ypey DL (1995) Tetrodotoxin sensitive fast Na^+ current in embryonic chicken osteoclasts. Pflügers Arch 430:596–598

Ince C, Van Dissel JT, Diesselhoff MMC (1985) A Teflon culture dish for high-magnification microscopy and measurements in single cells. Pflügers Arch 403:240–244

Ravesloot JH, Ypey DL, Vrijheid-Lammers T, Nijweide PJ (1989a) Voltage activated K^+ conductances in freshly isolated embryonic chicken osteoclast. Proc Natl Acad Sci USA 86:6821–6825

Ravesloot JH, Ypey DL, Nijweide PJ, Buisman HP, Vrijheid-Lammers T (1989b) Three voltage activated K^+ conductances and an ATP-activated conductance in freshly isolated embryonic chicken osteoclast. Pflügers Arch 414:166–167

Sims SM, Kelley MEM, Dixon JF (1991) K^+ and Cl^- currents in freshly isolated rat osteoclasts. Pflügers Arch. 419: 358–370

Weidema AF, Ravesloot JH, Panyi G, Nijweide PJ, Ypey DL (1993) A Ca^{2+}dependent K^+ channel in freshly isolated and cultured chicken osteoclast. Biochim Biophys Acta 1149:63–72

Wiltink A, Nijweide PJ, Scheenen WJJM, Ypey DL, Van Duijn B (1995) Cell membrane stretch in osteoclasts triggers a self reinforcing Ca^{2+} entry pathway. Pflügers Arch. 429:663–671

Ypey DL, Weidema AF, Hold KM, VanderLaarse A, Ravesloot JH, VanderPlas A, Nijweide PJ (1992) Voltage, calcium and stretch activated ionic channels and intracellular calcium in bone cells. J Bone Miner Res 7, supplement 2, S377–387

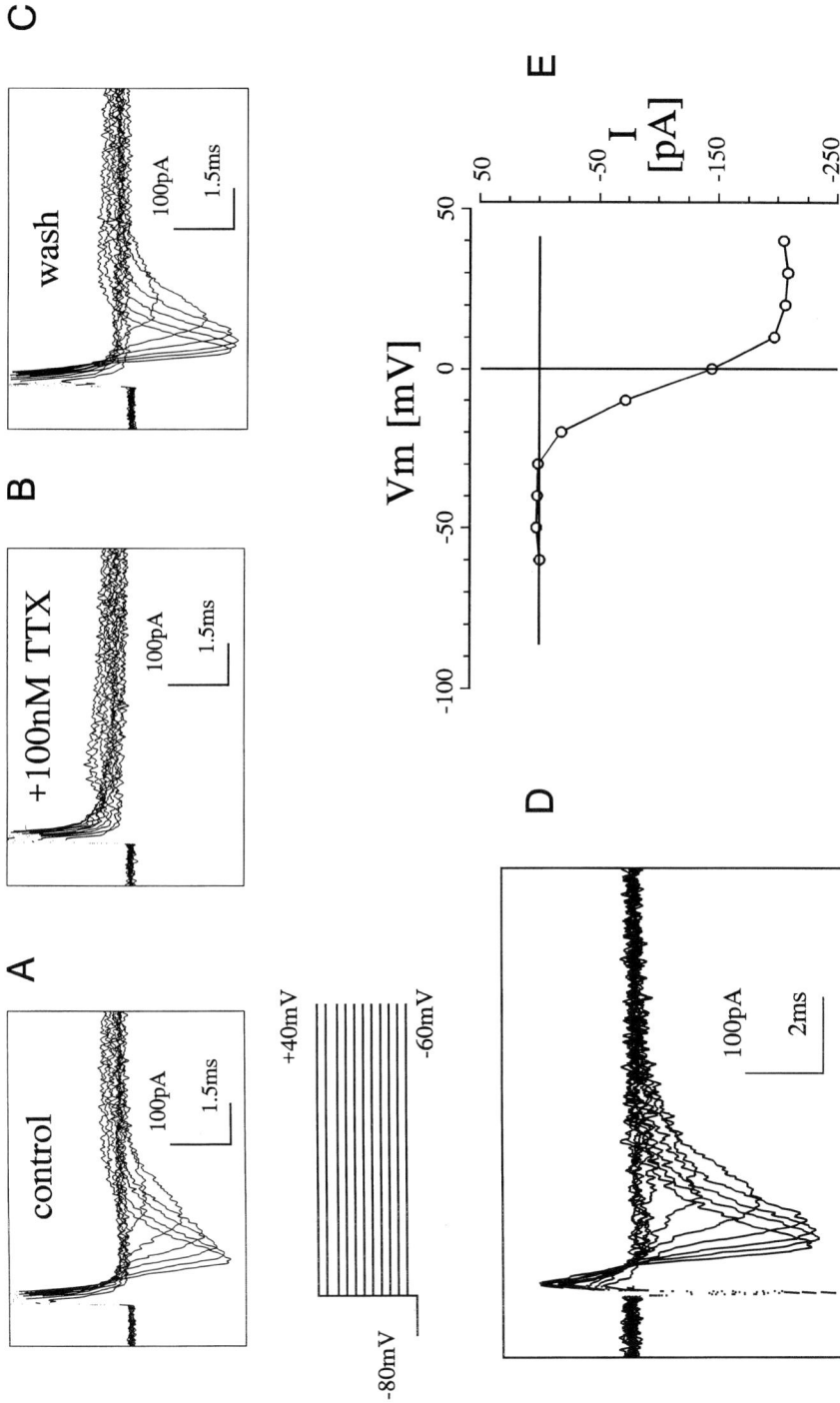

Fig. 18.2A–E. Voltage-dependent tetrodotoxin (TTX) blockable fast transient inward Na⁺ currents in chicken osteoclast measured in the nystatin whole cell configuration. The pipette was filled with 15-Na-Kglu-ICS and the bath contained 5 mM Ba²⁺ and 10 mM TEA to block potassium conductances in both inward and outward directions. Membrane potential was held at −80 mV and was stepped 10 ms in duration to potentials from −60 to +40 mV by 10 mV increments. **A** Currents under control conditions. **B** Addition of 100 nm TTX to the bath blocked I_Na completely. **C** The current recovered after the removal of TTX from the external solution. **D** The TTX blockable component (**B** substracted from **C**) of the current. **E** current voltage relationship of the peak I_Na of **D**.

Chapter 19

GABA$_A$ Receptor-Mediated Chloride Currents in Acutely Dissociated Hippocampal Neurons

Misa Dzoljic and Wytse J. Wadman

Background

γ-Aminobutyric acid (GABA) is the main inhibitory neurotransmitter substance in the central nervous system of practically all multicellular animals (Kaila 1994; Michelson and Wong 1991). The endogenous ligand GABA binds to two fundamentally different classes of receptors, GABA$_A$ and GABA$_B$. Despite their common sensitivity for GABA the two receptors are completely different in their sensitivity to agonists and antagonists. The classical GABA-gated receptors, or GABA$_A$ receptors, are sensitive to the agonist muscimol and blocked by picrotoxin or bicuculline. GABA$_A$ is a ligand-gated ion channel which is composed of several distinct polypeptide units, forming together a hetero-oligometric protein molecule with a specific conductance for chloride. GABA$_B$ receptors, by contrast, were found to respond to baclophen as an agonist and to be blocked by 2OH-saclophen or phaclophen. GABA$_B$ receptors are functionally coupled to certain K$^+$ and Ca^{2+} channels through GTP-binding proteins and/or other second messenger systems. Both receptors are interesting from a clinical point of view. Modulation of GABA$_B$ receptors by baclophen was shown to ameliorate the spasticity so common in paraplegic patients. GABA$_A$ receptor modulation was extensively studied in relation to sleeping disorders, epilepsy and anesthesia. An excellent introduction to the subject is given by Macdonald and Olsen (1994) and Kaila (1994).

The importance of drugs modulating the GABA$_A$ receptor in the clinic can hardly be exaggerated and much research is still being performed in this field. The functional aspects of the GABA$_A$ receptors can be explored using different techniques, including patch-clamp in various recording configurations. Here we describe a method to perform whole cell voltage-clamp measurements on acutely dissociated hippocampal cells (CA1 region) obtained from rats. The dissociated cells offer the advantage of perfect accessibility and immediate availability and absence of a culture environment which could influence receptor expression. Nonetheless, the dissociation process will limit extension of dendrites and possible interactions between the enzymes and the function of receptors should be kept in mind.

Our laboratories use hippocampal neurons for the measurement of GABA-induced currents based on a technique described by Kay and Wong (1986) . This preparation is also suitable for dissociation of neurons from other neuronal regions.

Materials

- Stock solutions   **Solutions**
  Three types of solutions are used in the experiments (Table 19.1): the dissociation solution used during the period of enzymatic digestion, the extracellular solution and the pipette solution. Stock solutions of some of the ingredients of the final solutions can be made or bought. CaCl$_2$, KCl and MgCl$_2$ can all be stored at 4 °C while the rest should be kept at −18 °C.

Stock	mM	Molecular weight (g/mol)	g/100 ml
CaCl$_2$ ·2H$_2$O	1000	147.02	14.702
KCl	1000	74.56	7.456
MgCl$_2$	1000	95.21	9.521

Stock	mM
NaGTP	100
TTX	0.5
Leupeptin	20
CP kinase	2500 U/ml
MgATP ·2H$_2$O	500

- Pipette solutions

	Molecular weight (g/mol)	mM	10 ml
CaCl$_2$	Stock	0.5	5 µl
CsF	151.9	100	152 mg
MgCl$_2$	Stock	2	20 µl
HEPES	238.3	10	24 mg

TEA-Cl ·H$_2$O	183.7	20	37 mg
EGTA	380.4	10	38 mg
P creatine	453.4	20	91 mg
MgATP	Stock	2	40 µl
NaGTP	Stock	0.1	10 µl
CP kinase	Stock	50 U/ml	200 µl
Leupeptin	Stock	0.1	50 µl

The pipette solution should be prepared on ice as a final 10 ml solution. MgATP, NaGTP, CP kinase and leupeptin should be added after all the other drugs. Finally the pH should be set to 7.3 with CsOH. Before dividing the pipette solution into micro-test tubes the pipette solution should be filtered once with a 0.22 µm pore (e.g., Millex-GS, Millipore S.A., 67 Molsheim, France). The micro-test tubes, each containing 0.5 ml, should be kept at −18 °C.

Tip. The patch pipette can be easily filled with a 1 ml syringe with a filter (Nalgene 4 mm syringe filters, Nalge Comp. Rochester, NY, USA) with a Microfil tip (MF34G, World Precision Instruments, Sarasota, Florida, USA) The syringe with the pipette solution should be kept on ice during the experiments.

- Extracellular solution

	Molecular weight (g/mol)	mM	100 ml
CaCl$_2$	Stock	5	500 µl
KCl	Stock	5	500 µl
MgCl$_2$	Stock	1	100 µl
NaCl	58.44	110	645 mg
TEA-Cl ·H$_2$O	183.7	25	459 mg
HEPES	238.3	10	238 mg
4-AP	94.12	5	47 mg
TTX	Stock	1	200 µl
D-glucose	180.16	25	450 mg

Finally the pH should be adjusted to 7.4 with NaOH. Usually 100 ml is sufficient for a few hours of experimentation, depending on the speed of perfusion through the experimental chamber.

● Dissociation solution

	Molecular weight	mM	100 ml
NaCl	58.44	120	701 mg
KCl	Stock	5	500 µl
CaCl$_2$	Stock	1	100 µl
MgCl$_2$	Stock	1	100 µl
PIPES	302.4	20	605 mg
D-glucose	180.16	25	450 mg

The pH should be adjusted to 7.0 using NaOH. Usually 100 ml is sufficient for cell isolation.

Tip. It is convenient to make 500 ml of the extracellular solution and dissociation solution. If D-glucose is omitted from the fluid it can be kept for several weeks at 4 °C without any bacterial or fungal contamination. D-glucose should then be added just before the experiment.

Procedure

Dissection of Brain Slices

Approval of the local animal experiments Ethics Committee is required for these experiments. After decapitation (with or without anesthesia) of the experimental animal (in our case a Wistar rat weighing approx. 200 g) the forebrain is cut free from the cerebellum, removed rapidly out of the skull and placed in well oxygenated dissociation solution at 0 °C (approx. 30 ml without trypsin). A few minutes are allowed before the brain is removed from this solution and placed on paper soaked in the dissociation solution and kept at 0 °C.

Note. The stress caused by the decapitation should be kept minimal. Advice and help should be sought from an experienced biotechnician. Very often animals are anesthetized before decapitation to reduce stress. Although we never could find any effect of inhalational anesthetics given before the decapitation on the cells we worked with, one should realize that anesthesia by itself can be very stressful to the animal.

Note. Tissue preparation should be fast (2 min), the temperature during the whole procedure must be kept low and care must be taken to prevent the tissue from drying out (use the dissociation solution).

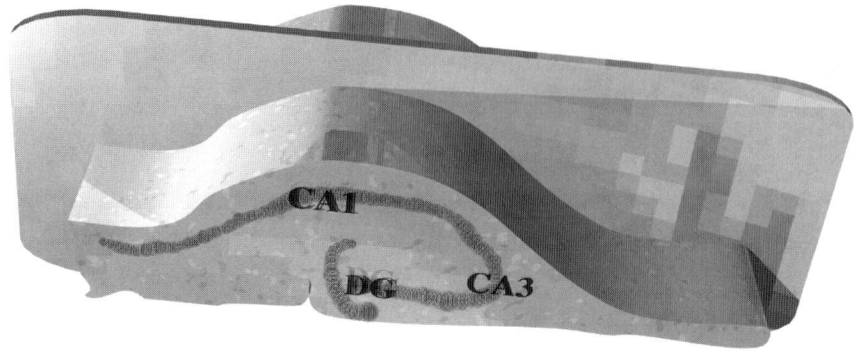

Fig. 19.1. The direction of the slicing device with respect to the hippocampus

The region of interest is blocked off with a scalpel using a soft paintbrush for handling of the brain. In our case the hippocampus was isolated from each hemisphere. It is important to isolate the region of interest without touching it or applying pressure, but rather by gently removing the surrounding tissue. Once the region is isolated from the rest of the brain tissue, it is cut into 500 μm slices. This can be done using a commercially available tissue chopper, but we have used a homemade set of razor blades with the same success. The slicing is done perpendicular to the longitudinal axis of the hippocampus (Fig. 19.1).

A device, able to hold several razor blades (in our case 7), is separated with either sheets of plastic (Fig. 19.2) or small rings that fit around the bars holding the razor blades ensuring a distance of 500 μm The set of blades is held together by a small block that fits over the two bars. Once the device is assembled a small amount of dissociation fluid (without trypsin) should be applied between the razor blades.

Tip. Razor blades are pretreated with oil or grease which might be helpful when shaving but which should be removed with acetone followed by 90 % ethanol before using the blades as a tissue chopper. Preferentially new razor blades should be used for each experiment.

A piece of silicone or rubber is used to support the tissue and to allow a full cut with the blades. The slices remain between the razor blades and can be removed, after opening the razor block, one by one with a paintbrush.

From these brain slices 1 mm^3 tissue blocks (in our case CA1 region) should be dissected using a scalpel and put in the dissociation chamber with dissociation solution containing 1 mg/ml trypsin.

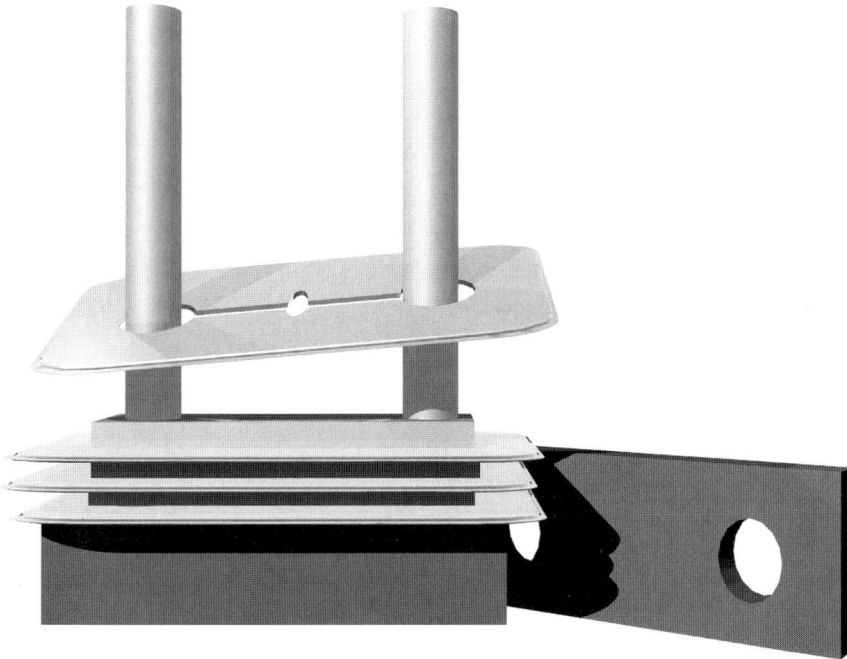

Fig. 19.2. The slicing device

Dissociation of Cells

Brain tissue is delicate and disintegrates easily when pressure is applied or when stirred vigorously. During enzymatic dissociation one should ensure oxygenation of the fluid (and the cells) as well as avoid that the cells are crushed between the stir bar and the dissociation chamber. Kay and Wong (1986) describe a chamber suitable for dissociation. Bellco spinner flasks (suspended bar type 1967 series 25 ml, Bellco Glass Inc. Vineland, New Jersey) have the advantage that tissue cannot be ground between the stir bar and the flask. During the enzymatic dissociation the cells are held at 32 °C for approx. 90 min and stirred slowly. A continuous flow of 100 % O$_2$ is kept in the dissociation chamber (but not bubbling through the dissociation fluid). The time allowed for dissociation is critical and strongly depends on the temperature. We used trypsin (Sigma Type XI, from bovine pancreas) for 90 min, but be prepared to experiment with the dissociation time. Usually less is needed. Older batches of trypsin, however, often require a higher concentration for the same effect.

Tip. Trypsin type XI from Sigma is preferred and should be used to start the technique and whenever problems are encountered. We also had excellent results with considerably cheaper bovine trypsin 1 mg/ml (trypsin 1–300, ICN, Nutritional Biochemicals, Cleveland, Ohio) if the dissociation time was reduced to 60 min.

After the Dissociation

After 90 min the cells are washed (exchange the fluid they are in) twice with the dissociation fluid (without trypsin) and kept in the dissociation fluid at room temperature until needed. Continuous flow of 100 % O_2 as well as gentle stirring is still needed to prevent the cells from settling. Under these conditions the tissue blocks yielded vital cells for the next 8–10 h. We replaced the dissociation fluid every 3 h.

Note. At this moment the tissue blocks should still look intact. If they look ragged or disintegrated they were probably stirred too vigorously, the temperature was higher than 32 °C or 90 min was too long.

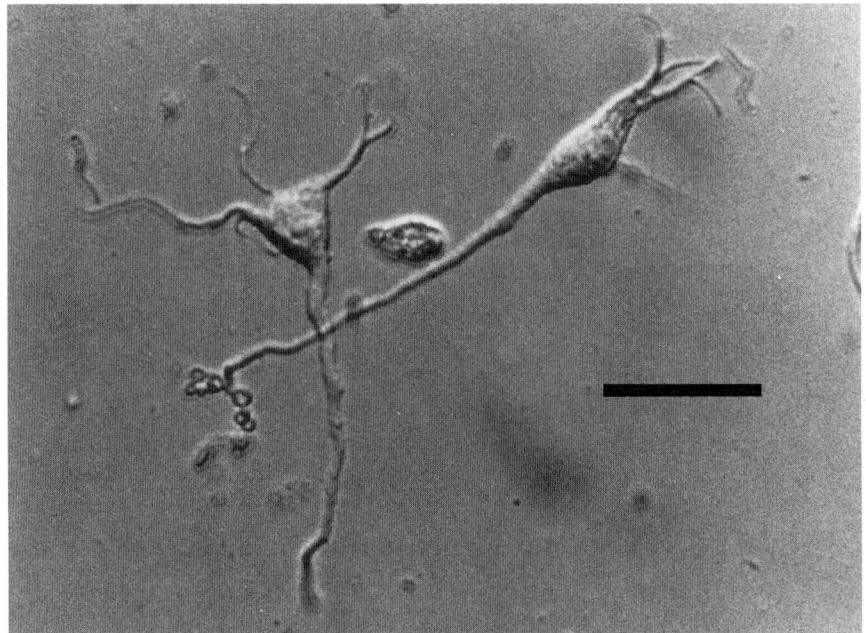

Fig. 19.3. Phase- contrast photograph of two hippocampal neurons after the dissociation. The *black bar* is approximately 30 μm

When cells are needed, a tissue block is removed from the dissociation chamber and triturated through a fire-polished Pasteur pipette in the extracellular solution. For this process we use 1 ml of extracellular fluid in a micro-test tube. When the slices are disintegrated and the extracellular fluid has an opalescent look, the solution is transferred to the recording chamber. Although different Pasteur pipettes with decreasing bore diameters can be used, we found that large bore alone was efficient. Finally, the neurons are allowed 5 min to settle on the bottom of the measurement chamber before the perfusion of extracellular fluid is started (Fig. 19.3).

Note. Neurons should be prepared before each measurement. Neurons will survive up to 1 h in the extracellular fluid; once patch-clamped a measurement should not last longer than 30 min to guarantee good quality of GABA-induced Cl$^-$ currents.

Patch-Clamping

Standard patch-clamp technique can be used to measure GABA$_A$ receptor-mediated Cl$^-$ currents under whole cell voltage-clamp conditions with a computer-controlled amplifier (APC-8, Medical Systems Corp., NY, USA). We perform the data acquisition, analysis and control of the amplifier with pClamp 6.0 in combination with a Digidata 1200 A/D converter (Axon Instruments Inc., Foster City, USA). Patch pipettes of 2–3 MΩ resistance, as measured in the extracellular solution, readily result in gigaseal formation. After membrane rupturing the whole cell configuration is entered and the membrane capacitance (usually between 8 and 10 pF) can be compensated. The input resistance is between 200 and 800 MΩ. Once the cell is in the whole cell configuration it is firmly attached to the pipette and lifted from the bottom. Rather fast flow can be tolerated (in our 1 ml measurement chamber, 4 ml/min of the extracellular fluid was the maximum tolerable flow). Application of the agonist is important and is dealt with in Chap. 4.

Results

The following experiments are examples of what can be expected when GABA$_A$ receptors on hippocampal cells are studied. All experiments were performed at room temperature.

Current-Voltage Relationship (I–V Relationship)

See Fig. 19.4.

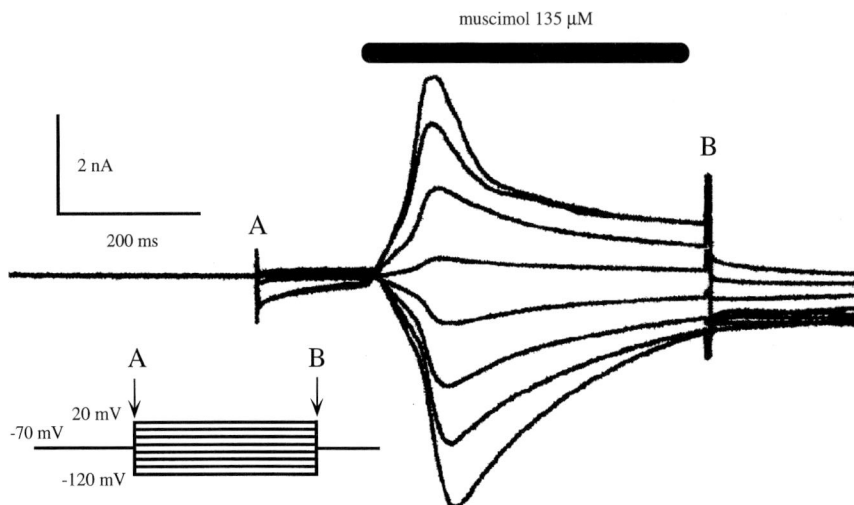

Fig. 19.4. In this experiment muscimol, a specific GABA$_A$ receptor agonist, is used. Muscimol is dissolved in extracellular fluid and applied through an application pipette. Although in this experiment high concentrations of muscimol are used, usually lower concentrations (up to approx. 30 µM) are advisable because of current rundown with higher concentrations. An I–V relationship between the membrane (clamped) voltage (between −120 mV and 20 mV in 20 mV steps), and the muscimol induced Cl$^-$ current is shown. Holding potential (V_h) is −70 mV. The *black bar* shows the period during which the muscimol (135 µM) is applied to the cell. The voltage protocol is depicted in the *left lower corner*. The reversal potential of peak currents in this cell is at approx. −50 mV which is in accordance with the predicted equilibrium potential for Cl$^-$ given by the Nernst equation. Additionally an inward current can be observed in the last depolarization step to 20 mV. This current is probably a Ca^{2+} current through voltage-gated Ca^{2+} channels

Effect of Pentobarbital

See Fig. 19.5.

Fig. 19.5. The cell (V_h = −70 mV) is exposed first to 20 μM muscimol and then to 20 μM muscimol in combination with 200 μM pentobarbital. A clear increase in Cl⁻ current can be observed. The *black bar* indicates the time at which the cell was exposed to the agonist.

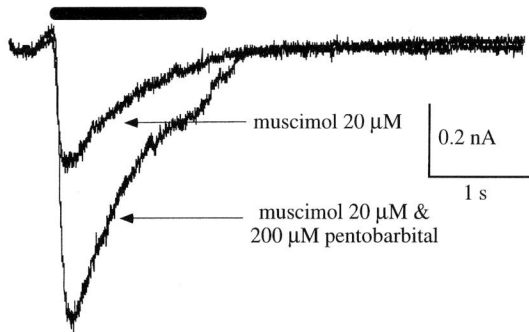

muscimol 20 μM

0.2 nA

1 s

muscimol 20 μM & 200 μM pentobarbital

References

Kaila K (1994) Ionic basis of GABA$_A$ receptor channel function in the nervous system. Progr Neurobiol 42:489–537

Kay AR, Wong RKS (1986) Isolation of neurons suitable for patch-clamping from adult mammalian central nervous systems. J Neurosci Meth 16:227–238

Macdonald RL, Olsen RW (1994) GABA$_A$ receptor channels. Annu Rev Neurosci 17:569–602

Michelson HB, Wong RKS (1991) Excitatory synaptic responses mediated by GABA$_A$ receptors in the hippocampus. Science 253:1420–1423

Measurement of Whole Cell Potassium Currents in Protoplasts from Tobacco Cell Suspensions

Bert van Duijn

Background

Fluxes of ions acros the plasma membrane are associated with the signal transduction of different intra- and extracellular stimuli such as the plant hormones auxin, abscisic acid and gibberellins. In these plant hormone signal transduction processes ion channels are involved. To be able to understand the significance and role of ion channels in plant hormone signal transduction an inventory and characterization of the plasma membrane ion channels in the plant cell of interest should be carried out.

The use of plant cell suspension cultures in plant hormone signal transduction research offers some methodological advantages as compared to cells isolated from tissue. Suspension cultures (cell lines) usually can be rather well controlled in growth, ensuring that cells are in a similar phase of growth and are able to respond to plant growth hormones. A cell suspension is a relative homogeneous population of cells, something which is in most cases difficult to achieve when isolating protoplasts from plant tissues. Furthermore, addition of drugs, growth regulators and genetic manipulation usually is less complex in cell suspension cultures. However, cells in suspension likely do not look like "normal" plant cells and do not show a specialization to perform a specific task in the plant. Although basic plant hormone signal transduction pathways will be present in many suspension cultures, any extrapolation of results obtained with suspension cells to normal plant cells must be done with great care.

We used tobacco cells from suspension cultures to study the signal transduction pathways of the plant growth regulator auxin. The cell suspension was dependent on auxin (or the auxin analog 2,4D) for growth, and showed expression of specific genes upon auxin stimulation (Boot et al. 1996; Van der Zaal et al. 1987). Patch-clamp experiments were performed on protoplasts from these cells to investigate the ion channels present (e.g., Thomine et al. 1994; Van Duijn et al. 1993; Van Duijn 1993). In this chapter the protocols used for protoplast preparation, patch-clamp experiments and analysis are described.

Materials

Cell Suspension Culture

Cell suspension cultures derived from several tissues and species are grown in different laboratories. In our experiments we made use of a tobacco (*Nicotiana tabacum* L., cv Bright Yellow) cell suspension culture derived from stem medulla parenchyma. The exact maintenance, growth, transfer and handling of this culture falls beyond the scope of this chapter. In short, cells are grown in 250 ml Erlenmeyer flasks containing 50 ml Linsmaier and Skoog medium (composition in Linsmaier and Skoog 1965) which is supplemented with aneurine-HCl (0.6 mg/l) and 180 mg/l KH_2PO_4. The growth of the cells in this suspension is plant growth hormone-dependent. Therefore, 880 nM of the auxin analogue 2,4-dichlorophenoxyacetic acid (2,4D) is added to the suspension culture. Suspensions can be subcultured every 7 days by transferring 6 ml of the 7-days-old culture to 50 ml fresh medium. The subculturing must be done under sterile conditions. The Erlenmeyer flasks are placed on a gyratory shaker at 120 rpm in the dark at 25 °C. A more detailed description can be found in An 1985.

Note. The properties of cell suspension cultures can change in time after a large number of subculture cycles. This may be noticed by changes in protoplasts preparation efficiency, changes in color, changes in growth, etc. In this case it is best to start a new culture from frozen starting material to obtain cultures with similar properties as initially used.

Protoplast Preparation

For preparation of protoplasts a method analogous to the one described by Dangle et al. (1987) for preparation of parsley protoplasts is used. We chose this method as protoplasts prepared in this way were viable (showing protoplasmic streaming) and allowed gigaseal formation (in contrast to protoplasts prepared in other ways we tried). In addition, these protoplasts showed normal expression of auxin inducible genes. Other methods, however, may be as good or even better (see also Chap. 9).

Note. After adopting a certain method for protoplast preparation it should be tested whether the obtained protoplasts are viable and still show some basic physiological responses (e.g., osmotic pressure-induced volume changes, cytoplasmic streaming, expression of specific genes).

Protoplast preparation

1. Wash the cells (50 ml suspension) three times in 0.24 M $CaCl_2$ solution by centrifugation (5 min) at about 200 g.

2. Incubate about 10 g (fresh weight) cells for 10–12 h at 25 °C in the dark in a 2-liter Erlenmeyer flask in 150 ml 0.24 M $CaCl_2$ solution supplemented with 0.05 g of cellulase (Onozaku RS, Yakult Honsha, Japan) and 0.005 g of macerozyme (Macerozyme R-10, Yakult Honsha, Japan). Do **not** put the flask on a shaker.

Note. The large Erlenmeyer is used to achieve a good surface area/volume ratio to ensure oxygenation of the cells without shaking. A large Petri dish may work as well in this case.

3. Wash the protoplasts at least three times with 0.24 M $CaCl_2$ solution by centrifugations (3 min) at 100 g.

4. Keep the protoplasts in 50 ml 0.24 M $CaCl_2$ solution before starting the experiment. The protoplasts can be stored at room temperature for about 4–5 h.

Patch-Clamp Experiments

Patch-clamp experiments on plant cell protoplasts can be performed with a standard patch-clamp setup and procedure (see other chapters in this volume). Our setup made use of a L/M EPC-7 patch-clamp amplifier (List Electronics, Darmstadt, Germany), in combination with the patch-clamp software pClamp on an A/D converter-equipped PC (Axon Instruments, Burlingame, USA). Protoplasts were observed under 40x objective magnification on an inverted microscope.

Solutions

Both extracellular and intracellular solutions have to be prepared before starting the experiments.

– Extracellular solution (ECS): 10 mM KCl, 2 mM $MgCl_2$, 1 mM $CaCl_2$, 1 mM KOH, 10 mM Mes (pH 5.5). Adjust the osmolarity of the solution to about 1010 mOsm with mannitol. The ECS can be stored in the refrigerator (5 °C) for about 1 week.

- Intracellular (pipette) solution (ICS): 100 mM K-gluconate, 2 mM MgCl$_2$, 1.1 mM EGTA, 0.45 mM CaCl$_2$, 4 mM MgATP, 6 mM KOH, 10 mM Hepes (pH 7.0). Adjust the osmolarity to about 708 mOsm with mannitol.

Note. Add the MgATP just before starting the experiment, or store 1.5 ml ICS in micro-reaction tubes at −20 °C. Keep the solution on ice during the experiment.

The protoplasts can be transferred to the experimentation chamber on the microscope with an automatic pipette using a plastic tip with the end cut off. Preferentially use an experimentation chamber with a thin (0.18 mm) glass bottom (e.g., the coverslip dish described in Chap. 1) to ensure an optimal view of the protoplasts. Clean the bottom of the chamber carefully with 96 % ethanol. Pipette 50 µl of cell suspension into 2 ml of ECS in the experimentation chamber (depending on the size of your chamber).

Handling of the protoplasts

Note. Pipetting 50 µl of cell suspension (0.24 M CaCl$_2$) in 2 ml ECS will change the calcium and chloride concentration of the ECS.

Allow the protoplasts to settle to the bottom of the dish for about 15 min. Use the firmly attached protoplasts for your experiments.

Note. Select protoplasts for your experiments that have intact cytoplasmic strands and cytoplasmic streaming. Reject protoplasts that have a "homogeneous" cytoplasm with brownian movements of particles.

Patch-clamp pipettes can be pulled from borosilicate glass (e.g., GC150TF-15 from Clark Electromedical Instruments, UK). Pipettes must be fire-polished before use. A seal success ratio of about 30 % can be achieved (ranging from 0 % to 80 % from culture to culture) using pipettes with a resistance ranging from 5 to 10 MΩ as measured in the ECS.

Patch pipettes and seal formation

Fill the pipette from the backside with a small syringe equipped with a nonmetallic needle (e.g., Microfil from WPI, USA).

Note. When using a steel needle, note that a reaction of the EGTA in the ICS with components (heavy metals) in the steel causes the release of protons thus acidifying the ICS. In this case a small amount of ICS (that part that was inside the needle for a longer time) should first be flushed through before pipette filling.

Position the pipette, while keeping a slight over-pressure in the pipette, onto the surface of the protoplast. Monitor the resistance constantly by applying small voltage pulses (1 mV) in the voltage-clamp mode. Before reaching the protoplast turn off the over-pressure and set the offset compensation of the amplifier in the right positions (current = 0).

Touching the protoplast should induce an increase in the pipette resistance.

Note. The position of the pipette on the protoplast surface seems to be important for the seal success ratio. Be careful not to position the pipette exactly on top or over the top of the protoplast.

By applying gentle suction to the lumen of the pipette, the pipette resistance should increase. Finally, a gigaseal ($2-10\,G\Omega$) will be formed. Now set the transient and capacitance compensation knobs of the amplifier in the right position.

Note. It is clear that the formation of a gigaseal is a crucial step in the patch-clamp technique. To establish the right conditions for gigaseal formation requires trial and error and "feeling" for your setup and protoplast. In most cases considerable training time is essential.

Note. The sealing process may be improved by applying a negative potential to the pipette (up to $-100\,mV$).

Once a gigaseal is formed the whole cell configuration can be obtained by applying short strong suction pulses to the pipette interior. Monitor the formation of the whole cell configuration by looking at the conductance and capacitance of the measurement configuration (both should increase upon obtaining the whole cell configuration). Adjust the capacitance compensation once the whole cell configuration is obtained.

Note. The capacitance of the whole cell membrane can later be used to normalize measured current values for cell size. In biological membranes the specific conductance is usually about $1\,\mu F.cm^{-2}$.

Procedure

Experiments designed to characterize (partly) two different potassium currents in the tobacco protoplasts will be described. The experiments are performed under voltage-clamp conditions in the whole cell configuration.

Holding potential   After obtaining the whole cell configuration a holding potential (V_h) must be applied to the pipette interior. Usually a value of about the "normal" resting membrane potential of the cell under study is chosen. For tobacco protoplasts from suspension culture a membrane potential of about $-40\,mV$ was found (Van Duijn and Heimovaara-Dijkstra 1994). In the experiments a V_h of $-50\,mV$ is used.

Outward Rectifying Potassium Conductance

In stomatal guard cells and pulvinar cells an outward rectifying potassium conductance functions as a potassium release channel during cell shrinking. In other cell types where this type of ion channel is present, its function is less clear. In protoplasts from tobacco suspension cells an outward rectifying potassium conductance is present as well (Van Duijn et al. 1993). A first characterization of the conductance can be easily carried out by using different voltage pulse protocols, different extracellular solutions and some ion channel blockers.

The outward rectifying potassium conductance in tobacco protoplasts is, using the solutions described above, depolarization activated with an activation potential of about $0\,mV$. This means that the ion channels underlying this conductance will open when the electrical potential across the membrane is made more positive and reaching values above $0\,mV$. The activation of the conductance can be studied using voltage pulse protocols. From the V_h the potential across the membrane is changed stepwise to a different value. Figure 20.1A illustrates the response of the conductance on depolarizing voltage steps of increasing amplitudes from a V_h of $-50\,mV$ (see the voltage protocol in the figure).

Current activation

Note. For presentation reasons only part of the applied voltages is shown in Fig. 20.1A. In your experiment voltage pulses over a large range (e.g., from -160 to $+160\,mV$) should be applied for initial characterization of a protoplast.

From the current response it can be clearly seen that the current increases with increasing depolarization and that the time constant of activation decreases (i.e., the current activates faster) with increasing depolarization. It is also clear that the current does not show inactivation (i.e., after activation the current does not decline as long as the potential is not changed).

Plotting the steady state current against the applied voltage results in the steady state I–V (current-voltage) relationship (Fig. 20.1B). The I–V

Steady state I–V relationship

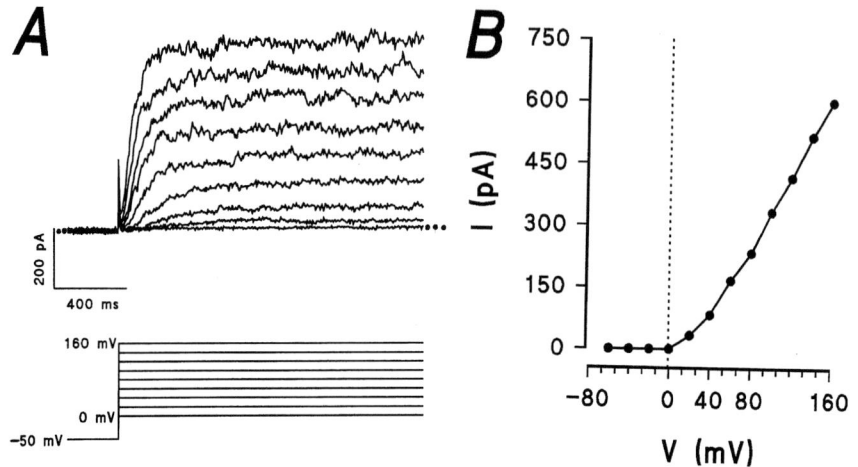

Fig. 20.1A, B. A Recording of activation of the outward directed K^+ current across the plasma membrane of a single tobacco cell suspension culture protoplast in the whole cell configuration of the patch-clamp technique. The extracellular solution contained 11 mM K^+ and the pipette solution contained 106 mM K^+. The *upper panel* shows superimposed records of the currents activated upon nine different voltage steps to depolarizing potentials from a holding potential of – 50 mV, as illustrated in the voltage protocol (*lower panel*). The current traces are corrected for the aspecific leak conductance (*dotted line* indicates zero current level). B Steady state I–V relationship from the currents shown in **A**, measured at $t = 1.2$ s. (Figure modified from Van Duijn 1993, with permission)

relationship shows more clearly that the current is only activated at depolarized potentials.

Leak correction   The ohmic membrane resistance (i.e., the membrane resistance when all ion channels are closed) can usually be measured only over a small voltage range or (in the presence of many different channels with different activation potentials) with the help of pharmacological blockage of the channels. For the protoplasts from tobacco cell suspension it seems that between −10 and −80 mV no active currents are activated and thus the ohmic membrane resistance can be determined from the I–V relationship in that region. The value varies around 5 GΩ. Current traces and I–V relationships can be corrected (leak correction) for the ohmic membrane resistance to show only the active currents, as is done for Fig. 20.1. The software program pClamp has a standard procedure for leak correction.

Note. Be careful with leak correction. Often the leak is not linear with the applied voltage and changes in time during the experiment as well.

Upon returning to a potential value more negative than the activation potential the open channels will close again (deactivate). The deactivation of the outward rectifying potassium conductance can be studied using a so-called backstep protocol. In this protocol the conductance is first activated (in this case by applying a depolarizing voltage step to, e.g., +110 mV) followed by a stepwise change in the applied voltage to less positive values.

Figure 20.2A illustrates the deactivation, seen as tail currents, of the outward rectifying potassium conductance and the voltage protocol used. The exponential decay of the current upon application of a deactivating voltage pulse can be clearly seen. The time constant of deactivation becomes smaller with more negative potentials.

Current deactivation

Tail currents

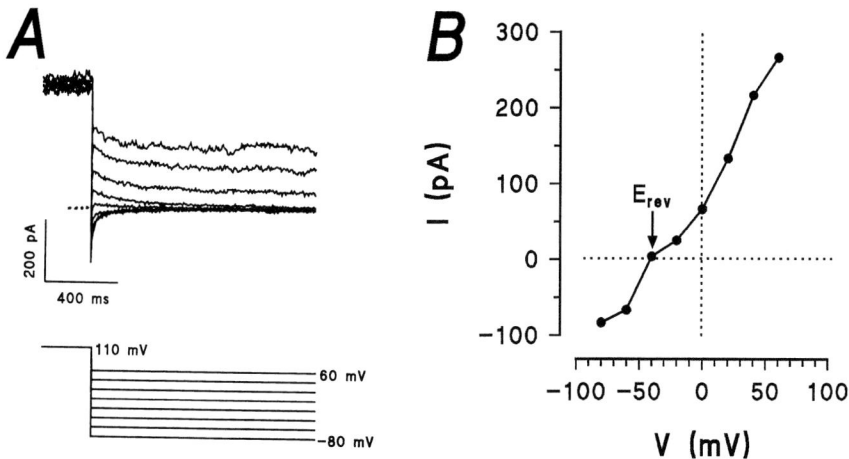

Fig. 20.2A, B. **A** Recording of deactivation of the outward directed K$^+$ current across the plasma membrane of a single tobacco cell suspension culture protoplast in the whole cell configuration of the patch-clamp technique. The extracellular solution contained 11 mM K$^+$ and the pipette solution contained 106 mM K$^+$. The *upper panel* shows superimposed records of currents, deactivating upon eight different voltage steps to less positive potentials from a current activating potential of +110 mV, as illustrated in the voltage protocol (*lower panel*). The voltage steps were made after the steady state current was reached after a voltage step to +100 mV from a holding potential of −50 mV. The current traces are corrected for the aspecific leak conductance (*dotted line* indicates zero current level). **B** I–V relationship of the instantaneous tail currents, measured just after stepping back to less positive potentials (as shown in **A**). (Figure modified from Van Duijn 1993, with permission)

Ion selectivity

The ion selectivity of the channels determines the current (I) through the conductance as the current is given by:

$$I = \gamma.(V - E_{rev}) \tag{1}$$

in which γ is the conductance (1/resistance), V the applied voltage and E_{rev} the reversal potential of the conductance. The E_{rev} is determined by the permeability of the channels for different ions and the concentrations of these ions inside and outside the cell (see Goldman-Hodgkin-Katz equation in Chap. 6). In case of perfect selectivity of the conductance for one type of ions the E_{rev} is equal to the Nernst equilibrium potential for this ion:

$$E_{rev} = \frac{RT}{nF} . \ln \frac{[I]_o}{[I]_i}$$

in which R is the gas constant, T the absolute temperature, n the valence of the ion, F the Faraday constant, $[I]_o$ and $[I]_i$ the concentrations of the ion in the extracellular and intracellular space, respectively.

To know more about the ion selectivity of the conductance, E_{rev} should be measured. The equation predicts that when the applied voltage is equal to the E_{rev} the current reverses its polarity. To measure this value the current must be measured over a voltage range including the E_{rev} value, preferentially at the same conductance value. As in many cases the channels are closed at potentials around the E_{rev}, the reversal of the current cannot be measured in the steady state. The backstep experiments offer the solution for this problem, as they measure current in a non-steady state situation.

In Fig. 20.2A it can be seen that the polarity of the current tails reverse at a certain backstep voltage.

Instantaneous I–V relationship

The instantaneous I–V relationship (i.e., the current amplitude just after stepping back to a less positive potential versus the backstep potential) shows this more clearly (Fig. 20.2B). This I–V relationship shows the current through the channels as a function of the applied voltage **independent** of the activation degree of the channels (so even at potential values where normally in steady state conditions the channels would be closed). This is because the channels are still activated to the same level, for a short moment, as they were just before stepping back to a less positive potential, as the deactivation is not instantaneous. The reversal potential (E_{rev}) as measured in this way can be used to identify the ion selectivity of the conductance. The value of E_{rev} is identical to the potential at which I = 0. From Fig. 20.2B it can be seen that this is about

−40 mV with ICS in the pipette and ECS in the bath. This negative potential indicates that the majority of the ions permeating the channel have a negative equilibrium potential (as has potassium). To test the involvement of different types of ions, different ECS and ICS solutions may be applied. For instance, an increase of the extracellular potassium concentration (resulting in a more positive potassium equilibrium potential) should result for a potassium selective conductance in a shift of the E_{rev} towards a more positive value.

For a complete analysis of the selectivity of the conductance different solutions with varying types of ions must be tested. From such tests it was established that the outward rectifying potassium conductance mainly carries potassium ions (Van Duijn et al. 1993).

Note. For the determination of E_{rev}, a good offset potential adjustment is essential as well as a careful leak correction. Be careful with changing solutions with different ion compositions as the offset potential is very sensitive to the ionic composition of the solutions. Corrections for these changes in offset potentials should be made.

From the steady state I–V relationship the conductance of the membrane can be determined according to the relationship:

Conductance

$$\gamma = \frac{I}{V - E_{rev}}$$

The E_{rev} value is determined from the backstep experiments. Figure 20.3 shows the relative conductance (relative to the maximal conductance of

Fig. 20.3. Steady state activation curve for the outward rectifying potassium conductance as shown in Fig. 20.1. The activation degree is expressed as the conductance γ normalized to the conductance at +160 mV (γ_{max}). γ was calculated with use of the E_{rev} indicated in Fig. 20.2

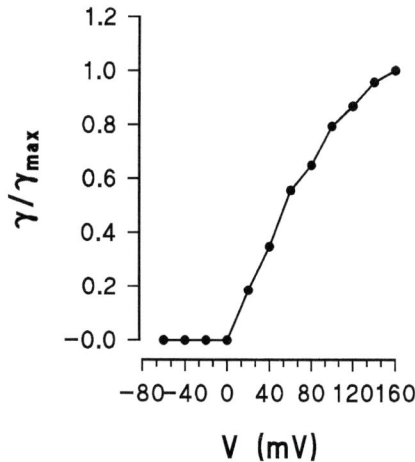

the membrane) as a function of the applied potential. From this curve the half-maximal activation potential and a slope factor can be determined. For a more detailed analysis see Van Duijn 1993.

Pharmacology A further characterization of the outward rectifying potassium conductance may be done by using different ion channel blocking agents. In our studies we used:

- Tetraethylamonium (19 mM): This concentration of TEA applied at the extracellular site only gives a slight inhibition of the current of about 20 % (van Duijn et al. 1993). In general TEA is not a very potent potassium channel blocker in plants. Note that when TEACl is used that the chloride concentration in the medium (and possibly offset potentials) changes.
- Quinidine (0.17 mM): This concentration results in a block of about 30 % (Van Duijn et al. 1993). Quinidine can be purchased in different forms. Note that the free acid form dissolves poorly in water dependent on the pH. Some pH manipulation and heating may be necessary.
- 4-Aminopyridine (5 mM): This concentration results in a block of about 30 % (Van Duijn et al. 1993). Note that this concentration of 4-AP induces a change in the pH of the solution. Adjust the pH to the original value after 4-AP addition.
- "Calcium" channel blockers: The IC_{50} of bepridil, verapamil and nifedipine are 1 μM, 5 μM and 5 μM, respectively (Thomine et al. 1994). These blockers should be prepared as stock solutions in DMSO (10 mM or 1 mM).

Note. From these data it is clear that the classical calcium channel blockers in animal cells are the most potent in blocking this plant potassium channel, while the classical potassium channel blockers are much less potent. So, be careful with the interpretation of the effects of calcium channel blockers in plants using different techniques.

Inward Rectifying Potassium Conductance

Besides the outward rectifying conductance (which is present in all the protoplasts), a hyperpolarization-activated inward rectifying potassium conductance can also be found in about 30 % of the protoplasts (Van Duijn et al. 1993). The measurements and analysis of this conductance can be performed in analogy with the description given for the outward rectifying potassium conductance. The same questions must be answered

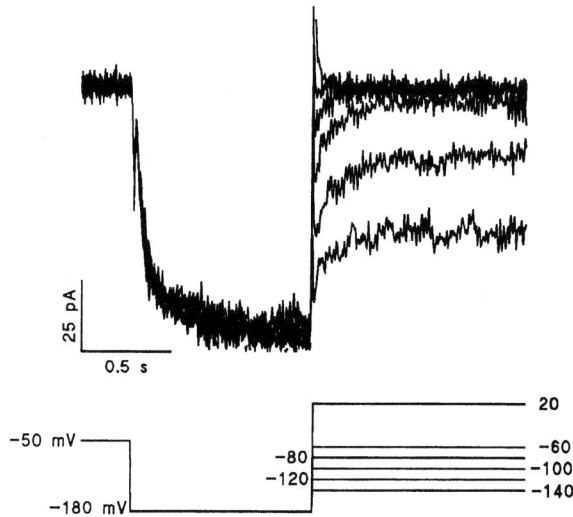

Fig. 20.4. Measurement of the inward rectifying potassium conductance. Current responses in the whole cell configuration upon the application of the voltage protocol shown in *lower part* of the figure. After application of activating hyperpolarizing pulses to −180 mV to the pipette interior, repolarizing pulses ranging from −160 to +40 mV (with a 20 mV interval) were applied. Because the current traces for repolarizing more positive than −60 mV show overlap, not all traces for these potential values are shown for reasons of clarity. The current responses on the repolarizing test pulse show a reversal of the instantaneous tail current at an applied potential of about −60 mV. Furthermore, voltage- and time-dependent deactivation of the conductance can be seen in the current responses upon repolarizing test pulses. The displayed currents are corrected for the aspecific leak conductance. The pipette contained ICS and the bath ECS. (Reprinted from Van Duijn et al. 1993, with permission)

for the characterization. Figure 20. 4 shows the activation and deactivation of the inward rectifying conductance as well as the reversal of the instantaneous tail current at about −60 mV.

Analysis of Kinetics

A complete characterization of the currents should involve fitting of models against the obtained data. A Hodgkin-Huxley analysis may be suitable for a more detailed description of the conductance. These procedures are beyond the scope of this chapter but can be found, e.g., in Van Duijn 1993. Note that for a good comparison of conductances between different cell types and species a detailed (mathematical) analysis of all aspects is inevitable.

References

An G (1985) High efficiency transformation of cultured tobacco cells. Plant Physiol 79:568–570

Boot CJM, Van Duijn B, Mennes AM, Libbenga KR (1996) Regulation of a class of auxin-induced genes in cell suspension cultures from *Nicotiana tabacum*. In: Smith AR, Berry AW, Harpham NVJ, Moshkov IE, Novikova GV, Kulaeva ON, Hall MA (eds) Plant hormone signal perception and transduction. Kluwer Academic, Dordrecht, pp 41–48

Dangle JL, Hauffe KD, Lipphardt S, Hahlbrock K, Scheel D (1987) Parsley protoplasts retain differential responsiveness to u.v. light and fungal elicitor. EMBO J 6:2551–2556

Linsmaier EM, Skoog F (1965) Organic growth factor requirements of tobacco tissue cultures. Physiol Plant 18: 100–127

Thomine S, Zimmermann S, Van Duijn B, Barbier-Brygoo H, Guern J (1994) Calcium channel antagonists induce direct inhibition of the outward rectifying potassium channel in tobacco protoplasts. FEBS Lett 340:45–50

Van der Zaal EJ, Mennes AM, Libbenga KR (1987) Auxin-induced rapid changes in translatable mRNA's in tobacco cell suspensions. Planta 172:514–519

Van Duijn B, Ypey DL, Libbenga KR (1993) Whole cell K^+ currents across the plasma membrane of tobacco protoplasts from cell suspension cultures. Plant Physiol 101:81–88

Van Duijn B (1993) Hodgkin-Huxley analysis of whole cell outward rectifying K^+ currents in protoplasts from tobacco cell suspension cultures. J Membrane Biol 132:77–85

Van Duijn B, Heimovaara-Dijkstra S (1994) Intracellular microelectrode membrane potential measurements in tobacco cell suspension protoplasts and barley aleurone protoplasts: interpretation and artifacts. Biochim Biophys Acta 1193:77–84

Chapter 21

Inward Rectifying Potassium Conductance in Barley Aleurone Protoplasts

MARCEL T. FLIKWEERT AND MEI WANG

Background

The mature barley grain consists mainly of embryo and starchy endosperm which is surrounded by three layers of cells, the aleurone. The aleurone cells are activated upon water imbibition. This activation is mainly due to plant hormones produced by the embryo during germination, resulting in production and secretion of hydrolytic enzymes. These enzymes will degrade the starchy endosperm to provide nutrients for further growth of the embryo.

Barley aleurone protoplasts are a model system to study diverse biological questions in plants, such as protein secretion, pH or ion traffic regulation and transient expression of foreign genes. Moreover, in dormant barley grains, not only the embryo but also the aleurone shows a differential responsiveness to plant hormones with regard to synthesis and production of hydrolytic enzymes. Barley aleurone protoplasts are thus a good system to study signal transduction of hormones such as abscisic acid (ABA) and gibberellins (GAs). Upon addition of ABA to barley aleurone protoplasts, there is activation of plasma membrane H^+-ATPases, Ca^{2+}-ATPases and K^+ fluxes (Van der Veen et al.1992: Wang et al. 1991; Heimovaara-Dijkstra et al. 1994). Intracellular Ca^{2+}, pH_i etc., have been considered as secondary messengers in plant hormone signal transduction as well. Therefore, to find out more about the mechanism and function of hormone signal transduction, it is very useful to use electrophysiological techniques.

Materials

Isolation of Aleurone Protoplasts from Mature Barley Grains

Protoplasts   Barley grains (*Hordeum vulgare L. cv.* Himalaya, harvest 1985) can be obtained from the Department of Agronomy, Washington State University, Pullman, WA, USA.

Barley dormant and nondormant grains (*Hordeum distichum L. cv.* Triumph, EBC trails 1988 at Carlsberg Plant breeding, Copenhagen, Denmark) are grown in a phytotron under the growing conditions described by Schuurink et al. (1992a). The mature grains must be stored at −20 °C to preserve dormancy (Schuurink et al. 1992b).

Note. Barley grains can be obtained from different research institutions or industrial and agricultural companies. There are two types of grains: naked grains, such as c.v. Himalaya, and grains with husks such as c.v. Triumph. The protocol described here is especially suitable for naked grains but can also be applied to husked grains. However, it is essential to remove the husk before cutting the grains in half.

Cellulase R-10 was from Yakult Honsha (Tokyo, Japan). Gamborg B5 was from Flow Laboratories (Irvine, UK). MES, nystatin and ampicillin were from Sigma Chemical (St. Louis, USA). PVP K25 was from Fluka Chemie (Tilburg, The Netherlands). All other chemicals were from Merck (Darmstadt, Germany).

Note. Cellulase R-10 can differ in quality and activity from batch to batch. Therefore, it is suggested to buy one small package and test its activity immediately; if it is good, it is advisable to buy and store more from the same batch (same batch number).

Solutions   Aleurone Protoplasts Isolation Medium (APIM) (800 mOSm)

- 0.385 % Gamborg B5
- 0.10 M D-glucose
- 10.0 mM L-arginine
- 20.0 mM $CaCl_2$
- 10.0 mM MES
- 1 % PVP K25
- 4.5 % Cellulase R-10
- 0.35 M Mannitol
- 50.0 µg/ml ampicillin (end concentration, dissolve in water)
- 5.0 µg/ml nystatin (end concentration, dissolve in DMSO)
- Adjust to pH 5.4 with KOH

Aleurone Protoplast Medium (AMP) (1080 mOSm)

- 0.385 % Gamborg B5
- 0.10 M D-glucose
- 10 mM L-arginine
- 20 mM $CaCl_2$
- 10 mM MES
- 0.65 M Mannitol
- Adjust to pH 5.4 with KOH

Extracellular Bath Solution (ECS)

- 1 mM $CaCl_2$
- 2 mM $MgCl_2$
- 10 mM MES
- 10 mM or 50 mM KCl
- Adjust pH to 5.6 with KOH
- Adjust solution to 1000 mOsm with mannitol

Pipette or Intracellular Solution (ICS)

- 1 mM EGTA
- 2 mM $MgCl_2$
- 10 mM HEPES
- 100 mM K-gluconate
- Adjust pH to 7.2, respectively 6.8, with KOH
- Adjust solution to 1000 mOsm with mannitol

Filter all solutions except for APIM and APM through a 0.2 μm Nucleopore filter (Pleasanton, CA, USA).

- Shaker   **Equipment**

- Sterile flow chamber
- Sterile Petri dishes (9 cm)
- Glass capillaries from hematocrit glass (type II glass ASTM E438, Dade Diagnostics Inc.)
- Narashige PB-7 or other type pipette puller
- Standard patch-clamp setup

Procedure

Sterilization and Incubation of Grains

1. After removal of the embryo by cutting off the front of the grain, cut the embryoless grain in half longitudinally.

2. Put the halved embryoless grains into water (25 grains in a 15 ml test tube).

3. Replace the water with 70 % ethanol for 30 s.

4. Wash intensively with distilled water three times.

5. Incubate the washed half-grains in a sodium hypochlorite solution (0.2 %) for 15 min. During incubation, the test tube should be placed horizontally on a shaker (100 rpm).

6. Wash the half-grains four times with 10 ml sterile distilled water and remove the residual hypochlorite by incubating with 4 ml 25 mM HCl solution (prepared with sterile water) for 10 min on the shaker.

7. Wash the half-grains intensively six times with 10 ml sterile distilled water.

8. Transfer, in a laminar flow chamber, the sterilized half-grains to a sterile Petri dish containing 2 ml sterile water.

9. Incubate the half-grains at 25 °C in the dark for 65 h.

Aleurone Layer Preparation and Enzyme Treatment

1. After incubating the grains as described above, the aleurone layers can be prepared by carefully removing the endosperm with a small spatula under sterile conditions (in a flow chamber).

2. Turn the aleurone layers from concave to convex and transfer the isolated aleurone layers to a new Petri dish containing 10 ml sterile water.

3. Incubate for about 2–3 h at room temperature.

4. Remove the water, and add 10 ml APIM (see above) to the isolated aleurone layers and incubate 3–4 h at room temperature.

5. Wash the aleurone layers carefully a few times with AMP in order to remove adhering starch.

6. Transfer the starch-free aleurone layers to a new Petri dish containing fresh 10 ml APIM and incubate in the dark at 25 °C for 16 h.

Protoplast Preparation

1. Add 5 ml APM to the incubation medium (aleurone layers + APIM after 16 h incubation).

2. To release the protoplasts from the aleurone layers, a transfer pipette should be used to swirl the incubation medium gently.

3. Filter the incubation mixture using a 100 μm filter. The protoplasts will pass through the filter and the cell wall residuals will stay on the filter screen.

4. Wash the protoplasts three times in 10 ml AMP by centrifugation at 550 rpm (50'g) for 5 min each time.

5. Add 5 μl of protoplast suspension (usually containing about 30 protoplasts, which proves useful when making high resistance seals) to 1 ml ECS buffer in a glass bottom Teflon culture dish. The protoplasts must be allowed to settle for 10 min.

Note. The bottom of the measurement chamber must be cleaned carefully with ethanol before use to ensure a clean surface and sticking of the protoplasts. It is important that the protoplasts stick to the glass and do not move freely. Free moving protoplasts should not be used because they have a very low seal success rate in patch-clamp experiments.

Electrophysiology

The conventional patch-clamp technique has been excellently described by Hamill et al. (1981). An often used amplifier is the L/M EPC-7 patch-clamp amplifier (List Electronics, Darmstadt, FRG). The signal is low-pass filtered at 3 kHz before it is sent to the analog-digital converter. The signal is acquired by the software program pClamp (Axon Instruments, Burlingame, CA), which also generates the voltage pulse protocols. Data can be analyzed and further processed with the software programs pClamp and FigP (Biosoft, Cambridge, UK).

The experiments are performed at room temperature (21 °C).

Preparation of the Glass Patch Pipette

To obtain a gigaseal it is very important to use the proper glass type and to make the proper patch pipette. Because of the diversity of plant membranes, the type of glass to be used should be decided empirically. We have had good experience with pipettes made from hematocrit glass (type II glass ASTM E438, Dade Diagnostics Inc.). The fabrication of a patch pipette consists of pulling the pipette and heat-polishing the pipette tip (Hamill et al. 1981).

Pipette fabrication procedure

1. Cut the glass capillaries from hematocrit glass (type II glass ASTM E438, Dade Diagnostics Inc.) to a length of about 8 cm. Each capillary is later on pulled into two patch pipettes of 4 cm. The glass capillaries should be stored in or cleaned with ethanol (98 %) before use.

2. After removing the glass capillary from the ethanol with tweezers pass it through a flame until the ethanol is vaporized. Avoid touching the middle of the capillary to keep the surface of the glass clean.

3. The patch pipettes are pulled on a pipette puller (e.g., Narashige PB-7 pipette puller). This is an important phase in the manufacturing of a patch pipette because the right pipette tip form and opening will arise during this proces. Make a batch of patch pipettes with different tip forms and openings to find the optimal shape. Use a microscope or have a microscope nearby with an ocular which has a micrometer scale. The tip should have an opening of about 3–5 μm.

4. After coating the patch pipette, heat-polish the tip by bringing the pipette under the microscope near a heat source. Heat-polishing the tip ensures that the tip will be clean and smooth and reduces the diameter of the opening. Place the finished patch pipettes tip upwards in a dust-free storage can.

5. After filling the pipette with ICS and removing the air in the tip by gently hitting the pipette with your fingernail, place the pipette in the suction pipette holder of the patch-clamp equipment. The pipette resistance should be between 5 and 15 MΩ, as measured in the extracellular solution. Adjust the offset when the pipette is in the bath.

Gigaseal Formation

An important property of the patch-clamp technique is the formation of a giga-seal. After the gigaseal is formed the whole cell configuration can be obtained by short powerful suctions to the pipette interior.

With a little overpressure, by blowing on the silicon tube connected to the outlet of the suction pipette holder, the glass pipette should be maneuvered through the ECS buffer to the vicinity of the protoplasts. Applying overpressure to the patch pipette avoids debris, etc. which floats around in the bath solution and can stick to the pipette tip. Observation of protoplasts and maneuvering of the pipette is done using 40x objective magnification. The pipette must then be pressed gently against the plasma membrane where immediatly an electrical seal of approximately $50\,M\Omega$ will be formed. After attaining this seal apply a negative potential to the pipette of about $-50\,mV$. It is now possible to get a seal resistance which is in the order of some giga-Ohms, the so-called gigaseal. This high resistance reduces the background noise. The formation of the gigaseal is monitored by measuring the current induced by a $+1\,mV$ pulse while creating an underpressure in the glass pipette through suction. In our experiments with Himalaya protoplasts the formation of a $5–20\,G\Omega$ seal took between 3 and 10 min to form. In Triumph nondormant and dormant protoplasts a $5–20\,G\Omega$ seal was formed between 1 and 5 min.

Note. The high resistance of the gigaseal is of twofold importance: (1) a high resistance reduces the background noise and (2) it allows the current in the membrane patch to pass through the pipette. To more easily form a gigaseal, a clean (enzymatically) membrane and a clean pipette tip surface are required. Using Ca^{2+} in the ICS buffer ($0.1–1\,mM$) or an increased concentration of Ca^{2+} in the ECS buffer ($1–10\,mM$) can help to increase the formation of seals. A bath perfusion system can be used to change the extracellular solution during the experiments. Use a patch pipette only once, because a used pipette will not have a clean tip anymore, resulting in a decreased likelihood of forming gigaseals.

Whole Cell Configuration

After formation of a gigaseal the whole cell configuration can be established by applying additional suction to the patch electrode interior. Beware that this is a short powerful suction which only removes the patched membrane inside the electrode. Directly after the whole cell configuration is formed the membrane potential should be clamped with a holding potential (-50 mV). The formation of the whole cell can be observed by an increase of the membrane capacitance on the oscilloscope as seen by a transient membrane current. The transient can be eliminated by whole cell capacitance control on the patch-clamp amplifier. The settings (values) of the capacitance and access resistance on this control are unique for each cell being clamped and should be noted. This value of the membrane capacitance can be used to estimate the size (surface area; membrane capacitance per unit area is about $1 \, \mu F/cm^2$) of the patched cell, used later for cell size correction of the currents.

Applications

In this section an example is given of patch-clamp measurements and data analysis of barley aleurone cells (see Van Duijn et al. 1996).

Whole Cell Current-Voltage Characteristics

After the formation of the whole cell configuration, the membrane potential (V_m) is clamped at a holding potential (V_H) of -50 mV. Upon applying a varying hyperpolarizing potential, by means of a voltage step

---

Fig. 21.1A–D. Whole cell inward rectifying K^+ current response, upon a voltage step ▶ protocol, across the plasma membrane of barley aleurone nondormant Triumph cultivar protoplasts. **A** Time and voltage dependent whole cell inward current response upon a stepwise changing of the membrane potential. The current was induced by a voltage step protocol (*inset*) with a holding potential of -50 mV. A series of tests voltages was applied from -180 mV to 160 mV. **B** Steady state current-voltage (I–V) relationship of the responses shown in **A**. The bath solution contained ECS with 10 mM potassium chloride and the pipette solution contained ICS with 100 mM potassium gluconate. **C** Determination of the reversal potential (E_{rev}) of the inward rectifying K^+ current upon application of a backstep protocol (*inset*) in the whole cell configuration in a barley aleurone nondormant Triumph cultivar protoplast. Backstep protocol which started at a holding potential of -50 mV stepped to hyperpolarizing voltage pulses of -180 mV to the pipette interior then repolarizing from -180 mV to 20 mV with steps of 20 mV. **D** Steady state current-voltage (I–V) relationship of the instantaneous tail currents shown in **A**. The E_{rev} is the potential at which the instantaneous current response reverses. The bath solution contained ECS with 50 mM KCl and the pipette solution contained ICS with 100 mM K-gluconate. The current response at which the instantaneous current reverses is close to the Nernst potential of potassium in this measurement condition.(Van Duijn et al. 1996; reprinted by permission of Kluwer Academic Publishers)

protocol across the plasma membrane, the activation kinetics shows a time- and voltage-dependent inward-directed ion current through the membrane as a response (Fig. 21.1A).

The current traces in the example experiments were corrected for the leak resistance, which was about $1.5\,G\Omega$. With these measurements the relationship between the applied potential and the current response in the stationary phase, the current-voltage curve (I–V curve, Fig. 21.1B), can be determined. Considering the size of the protoplasts, there was a similar current magnitude in protoplasts derived from aleurone of Himalaya and Triumph nondormant grains. Protoplasts derived from aleurone of Triumph dormant grain show in general a smaller inward rectifying current.

Current-voltage relationship

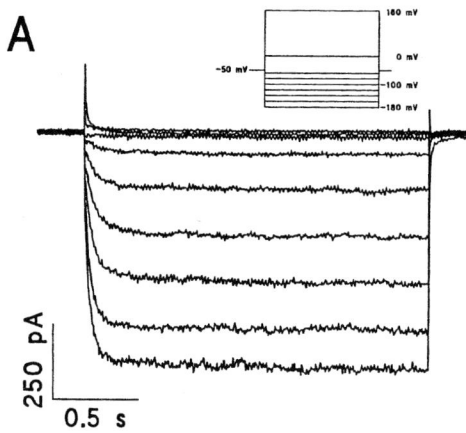

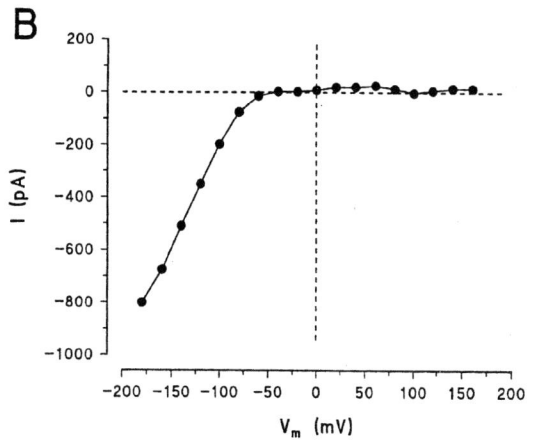

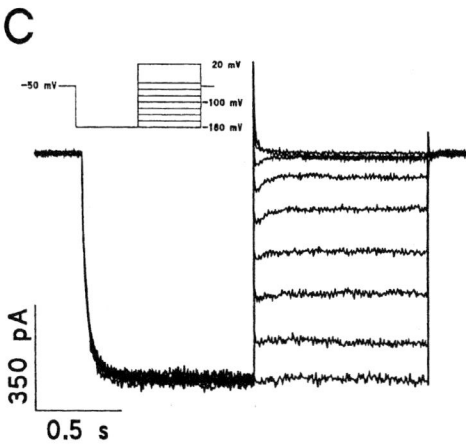

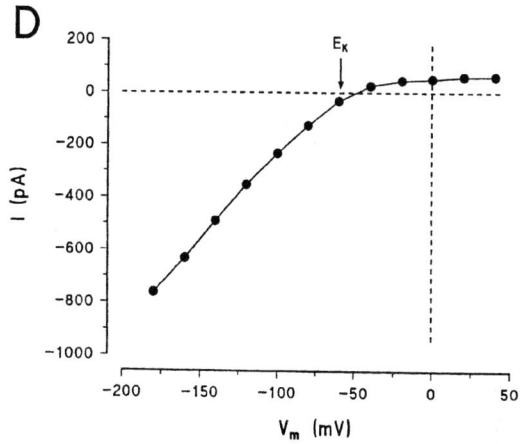

Note. In aleurone protoplasts from both the Himalaya and the Triumph (dormant and nondormant) grains a depolarization-induced current was seen in a small portion (about 10 %) of the protoplasts. This outward current, which was likely the outward rectifying potassium current described by Bush et al. (1988), was not investigated since the amplitude always ran down during the whole cell recording.

Reversal Potential

To determine the reversal potential of the ion channels, activating hyperpolarizing potentials are applied long enough to force the channels to maximal activation. Voltage- and time-dependent deactivation of the conductance can be elucidated using a backstep protocol giving as a result the reversal potential (E_{rev}). Reversal potentials in the barley aleurone protoplasts can be measured by hyperpolarization of the membrane to $-180\,mV$ from a holding potential of $-50\,mV$ and then increasing stepwise to less negative potentials (Fig. 21.1C), the so-called tail current method.

Tail current method The E_{rev} is equal to the tail current at the potentials where the instantaneous current reverses, since at this potential there is no ion flux through the still open channels (Fig. 21.1D).

The inward directed current in the example experiment has a reversal potential (E_{rev}) which is close to the equilibrium (Nernst) potential of potassium (E_K). The Nernst potential can be calculated by Eq. (1) when the channels are selective for only one ion (see also Chap. 6).

$$E_{rev} = \frac{RT}{F} \ln \frac{[I]_o}{[I]_i}$$

In our experiments the E_{rev} is related to the concentration of K^+ in the intracellular and extracellular solutions. Using different ion compositions in the extracellular solution gives an idea of how the E_{rev} depends on the type and concentration of the different ions. Increasing the extracellular potassium concentration (up to $20\,mM$) should give a shift towards the new E_K for potassium channels.

In all three types of barley protoplasts (Himalaya, Triumph nondormant and Triumph dormant) the reversal potentials are around the equilibrium potential of potassium.

Note. Before analyzing the currents achieved by the tail current method it is very imoprtant to correct the data for leakage current which can occur due to a change of the offset in the control circuitry by temperature and time during the experiment.

Conductance

The maximal steady state conductance (G_K) for the inward directed K^+ current can be determined with Eq. (2) from the steady state current-voltage relationship for the different types of protoplasts:

$$G_{K, max} = \frac{I}{E_m - E_{rev}}$$

If a line is drawn from the reversal potential towards the current at an applied potential in the current-voltage curve (Fig. 21.1B), the conductance is equal to the slope.

This conductance is called the chord conductance (G_K). With these different conductances and the maximal conductance the activation degree ($G_K/G_{K,max}$) can be calculated. Figure 21.2 shows the normalized steady state conductance as a function of the applied potential for the three types of barley aleurone protoplasts. The calculated activation degrees can be fitted according to the Boltzmann distribution as stated in Eq. (3). **Chord conductance**

$$\frac{G_K}{G_{K, max}} = \frac{I}{I + e \left(\frac{V_{0,5} - V_m}{S} \right)}$$

The mean half-maximal activation potentials ($V_{0.5}$) and the slope (S) of the different aleurone protoplasts can be determined in this way.

Modulation of the K^+ Inward Rectifier Current

In order to get more insight into the properties of the K^+ inward rectifier additional experiments can be done. For instance, the well known K^+ inward rectifier blocker Ba^{2+}(500 μM), when added to the measurement chamber, completely blocks the inward current. In addition, to study the role of plant hormones in signal transduction the controlling effect of the hormone abscisic acid (ABA) on this ion channel can be investigated.

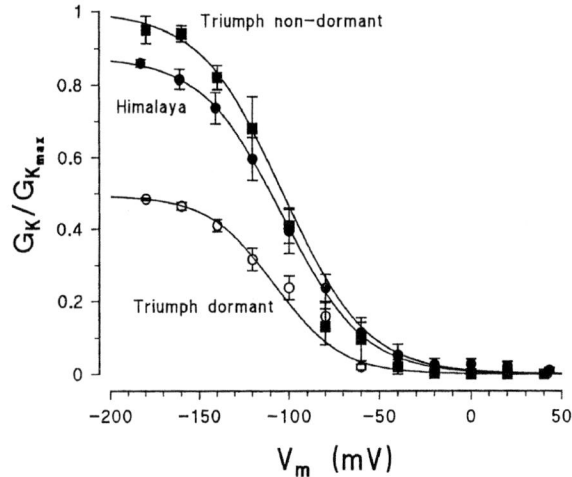

Fig. 21.2 Steady state activations of the inward K^+ currents of barley aleurone protoplasts from Himalaya (*filled circles*), Triumph nondormant (*filled squares*) and Triumph dormant (*open circles*) grains. The activation degree ($G_K/G_{K,max}$) was calculated with respect to the determined E_{rev}. $G_{K,max}$ is the maximal conductance as found in protoplasts from Triumph nondormant grains. K^+ concentration in the extracellular solution (ECS) was 10 mM. The data points are fitted according to the Boltzmann distribution. (Van Duijn et al. 1996; reprinted by permission of Kluwer Academic Publishers)

References

Bush DS, Hedrich R, Schroeder JI, Jones RL (1988) Channel-mediated K^+ flux in barley aleurone protoplasts. Planta 176:368–377

Hamill OP , Marty A, Neher E, Sakmann B, Sigworth FJ (1981) Improved patch-clamp techniques for high-resolution current recording from cell-free membrane patches. Pflügers Arch 391:85–100

Heimovaara-Dijkstra, S, Van Duijn, B, Libbenga, KR, Heidekamp, F, Wang M (1994) Abscisic acid-induced membrane potential changes in barley aleurone protoplasts: a possible relevance for the regulation of rab gene expression. Plant Cell Physiol 35; 743–750

Hille B (1993) Ionic channels of excitable membranes. 2nd ed. Sinauer Associates Inc Sunderland, MA

Ince C, Dissel JT, Diesselhoff MC (1985) A Teflon culture dish for high-magnification microscopy and measurements in single cells. Pflügers Arch 403:240–244

Sakmann B, Neher E (Eds) (1995) Single channel recording. 2nd ed. Plenum, New York

Schuurink RC, Van Beckum JMM, Heidekamp F (1992a). Modulation of grain dormancy in barley by variation of plant growth conditions. Hereditas 117: 137–143

Schuurink RC, Sedee NJA, Wang M (1992b) Dormancy of the grain is correlated with gibberellic acid responsiveness of the isolated aleurone layer. Plant Physiol 100: 1834–1839

Van der Veen R, Heimovaara-Dijkstra S, Wang M (1992) The cytosolic alkalinization mediated by abscisic acid is necessary but not sufficient, for abscisic acid-induced gene expression in barley aleurone protoplasts. Plant Physiol 100: 699–705

Van Duijn B, Flikweert MT, Heidekamp F, Wang M (1996) Different properties of the inward rectifying potassium conductance of aleurone protoplasts from dormant and non-dormant barley grains. Plant Growth Regulation 18:107–113

Wang M, Van Duijn B, Schram A (1991) Abscisic acid induces a cytosolic calcium decrease in barley aleurone protoplasts. FEBS Lett 278: 69–74

Fluorescence To Measure Intracellular Ions

Introduction to the Measurement Technique

An Introduction to the Use of Fluorescent Probes in Ratiometric Fluorescence Microscopy

PIET VIS AND JOLANDA LEMMERS

Background

The possibilities of measuring the behavior of intracellular ionized calcium in living cells have increased considerably with the introduction of the ratiometric calcium probe Fura-2 in 1985. Before that, the luminescent calcium probe aequorin was used widely. This probe however has two important drawbacks: the concentration must be known in order to make absolute measurements and the aequorin is consumed during the process. The new ratiometric probes overcome these problems. In this chapter the principles of the **ratiometric method** will be discussed using the calcium probe Fura-2 as an example. The principles and practical methods however also apply to most other ratiometric probes.

The central principal of the method is a specific change in the fluorescence spectrum of the probe (also named a fluorochrome) upon binding to the target molecule (calcium in the case of Fura-2). That specific change is a **shift in the wavelength of maximal fluorescence intensity** either in the emission spectrum or in the excitation spectrum (spectral shift). After adding the probe to the target molecule, an equilibrium is established between bound and unbound probe molecules. Hence there is a mixture of two fluorescence spectra, the shifted and the unshifted one. The ratiometric method makes it possible to unravel the two spectra and to calculate the concentration of the target molecule from the relative contributions of the shifted and unshifted spectra.

The Calcium Probe Fura-2

Fura-2 (Fig. 22.1) was first described by Grynkiewicz et al (1985) and is based on the calcium chelator EGTA. A major modification is the addition of a fluorescent furan group (hence its name, with capital F and a hyphen). Its four ionized carboxylate arms constitute a potential well

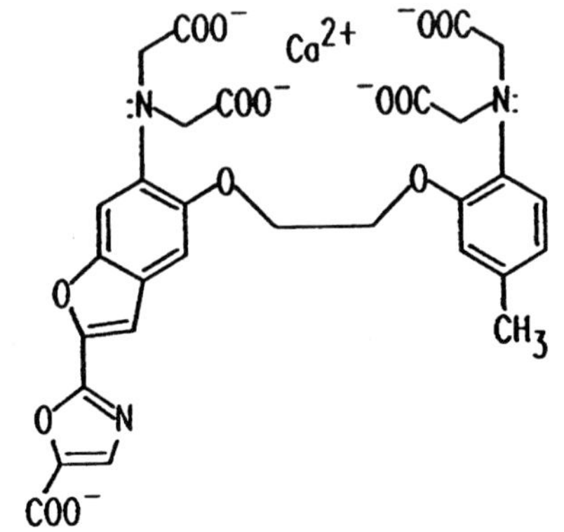

Fig. 22.1. The Fura-2 molecule. (Reprinted with permission from Molecular Probes)

into which the calcium ion is attracted. When unbound Fura-2 is excited with ultraviolet (UV) light, it fluoresces at around 510 nm. Figure 22.2 shows the fluorescence intensity as a function of the excitation wavelength (excitation spectrum). There is an optimal excitation wavelength at 362 nm for maximum fluorescence. This optimal excitation wavelength changes to 335 nm when calcium is bound. Thus binding of calcium shifts the optimal excitation wavelength whereas the emission wavelength remains 510 nm. There also exist dual emission probes that

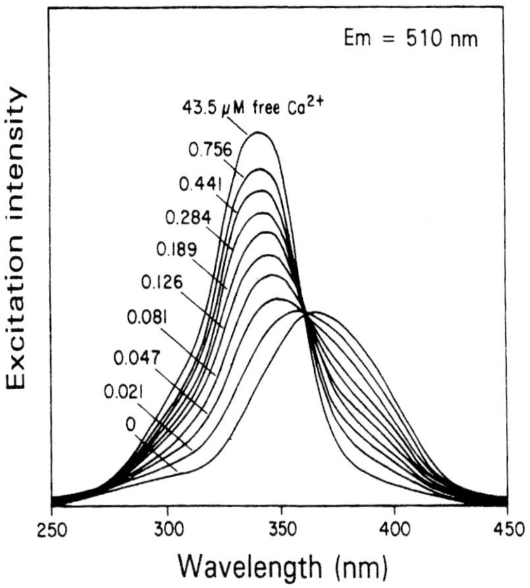

Fig. 22.2. The excitation spectrum of Fura-2. (Reprinted with permission from Molecular Probes)

exhibit a shift in the emission spectrum. The theoretical and practical approaches are similar.

In a mixture of bound and unbound Fura-2, the excitation spectrum will be a mixture of the two basic spectra. When drawn in one graph all these spectra cross at the same point: the "isosbestic" or "isoemissive" point (360 nm). The fluorescence intensity at this wavelength is independent of the calcium concentration. This can be used to monitor the Fura-2 concentration in a cell.

The ratio of bound to unbound Fura-2 can be calculated from the chemical equilibrium

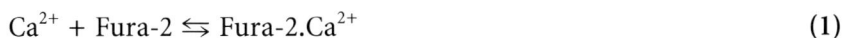

$$Ca^{2+} + Fura\text{-}2 \leftrightarrows Fura\text{-}2.Ca^{2+} \tag{1}$$

$$[Ca^{2+}] = K_d.c_b/c_f \tag{2a}$$

$$Ca' = [Ca^{2+}]/K_d = c_b/c_f \tag{2b}$$

where K_d is the dissociation constant of Fura-2 for calcium, c_b the concentration of Fura-2 bound to calcium and c_f the unbound concentration. The dissociation constant reported by Grynkiewicz et al is $K_d = 224$ nanomole (37 °C, pH 7.1, $[Mg^{2+}] = 1$ mM). However this value changes with circumstances, notably the magnesium concentration (see below). The cytosolic calcium concentration is about 100 nM ($0.5K_d$) at rest and can rise to about 1000 nM ($5K_d$) when the cell is activated. The K_d of Fura-2 for calcium lies nicely between these extremes making it a suitable probe for cytosolic free calcium.

Fura-2 also is a chelator for Mg^{2+} with $K_d(Mg) = 6$–10 mM. The cytosolic free magnesium level however is only about 0.7 mM, approximately 0.1 $K_d(Mg)$. Magnesium induces the same spectral shift as calcium. So in a calcium-free environment Fura-2 can also be used as a magnesium probe. When comparing different competing ions like Ca^{2+} and Mg^{2+} it is convenient to express their concentration in their respective K_d units (see Eq. 2b). In this way their relative strengths are directly comparable.

Also manganese (Mn^{2+}) and iron (Fe^{2+}) bind to Fura-2. In this situation there is no spectral shift but a complete quenching of the fluorescence. This can be used to measure the background fluorescence at the end of an experiment or to determine leakage of Fura-2 out of the cell.

The Ratio Method

Consider a solution containing a certain concentration of free ionized calcium. Fura-2 is added and an equilibrium is established according to Eqs. (1) and (2). This solution is excited by a beam of light (intensity I_1) with a wavelength λ_1 that is shorter than the isosbestic wavelength. For Fura-2 this is usually 340 nm. The fluorescence F_1 emanates from two sources: the free Fura-2 and the bound Fura-2. Per unit volume this is

$$F_1 = I_1(\alpha_1 c_f + \beta_1 c_b) \tag{3}$$

The parameters α and β are proportionality constants comprising the quantum efficiency of Fura-2 and optical characteristics of the measuring system. In addition, the sample is excited by a beam of light (intensity I_2) with a wavelength λ_2 that is longer than the isosbestic wavelength. For Fura-2 this is usually 380 nm. Likewise the fluorescence now is

$$F_2 = I_2(\alpha_2 c_f + \beta_2 c_b) \tag{4}$$

The ratio R after which the method is named, is defined by

$$R = F_1/F_2 \tag{5}$$

Now consider the special situation of a calcium-free solution. Then $c_b = 0$, $c_f = c_{tot}$ and

$$F_1 = F_{1,min} = I_1.\alpha_1.c_{tot} \tag{6}$$

$$F_2 = F_{2,min} = I_2.\alpha_2.c_{tot} \tag{7}$$

$$R_{min} = F_{1,min}/F_{1,min} = I_1.\alpha_1/I_2.\alpha_2 \tag{8}$$

In the special situation when the solution is fully saturated we have $c_f = 0$, $c_b = c_{tot}$ and

$$F_{1,max} = I_1.\beta_1.c_{tot} \tag{9}$$

$$F_{2,max} = I_2.\beta_2.c_{tot} \tag{10}$$

$$R_{max} = F_{1,max}/F_{2,max} = I_1.\beta_1/I_2.\beta_2 \tag{11}$$

We can now calculate the calcium concentration from a measured ratio by using the above definitions in the following way:

$$R = F_1/F_2 = (I_1\alpha_1\, c_f + I_1\beta_1.c_b)/(I_2\alpha_2\, c_f + I_2\beta_2.c_b) \tag{12a}$$

From Eq. (2a) and $c_f + c_b = c_{tot}$ we calculate that

$$c_f = c_{tot}.K_d/(K_d + [Ca^{2+}]) \tag{12b}$$

$$c_b = c_{tot}.[Ca^{2+}])/(K_d + [Ca^{2+}]) \tag{12c}$$

Substituting Eqs. (12b) and (12c) in (12a), using definitions from Eqs. (6)–(10) and then inverting the result yields

$$[Ca^{2+}] = K_d.\{F_{2,min}/F_{2,max}\}.(R - R_{min})/(R_{max} - R) \tag{12d}$$

This is the forward equation of the ratio method: from the ratio of two measured fluorescence intensities one can calculate the calcium concentration in the solution. The parameters in Eq. (12d) are determined in a calibration procedure (see below). As a consequence, the absolute values of the emission intensities F_1 and F_2 and hence the Fura-2 concentration, need not be known. The Fura-2 concentration may even change during the measurement (provided F_1 and F_2 are corrected for background radiation). This is the great advantage of this method over nonratiometric methods.

When the calcium concentration changes, the ratio c_f/c_b will change. This means that some Fura-2 molecules will shift their excitation spectrum. F_1 then will increase and F_2 will decrease or visa versa. Essential to the method is that, whatever the change in the calcium concentration, the emission intensities F_1 and F_2 should always change in opposite directions. For a variety of reasons the ratio may change without a change in the calcium concentration. Therefor the individual signals F_1 and F_2 should always be inspected to see whether a change in the ratio reflects an actual change in the calcium concentration or is some artefact.

Dual Excitation Fluorescence Microscopy

There are several types of instruments that can be used with the ratio **Principles** method. The simplest is a suspension of loaded cells in a dual wavelength spectrometer. The individual behavior of the cell may vary considerably. A single cell measuring technique like flow cytometry or fluorescence (confocal) microscopy then is better suited to study the details of intracellular calcium dynamics.

In the latter technique an inverted epifluorescence microscope is used as shown in Fig. 22.3 (e.g., see Methods in Cell Biology, vols. 29 and 30: Fluorescence Microscopy of Living Cells in Culture). The excitation beam

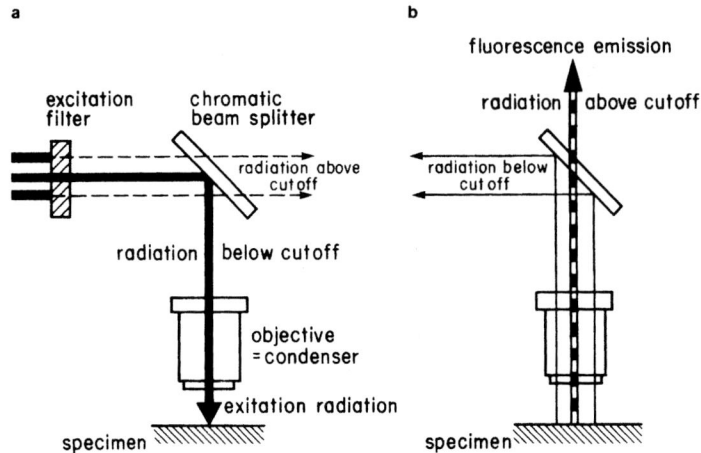

Fig. 22.3a, b. Epifluorescence microscope. Diagram of an incident or epiillumination fluorescence microscope. **a** The excitation of the spectrum using the objective as condensor. The chromatic beam splitter or dichroic mirror has the characteristics that light below a designed cutoff wavelength is reflected through the objective (condensor) and excites the specimen (*heavy black line*). **b** The emission of light from the specimen. Emitted light (*heavy dashed line*) that enters the objective above the designed cutoff is not reflected, but passes to the detector. Wavelengths below the designed cutoff are reflected away from the detector. (Reprinted with permission from Methods in Cell Biology vol. 29: Fluorescence microscopy of living cells in culture)

emanates from an UV light source, usually a xenon arc lamp. A xenon arc lamp produces a smoother UV spectrum than a mercury arc lamp, making it better suited for measuring continuous spectra in the UV range. In between the light source and the UV input port of the microscope is a filtering device that produces a monochromatic beam of light with a wavelength alternating between 340 nm and 380 nm.

The filtering device may use either fixed filters or one or two monochromators. The most common type uses fixed filters. These are interference filters with a central wavelength of 340 nm and 380 nm and a bandwidth of 5–15 nm. They are placed in the UV lightpath in a computer controlled rocking or rotating filter wheel. The changing of the filters or monochromators is synchronized with the emission measuring system.

Since the two excitation beams alternate, there is a maximum sampling speed. The two fluorescences from which the calcium concentration is calculated are shifted in time. This impairs the resolution in time still further. In a dual emission system the two emission beams can be split by a dichroic mirror and measured simultaneously, enabling a higher sampling speed and a better time resolution.

The excitation beam is fed into the normal lightpath of the microscope via a dichroic mirror. The dichroic wavelength should be between 380 nm and the emission peak. The objective of the microscope focuses the excitation beam onto the cells on the stage of the microscope. In the case of Fura-2 quartz lenses must be used since the objective must be transparant to UV light.

The emission is collected by the same objective and the fluorescent image of the cells is then passed through the emission filter onto the detector. The emission filter also is a bandpass filter but can have a greater bandwidth (30–40 nm) to collect more energy. Light levels are very low. Care should be taken to collect as much fluorescence as possible. A high-quality objective with a high numerical aperture (NA) is crucial as is proper alignment and maintenance of the microscope.

Two types of detectors are possible: a combination of a pinhole and a photomultiplier or some type of intensified video camera (see below). A photon multiplier is more sensitive than a video camera and is able to count single photons. The fluorescent image is blocked out by a diaphragm with an adjustable aperture. The image is centered such (by moving the stage of the microscope) that only a small selected portion of the field of view is seen by the photon multiplier. In this way just a single cell or a group of cells can be studied. All cells are studied at the same central position in the field of view (the light axis of the microscope) so there is no positional variance in the calibration parameters. This is also an advantage of the flow cytometer. The fluorescence is averaged over the pinhole. There is no spatial resolution. With a camera many cells and different regions inside cells can be studied simultaneously. However positional variance in the calibration parameters must be checked and accounted for. This makes the data processing more complicated.

Choice of wavelength

There is one excitation wavelength at either side of the isosbestic point. The best choices are those wavelengths at which the difference in intensities of the bound and unbound spectra are maximal (substract the spectra and find the maximum difference left and right from the isosbestic point)

The absorption peak of calcium bound Fura-2 is at 335 nm. This in theory is a good excitation wavelength. In practice however a wavelength of 340 nm is used due to the availability of filters in the early days of this technique.

The absorption peak of calcium free Fura-2 is at 362 nm. This is very close to the isosbestic point (less then the usual bandwidth). Here the intensity difference between bound and unbound Fura-2 is only small. At a longer wavelength (380 nm) this difference and hence resolution is much greater.

It should be emphasized that these short wavelengths are damaging to the cells and to the eyes. Great care should be taken and exposure restricted to the minimum.

The emission peak does actually shift but only from 505 (bound Fura-2) to 512 (unbound Fura-2). The bandwidth of the emission filter should accommodate this shift and can even be wider to collect more photons, e.g., a 490–530 nm bandpass filter makes a good choice. It is advised that the spectra be measured again using the solutions and equipment employed in your experiment. The spectra may change slightly with circumstances and equipment (like type and age of light source).

Both a photomultiplier and a video camera are very sensitive and must be guarded against overexposure. The system is best used in a dark room with a shutter mounted on the detector. The detectors also are sensitive to the infrared radiation that shines into the objective and that is generated inside the microscope. This can be dealt with by placing an infrared filter as close to the detector as possible. If so, red darkroom lighting can be used.

22.1
Loading of the Cells

Loading

There are several methods for loading cells with Fura-2. Fura-2 is a hydrophilic membrane-impermeant molecule. One method for loading is by incubating the cells in a solution containing the lipophilic, membrane-permeant acetoxymethyl (AM) ester Fura-2-AM. This fluorescent ester contains 5-AM groups and is insensitive to calcium. After crossing the cell membrane Fura-2-AM is hydrolyzed by intracellular esterases. Byproducts of this hydrolysis are acetate, formaldehyde and protons, which are toxic to living cells. Therefore it is important not to overload the cells with Fura-2, but to keep the Fura-2 concentration as low as possible and to use very sensitive measuring equipment. This is also of importance because of the buffering capacity of Fura-2 for calcium. It is advised to add Fura-2-AM to the loading medium in a concentration of 1–10 μM. Dispersing agents, such as pluronic F-127 and fetal calf serum (FCS), can be used to improve the solubility of Fura-2-AM and avoid binding of Fura-2-AM to the cell membrane (Haugland 1992). Since the intracellular Fura-2-AM is removed because it is hydrolyzed, the loading gradient remains intact. The intracellular concentration of Fura-2 may therefore increase well above the concentration of the extracellular Fura-2-AM (Tran et al. 1995).

Fura-2-AM can be hydrolyzed by enzymes on the outside of the cell. Loading then is not successful. This can be checked by placing the supernatant of the loading medium in a dual wavelength spectrometer. The supernatant will be rich in Fura-2 and calcium sensitive.

Fura-2 is also available coupled to dextran molecules which have to be microinjected. The advantage of this technique is that compartmentalization and leakage do not occur and there is less binding to cellular proteins than with regular Fura-2-AM. Also there is no need to de-esterize (no toxic byproducts, no intermediate forms, no incubation time).

Incubation Time and Temperature

Several factors determine the incubation time and temperature for loading the cells with Fura-2-AM. Firstly it is important to be sure that all Fura-2-AM molecules are hydrolyzed completely. Incomplete hydrolysis results in several different, highly fluorescent, but Ca^{2+}-insensitive forms. Secondly, Fura-2-AM will also cross intracellular membranes because of its lipophilic character or by endocytosis. This process of compartmentalization is temperature-dependent. Therefore the temperature should be low enough to inhibit or retard endocytosis and high enough to hydrolyze all Fura-2-AM molecules (Malgaroli et al. 1987). Depending on the cell type used, the incubation temperature may vary from room temperature to 37 °C. The incubation time may vary from 10 min to 1 h. The optimal conditions are found by trial and error.

Material and Procedure

Several culture media are commercially available. Also there are several recipes for making your own media, e.g., HBSS (Hanks balanced salt solution) or DME. The storing and loading medium can contain vitamins and may even be necessary for keeping the cells viable during loading. The washing and measuring medium should not contain vitamins or indicators because of their high fluorescence. A possible loading protocol is:

1. Add the loading medium to the cells and load the cells for 30 min at room temperature in the dark while shaking gently.

2. Remove the medium and wash the cells twice with washing medium.

3. Add washing medium to the cells and leave them 30 min in the dark while shaking gently. During this time the esterases can hydrolyze all the Fura-2-AM.

4. Remove the medium and wash the cells again with washing medium.

5. Add washing medium at 37 °C to the cells to perform the measurement.

Media
- Loading Medium. The following recipe is for 2 ml. The chemicals should be added in the following order and the mixture vortexed after each addition:

1. 2 ml culture medium (with vitamins) +5 % FCS

2. 3.1 µl 20 % w/v pluronic (final concentration: 0.03 % w/v)

3. 6.4 µl 3 mM Fura-2-AM (final concentration: 9.4 µM)

4. 4 µl 0.5 M probenecid (final concentration: 1 mM)

- Washing and Measuring Medium. Two ml culture medium (without vitamins) +5 % FCS and 4 µl 0.5 M probenecid

Buffer solutions
For buffering Ca^{2+} one needs a chelator with high affinity and specificity for Ca^{2+}, such as EGTA. The dissociation constant of EGTA is extremely sensitive to pH near 7.00, ionic strength, temperature and ionic composition (Bers 1982). It is very important to know the right value of $K_{d(EGTA-Ca2+)}$, because the wrong value will result in a wrong value for $K_{d[Fura-2-Ca2+)}$ and consequently the wrong calculated $[Ca^{2+}]$.

Durham (1983) determined the K_d values of the Ca^{2+} buffers EGTA, EDTA, HEDTA, DPA, NTA, ADA and citrate. In calculating the dissociation constant he used the thermodynamic activity of calcium a_{Ca} and the concentrations of the two forms of the chelator:

$$K_{Ca} = a_{Ca}.[EGTA^{4-}]/[EGTA\text{-}Ca^{2+}]$$ (13)

$$pK_{Ca} = -\log_{10}(K_{Ca})$$ (14)

The thermodynamic activity can be calculated with the formula (analogous to the theoretical ideal to which the practical pH scale approximates):

$$pCa = -\log_{10}(a_{Ca}) = -\log_{10}[Ca^{2+}] - \log_{10}(\gamma_{Ca}) \tag{15}$$

γ_{Ca} is the activity coefficient of the Ca^{2+} ions: $\log_{10}(\gamma_{Ca}) = -0.48$ at 37 °C and ionic strength 0.15.

For a given set of pH, pCa and pMg values it is a straightforward procedure, using equations like these, to calculate the concentrations of all the species present relative to the concentration of the fully ionized form of the chelator. The total concentration of chelator, or that of magnesium, or that of calcium then defines the total concentrations of the other two. Working back in the opposite direction, i.e., calculating the pCa and/or pMg values in a solution with known total metal and chelator concentrations, is more difficult: an iterative routine is needed to evaluate the concentration of the fully ionized form of the chelator.

It is often convenient to ignore the multiple protonated forms of a chelator present in solution and to define an apparent or "conditional" stability constant, pK'_{Ca} or pK'_{Mg}, for any particular pH. Thus:

$$pK'_{Ca} = -\log_{10} \cdot \frac{[\text{all calcium-free forms of } E].a_{Ca}}{[\text{all calcium-bound forms of } E]} \tag{16}$$

This value of any chelator at a particular pH value represents the middle of its calcium buffering range, i.e., the pCa value at which half of the chelator is calcium-bound and half is calcium-free.

At 37 °C, ionic strength of 0.16 M and "physiological" pH 7.35:

	EGTA	EDTA
pK_{Ca}	11.00	10.01
pK_{CaH}	4.8	3.6
pK_{Mg}	5.4	8.50
pK_{MgH}		3.3
pK'_{Ca}	7.96	7.38
pK_{Mg}	1.9	6.35

There are several computer programs available to determine the free calcium concentration at any concentration EGTA and EDTA at any pH and ionic strength.

22.2
Characteristics of Intracellular Fura-2

Aberrant Forms

Incomplete hydrolysis of Fura-2-AM will result in the presence of several different aberrant forms of partially hydrolyzed Fura-2-AM. They are highly fluorescent, but insensitive to changes in Ca^{2+} concentration and contribute greatly to the background fluorescence. Sufficient time should be allowed for hydrolysis to take place.

The Dissociation Constant K_d

For accurate measurement and calibration, a proper determination of the dissociation constant K_d is necessary. The K_d is the ionic concentration at which 50 % of the probe is bound to calcium (see Eq. 2). It depends on pH, temperature, ionic strength, other ions like Mg^{2+} and Mn^{2+}, viscosity and protein binding (Grynkiewicz et al. 1985).

As mentioned before the $K_d = 224\,nM$ at 37 °C, pH 7.1 and $[Mg^{2+}] = 1\,mM$. With lower pH or higher ionic strength, the binding of Fura-2 to Ca^{2+} will decrease (see Catalog of Molecular Probes).

The effect of viscosity on the fluorescence intensity, spectrum and the K_d has been investigated by many researchers. The viscosity of the cytoplasm of most cells is 10–15 cP (centiPoise), while the viscosity of water is 0.7 cP. David-Dufilho et al. (1988) found that a higher viscosity raised, but did not shift, the total spectra of Fura-2 and thus decreased the ratio F_{340}/F_{380}. This was also found by Owen (1991), who showed that changes in temperature (and resulting viscosity) had no effect on the shape of Fura-2 excitation spectra. He concluded that polarity effects are the principal mechanism for the shifts in excitation wavelengths and that viscosity could affect fluorescence intensity. Viscosity and light absorption/ scattering together may be responsible for a change in relative peak heights of calcium-bound and free Fura-2 in the cell. The change in relative peak heights directly influences the actual calibration of the system for measurement of intracellular calcium activity, so an intracellular calibration is necessary.

Leakage

Once inside, Fura-2 will leak out of the cells, because the cell membrane is not completely impermeable to Fura-2. Inhibitors of anion transport, such as probenecid or sulfinpyrazone, can be used to inhibit sequestration and secretion of Fura-2. In a constantly perfused system secretion of Fura-2 will only cause a small decrease of fluorescence intensity, which will have no effect on the ratio as long as the intensity is high enough with respect to background fluorescence. In static systems with no perfusion, long-term leakage may increase the background fluorescence slightly, but this effect is usually minimal (Haugland 1993). The use of inhibitors is advised; however, depending on the cell type, inhibition may alter cell function (Di Virgilio et al. 1989).

22.3
Calibration

The Calibration Curve

The theoretical relationship between $[Ca^{2+}]$ and the ratio R is described by Eq. (12d). In practice this is only useful when numerical values are substituted for the parameters in Eq. (12d). These values are determined in a calibration procedure. In such a procedure the calibration curve is constructed experimentally. Theoretically this curve is the inverse of Eq. (12d) and is given by

$$R = ([Ca^{2+}].R_{max}+K'.R_{min})/([Ca^{2+}]+K') \tag{17}$$

$$K' = K_d.(F_{2,min}/F_{2,max}) \tag{18}$$

A graphical presentation of Eq. (17) is shown in Fig. 22.4. The ratio R (dimensionless) is plotted against the calcium concentration (nanomoles).

Two major approaches are possible here. In the first approach, called here the R_{min}/R_{max} calibration, the numerical values of the parameters are measured directly. In the second approach, the full calibration method, many points $[Ca_i^{2+}, R_i]$ on the curve are measured. The parameter values can then be estimated via a curve fitting procedure.

Each calibration method can be executed either intracellularly (in vivo calibration) or in a cytosol-like cell free solution (in vitro calibration). One of the parameters is the dissociation constant K_d. From the preced-

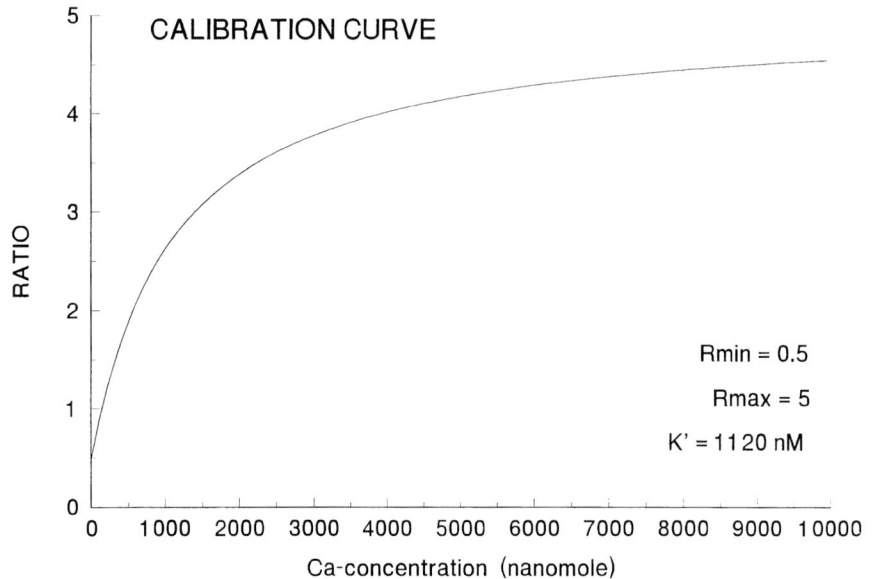

CALIBRATION CURVE

Rmin = 0.5

Rmax = 5

K' = 1120 nM

RATIO

Ca-concentration (nanomole)

Fig. 22.4. Calibration curve

ing paragraph it is clear that K_d varies with circumstances. Hence the best calibration procedure is an intracellular one. Only this option will be discussed here (Williams and Fay 1990).

Setting Intracellular Calcium Levels

Whatever calibration method is applied, the foremost important and difficult part is setting the calcium level at a predetermined well known value. To do an intracellular calibration one must manipulate the cytosolic calcium level. However the cell membrane is rather impermeable for calcium and any excess calcium is readily expelled from the cell. Remember that animal cells are capable of maintaining a 1:10 000 calcium gradient across their membrane.

One way of getting calcium into the cell is by "shooting holes" in it with ionophores. These are molecules that settle in the membrane and form ionselective pores (Pressman 1976). The most widely used calcium selective ionophores are ionomycin (Liu and Hermann 1978) and 4-bromo-A23187 (Deber et al. 1985). These ionophores are capable of transporting monovalent and divalent cations across the membrane with a high selectivity for calcium. After incubating the cells with the iono-phore one can set the intracellular calcium level by simply bathing the cells in a calcium buffer having the desired concentration of ionized cal-cium. Different levels can be set by changing the bathing fluid or by add-ing calculated amounts of calcium. When changing the bathing fluid it is possible to wash out the ionophore. To prevent this, ionophore should be added to the buffer solution. Also manganese is passed by these iono-phores so they can be used to determine the background fluorescence by intracellularly quenching the Fura-2. There are other ways of permeabi-lizing a cell such as using digoxin. However the membrane can be dis-rupted to such an extent that the Fura-2 will leak out of the cell. The use of ion selective ionophores is thus preferred.

During the calibration procedure active cellular processes may be very much in the way. When the calcium level of the calibration buffer is not too far away from the resting level of intracellular calcium, then the extrusion capacity of the cell may readily compensate for the enhanced calcium influx. This can be overcome by depriving the cells of energy by blocking their ATP production with KCN (0.1–1 mM). In addition the bathing fluid can have an intracellular composition to stop the Na^+- and K^+-driven membrane pumps. An additional advantage is that the cyto-solic composition is the least disturbed. When a calibration is done at the end of an experiment the cell no longer needs to function or even be alive as long as the membrane is still intact. All you need is "just a bag of cytosol", provided the pH, viscosity, etc., have not deviated too much from their normal values.

The R_{min}/R_{max} Calibration

This is a two-point calibration at the extremes of the calibration curve. First the calcium ionophore is added, followed by a calcium chelator, e.g., EDTA to extract all the calcium from the cytosol. We then have a calcium-free situation as in Eqs. (6)–(8). The measured fluorescences are $F_{1,min}$ and $F_{2,min}$ from which R_{min} is calculated.

Next an overdose of calcium is added to saturate the Fura-2. This situ-ation is described by Eqs. (9)–(11) and yields $F_{1,max}$ and $F_{2,max}$ and hence R_{max}.

We now have all the parameters in Eq. (2) except for K_d. The affinity constant must be measured separately or be taken from the literature. The latter method of course introduces great uncertainty.

This fairly simple procedure does have its limitations however. To measure R_{min} the calcium concentration must be $[Ca^{2+}] = 0$. This is practically impossible because calcium is abound everywhere, and the buffer capacity of the added chelator is limited. To reach a situation of 99 % unbound Fura-2, the calcium concentration must be as low as $0.01K_d$ (approx. 2 nM) according to Eq. (2).

To measure R_{max} the calcium concentration must, in theory, be infinite. This also is practically impossible. To reach a 99 % binding, a calcium concentration of $99K_d$ (approx. 20 µM) is needed.

The Full Calibration

This is a multipoint calibration. Several points $[Ca_i, R_i]$ along the calibration curve are measured. In both Eqs. (12) and (17) the expression $K' = K_d.(F_{2,min}/F_{2,max})$ can be treated as one single parameter. In a curve fitting procedure only three parameters (R_{min}, R_{max} and K') need to be estimated.

The advantage of a full calibration is that the affinity constant K_d is included in the calibration (via K'). However the procedure takes a longer time to execute and the cells are subjected to many changings of the bathing fluid. There is thus a fair chance they will detach or burst.

The Partial Calibration

The level of cytosolic calcium ranges from resting level (around 50 nM) to the highest level during spikes and oscillations (around 1000 nM). The calibration curve need not extend beyond that range. It suffices to perform only a partial calibration in the working range. This is less work and the cells are less likely to become damaged. A good series of calcium levels for a partial calibration would be 50, 100, 200, 500 and 1000 nM.

Parameter values estimated in a series of partial calibrations may vary considerably though the curves are very close together. This is partially due to noise and real differences, but mostly due to the fact that no calibration was done at or near R_{min} and R_{max}. This is no reason for concern. The curve counts, not the parameter values.

Comments and Troubleshooting

System Dependence

In a dual excitation system, the two fluorescence intensities F_1 and F_2 have the same wavelength and follow the same optical pathway. So the ratio is not influenced by changes in this part of the equipment but is influenced by the signal to noise ratio. The fluorescence intensities are proportional to the excitation intensities I_1 and I_2 (see Eqs. 3 and 4). The excitation beams do have different wavelengths and partially follow a different optical path. So the ratio depends on the actual construction of the equipment. The ratios may even change with aging of the UV light source. So, when changes have been made in the excitation pathway, the system needs to be calibrated again.

Beam Balancing

When the fluorescences F_1 and F_2 differ markedly in intensity (either a very small or a very large ratio) the lower intensity may be barely detectable and/or the higher intensity may be saturating the detector. In either case an incorrect ratio is measured. This problem can be avoided by balancing the excitation intensities such that within the working range the ratio varies around 1. This can be achieved by placing a neutral density filter in the light path of the excitation beam that produces the strongest fluorescence. If this is not possible a suitable color filter can be placed in the common pathway of the excitation beams (for example, in front of the dichroic mirror) such that the one beam is attenuated more than the other.

Background Fluorescence

Apart from the intended fluorescence from the fluorochrome, there also is unwanted fluorescence. This can be caused by specks of dust on the coverslip, chemicals in the bathing fluid (like certain vitamins) or chemicals inside the cell (like NADH). Furthermore there is dark current in the detector and the ever present infrared radiation. All these sources contribute an amount B_1 and B_2 to the fluorescences F_1 and F_2. So the basic equations now become

$$F_1^* = I_1(\alpha_1 c_f + \beta_1 c_b) + B_1 \tag{19}$$

$$F_2^* = I_2(\alpha_2 c_f + \beta_2 c_b) + B_2 \tag{20}$$

$$Ca' = [Ca^{2+}]/K_d = c_b/c_f \tag{21}$$

$$c_{tot} = c_f + c_b \tag{22}$$

The forward equation is again calculated using the same definitions of the parameters, however now with the background radiation included (indicated by the asterisk)

$$R^* = F_1^*/F_2^* \tag{23}$$

$$R^* = F_{1,min}^*/F_{2,min}^* = (I_1\alpha_1 c_{tot} + B_1)/(I_2\alpha_2 c_{tot} + B_2) \tag{24}$$

$$R^* = F_{1,max}^*/F_{2,max}^* = (I_1\beta_1 c_{tot} + B_1)/(I_2\beta_2 c_{tot} + B_2) \tag{25}$$

Combining Eqs. (21) and (22) gives

$$c_f = c_{tot}/(Ca' + 1) \tag{26}$$

$$c_b = Ca'.c_{tot}/(Ca' + 1) \tag{27}$$

Substituting Eqs. (26) and (27) in Eqs. (19) and (20) gives the calibration equation:

$$R^* = F_1^*/F_2^* = \{I_1\alpha_1 + I_1\beta_1 Ca' + B_1(Ca' + 1)/c_{tot}\}/\{I_2\alpha_2 + I_2\beta_2 Ca' + B_2(Ca' + 1)/c_{tot}\} \tag{28}$$

By inverting this equation we find the forward equation:

$$Ca' = \{R^*(I_2\alpha_2 c_{tot} + B_2) - (I_1\alpha_1 c_{tot} + B_1)\}/\{I_1\beta_1 c_{tot} + B_1 - R^*(I_2\beta_2 c_{tot} + B_2)\} \tag{29}$$

Using the above definitions (24) and (25) we find

$$[Ca^{2+}] = K_d.\{F_{2,min}^*/F_{2,max}^*\}.(R^* - R_{min}^*)/(R_{max}^* - R^*) \tag{30}$$

This equation is just the same as Eq. (12d) which is the forward equation without background radiation. So, with or without background radiation, we can use the same procedures and equations. However this is only true if the background radiation and the Fura-2 concentration remain constant.

The Pseudo-isosbestic Point

The fluorescence intensity at the isosbestic wavelength is independent from the calcium concentration and directly proportional to the Fura-2 concentration. During an experiment any loss of Fura-2 can be checked by measuring the fluorescence at the isosbestic wavelength. This requires an additional measurement and an additional excitation wavelength.

It is possible however to check the Fura-2 concentration without additional hardware by a simple weighted summation of the two fluorescences F_1 and F_2. This can be done in real time during the experiment or later during data analysis.

$$F^\star{}_{tot} = F^\star{}_1 + \gamma\, F^\star{}_2 \tag{31}$$

is constant.

Substituting Eqs. (19), (20) and (22) gives

$$F^\star{}_{tot} = c_{tot}(I_1\alpha_1 + \gamma I_2\alpha_2) + c_b(I_1\beta_1 + \gamma I_2\beta_2 - I_1\alpha_1 - \gamma I_2\,\alpha_2) + B_1 + \gamma B_1 \tag{32}$$

All terms in Eq. (32) are independent from the calcium concentration except c_b. So $F^\star{}_{tot}$ can only be constant if the term associated with c_b equals zero.

$$(I_1\beta_1 + \gamma I_2\beta_2 - I_1\alpha_1 - \gamma I_2\alpha_2) = 0 \tag{33}$$

From this we can solve the weighing factor γ.

$$\gamma = (I_1\alpha_1 - I_1\beta_1)/(I_2\beta_2 - I_2\alpha_2) = (F_{1,min} - F_{1,max})/(F_{2,max} - F_{2,min}) \tag{34}$$

So the weighing factor can be calculated using the fluorescence intensities (corrected for background radiation) from the R_{min}/R_{max} calibration.

Bleaching

When fluorescent probes are excited, they are known to bleach. To retard this fading of fluorescence anti-fading agents, such as p-phenylenediamine (PPD), n-propylgallate (NPG) or 1,4-diazobicyclo[2, 2, 2]-octane (DABCO) or Slowfade (Molecular Probes) can be used (Krenik 1989). These agents inhibit generation and diffusion of free radicals, which increase fading, but are only to be used for nonliving specimens. Brief excitation, low light levels and sensitive detectors and a microscope with high light collecting power all help to prevent bleaching.

Changes in Cell Volume

When the cell is entirely in the field of view of the microscope, the fluorescence emanating from it is proportional to the total Fura-2 content. Changes in the isosbestic fluorescence F^{iso} reflect changes in the Fura-2 content.

When only a part of the cell's image is analyzed, say by studying only a limited number of pixels of a CCD-camera image, then the isosbestic fluorescence is proportional to the Fura-2 concentration. Changes in the isosbestic fluorescence not only reflect possible loss of Fura-2 but also changes in cell volume or a redistribution of the fluorochrome.

This feature can be used to study concomitant changes in cell volume when this is not possible morphologically. This would be the case in an epithelial layer where the volume changes in the line of sight. If the axial resolution of the imaging system is smaller than the thickness of the cell layer then the volume of the cell is proportional to:

$$\text{cell volume} \approx F^{iso} \text{ (part of cell)}/F^{iso}(\text{whole cell})$$

irrespective of loss of fluorochrome.

Working Range

The sensitivity of the system is proportional to dR/dCa, the tangent of the calibration curve. As can be seen from Fig. 22.4, the sensitivity goes to zero at increasing calcium levels. Sensitivity is highest when $[Ca^{2+}] = 0$. However the relative error $\Delta Ca/Ca$ then is infinite. As a rule of thumb, do not go beyond 90% saturation of the fluorochrome. According to Eq. (2), the working range then is $0.1K_d < [Ca^{2+}] < 9K_d$.

Storage and Handling

Fura-2-AM will slowly hydrolyze spontaneously in the presence of water. On arrival from the supplier it is best dissolved in DMSO, divided in small aliquots and stored in separate vials in a freezer. When a vial is retrieved from the freezer, it is best to wait until it is fully equilibrated with room temperature before opening it. This way, the water vapour in the ambient air will not condensate inside the vial.

Calcium

The cytosolic calcium level is in the nanomole range. Outside the cell levels are much higher and calcium is lurking everywhere. Even laboratory grade chemicals contain trace amounts of calcium as contamination. The purest water may contain calcium in the micromolar range. Calcium will leach from the glassware into the contents. The buffer capacity of the calibration solutions must be able to accommodate this environmental calcium. Critical glassware is best rinsed with and stored in water with some EDTA.

References

Bers DM (1982) A simple method for the accurate determination of free [Ca] in Ca-EGTA solutions. Am J Physiol 242: C404-C408

David-Dufilho M, Montenay-Garestier T, Devynck M (1988) Fluorescence measurement of free Ca^{2+} concentration in human erythrocytes using the Ca^{2+} indicator Fura-2. Cell Calcium 9: 167–179

Deber CM, Tom-Kun J, Mack E, Grinstein S (1985) Bromo-A23187: a nonfluorescent calcium ionophore for use with fluorescent probes. Anal Biochem 146: 349–352

Di Virgilio F, Steinberg TH, Silverstein SC (1989) Organic-anion transport inhibitors to facilitate measurement of cytosolic free Ca^{2+} with Fura-2. Methods Cell Biol 31: 453–462

Dix JA, Verkman AS (1990) Mapping of fluorescence anisotropy in living cells by ratio imaging; Application to cytoplasmic viscosity. Biophys J 57: 231–240

Durham ACH (1983) A survey of readily available chelators for buffering calcium ion concentrations in physiological solutions. Cell Calcium 4: 33–46

Grynkiewicz G, Poenie M, Tsien RY (1985) A new generation of Ca^{2+} indicators with greatly improved fluorescence properties. J Biol Chem 260: 3440–3450

Haugland R (1992) Handbook of fluorescent probes and research chemicals, Molecular Probes Inc., Eugene, OR USA

Haugland R (1993) Intracellular ion indicators. In: Mason WT (ed) Fluorescent and luminescent probes for biological activity. A practical guide to technology for quantitative real-time analysis. Academic Press, London, pp 34–43

Krenik KD, Kephart GM, Offord KP, Dunette SL, Gleich GJ (1989) Comparison of antifading agents used in immunofluorescence. J Immunol Methods 117: 91–97

Liu C, Hermann TE (1978) Characterization if ionomycin as a calcium ionophore. J Biol Chem: 253:pp 5892–5894

Malgaroli A, Milani D, Meldolesi J, Pozzan T (1987) Fura-2 measurement of cytosolic free Ca^{2+} in monolayers and suspensions of various types of animal cells. J Cell Biol 105: 2145–2155

Owen CS (1991) Spectra of intracellular Fura-2. Cell Calcium 12: 385–393

Pressman BC (1976) Biological applications of ionophores. Annual Rev Biochem 45: 501–530

Tran NNP, Leroy P, Bellucci L, Robert A, Nicolas A, Atkinson J, Capdeville-Atkinson C (1995) Intracellular concentrations of Fura-2 and Fura-2/AM in vascular smooth muscle cells following perfusion loading of Fura-2/AM in arterial segments. Cell Calcium 18: 420–428

Williams DA, Fay FS (1990) Intracellular calibration of the fluorescent calcium indicator Fura-2. Cell Calcium 11: 75–83

Chapter 23

How To Use Video Cameras To Acquire Images

JOCHEM HERRMANN

Background

Accurate measurements can only be made if you know your measurement setup, and understand its limitations. This chapter deals with an important part of the measurement setup: the video camera. There are thousands of cameras on the market, but it is often very difficult to understand the specifications of the various manufacturers and to discriminate important from unimportant parameters.

This chapter provides the necessary background for understanding the most important characteristics of a charge-coupled device (CCD) camera and helps to select the appropriate type of camera for your application. Also, recommendations are given to get the most out of your measurement setup.

Common pitfalls and the methods to prevent them from crossing your path conclude this chapter.

Materials and Procedure

CCD Cameras: Video Standards

CCIR and RS-170 Video Standards

Here, we will concentrate on two video standards that are widely found and which are designed to be displayed using an ordinary video monitor: CCIR and EIA RS-170. The main difference between the two standards are the number of lines in an image and the number of images per second (frames), as can be seen in Table 23.1.

Regardless of the video standard, a complete video image (frame) is composed of two interlaced fields, each with half the number of lines but scanned at twice the frame rate. After the first field is scanned, the second field is scanned with an offset of one (half) line: effectively, the lines of the second field are scanned just in between the lines of the first field (Fig. 23.1).

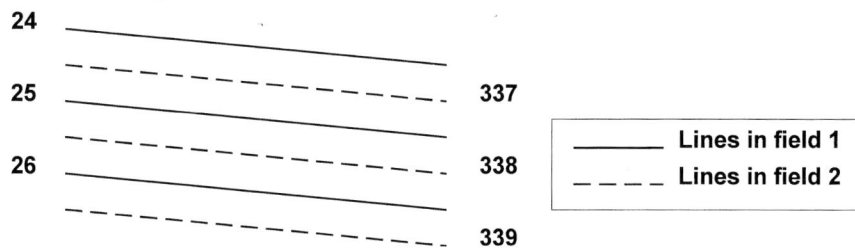

Fig. 23.1. Interlaced scanning (CCIR format)

The advantage of this somewhat complex scanning method is that it combines a high refresh frequency on the monitor (50 or 60 Hz) with a high vertical resolution. Should noninterlaced scanning be used, the refresh frequency on the monitor would be half (25 or 30 Hz). This frequency is too low for display on a standard monitor: the human observer would see severe flickering in the image.

The standard that is used depends on the country where you are living. In Europe, the CCIR standard is used, the USA and Japan use the RS-170 standard.

We advice you to use the video standard that is common in your country. Not only is the majority of the equipment locally available according to that standard, but you will also prevent problems due to interference caused by the difference between the main frequency and the field frequency of the video signal.

There are dozens of other video standards on the market, each with its specific application in mind. The most common is perhaps the noninterlaced (also called progressive scan) video standard. It is used for captur-

Other video standards

Table 23.1. Main characteristics of video standards

	Fields/s	Frames/s	Lines/image	Active lines
CCIR	50	25	625	575
RS-170	60	30	525	485

ing images of moving targets with full vertical resolution. Also, there are standards for high resolution cameras, e.g., with 1249 or 1125 lines (HDTV).

Cameras and other equipment using these special standards are (much) more expensive than those according to the CCIR and RS-170 standards. Also, you will experience more interfacing problems using nonstandard equipment. These special video standards should only be considered if the normal video standards cannot be used for your application (e.g., you need a higher vertical resolution or a camera for long integration times).

CCD Cameras: Types

Below we will discuss three different types of monochrome matrix CCD cameras: a standard camera, a long integrating camera, and an intensified camera.

Standard CCD camera

Actually, there is no standard camera. Instead, what we mean is a camera using one of the major video standards CCIR or RS-170, equipped with a normal CCD sensor and without special features such as an image intensifier or long integration.

Although this type of camera is too insensitive to use for most fluorescent imaging, we start with it because it serves as a basis for the two other cameras we will describe later on.

Figure 23.2 shows a simplified block diagram of a standard camera: we left out all parts that are not of direct importance for the description of the signal path.

Starting from the left side: the CCD sensor (together with the clock pattern generator that is used to scan the picture elements of the CCD) transforms a light pattern on its surface to an analog output signal.

As long as the CCD is operating within its dynamic range, the signal level on the CCD output is linear with the product of the light level at the CCD surface and the integration time. The sensitivity of the CCD is further a function of the spectral distribution of the light at the input: most CCDs are sensitive in the range of 450–700 nm with the peak sensitivity in the 500–550 nm range (green). There are however also CCDs which are peaked more towards the red and near infrared spectrum.

The pre-amplifier conditions the video signal for use by the AGC (automatic gain control) amplifier. The gain of the AGC amplifier can in most cases be set by changing a voltage level at the control input. In the block diagram one can choose to set the gain either manually via an (external) voltage or let the camera set its gain automatically by means of its AGC loop. For a limited light level range at the CCD input, this loop keeps the video signal at the camera output at the same level by adjusting the camera gain automatically. If the camera is used for quantitative

CCD
sensor

Pre
amplifier

AGC
amplifier

output
circuit

video output

gamma
on/off

video
gain input

AGC detector

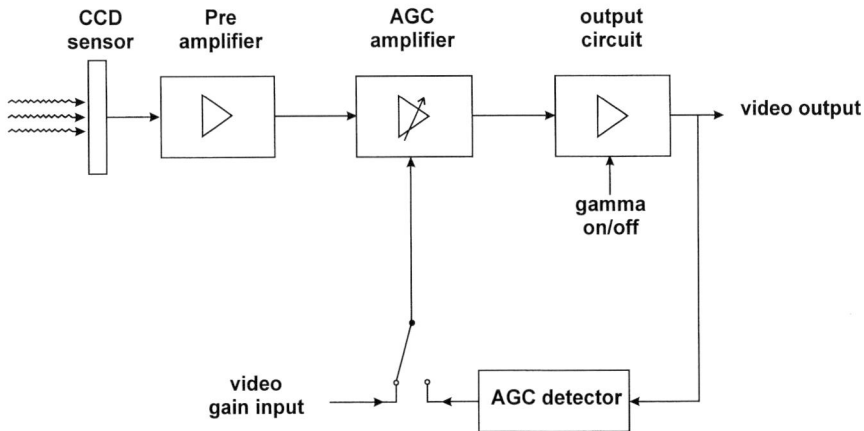

Fig. 23.2. Block diagram of signal path in standard CCD camera. The different components are discussed in more detail in the text

measurement of light levels, the camera is normally set at a fixed gain level.

The output circuit of the camera processes and buffers the signal so that it complies to the relevant video standard. In most cameras, the output circuit is also capable of performing a nonlinear transformation on the video signal called the γ function. With a γ value of 1 the function is linear, whereas lower γ values will give a higher gain in dark areas of the image than in light areas. In most cameras, the users can set the γ value to 1 and a lower value like 0.45 or 0.5. For measurement purposes, the γ should be set to 1 for a linear relationship between the light level at the CCD input and the video level at the output of the camera.

Long-integrating CCD camera

In cases in which the sensitivity of the standard CCD camera is not high enough, a long integrating CCD camera can be of use. In contrast to the standard CCD cameras, in which the maximum integration period is limited to 20 or 40 ms due to the video standard, a long integrating camera can integrate for much longer periods.

Technically speaking, the maximum integration time is limited by what is called the dark current of the sensor. The dark current is caused by the fact that each picture element (pixel) of the CCD has a small leak current. The generated charge in a pixel caused by the dark current of the sensor is linear with the integration time and doubles every 8 °C raise of CCD temperature. Cameras that are made for long integration periods (seconds to even hours) have therefore integrated coolers that keep the CCD temperature low.

In practice, the maximum integration time is often limited by factors such as available time per measurement and movements in the image.

Note that it is very difficult to adjust your microscope if you are using long integration times (more than 1 s); moving images get blurred and focusing of the image is difficult because of the image lag. Also, keep in mind that you need a frame buffer to be able to display an image on a monitor. The frame buffer may, or may not, be integrated in the camera.

Intensified CCD camera

Another alternative for achieving a high sensitivity is to use an intensified CCD camera. In this type of camera, an image intensifier type is used to raise the light level before it reaches the CCD sensor.

There are several types of image intensifiers available:

- First generation. This type of intensifier has a moderate fixed gain of, say, 100x with good resolution.
- Second generation. This type of intensifier is compact, has a high and adjustable gain and can be gated. (Gating is the process of switching the intensifier on and off by an external signal. It can be used to reduce the sensitivity and for life time imaging). The gain range is approximately 10 000–100 000x. We will describe this type of intensifier in more detail.
- Third generation. In construction similar to the second generation, but with a higher sensitivity in the (near-) infrared range and a much higher price.
- Hybrid. This intensifier is a combination of a first generation and second generation intensifier and has the highest gain of the described intensifiers. The gain can be up to 1 000 000x. It is the right choice if the light levels are extremely low.

A second generation image intensifier integrated in a CCD camera is a popular combination that has a high sensitivity and a high dynamic range. Figure 23.3 shows the most important parts in this type of intensifier.

Photons falling on the photo cathode generate electrons that are accelerated by the strong electric field in the space between the photo cathode and the microchannel plate (MCP). The MCP is a wafer with a large number of channels (running from left to right in the diagram) with a diameter in the range of 5–10 μm. Each time an electron hits the wall of a channel in the MCP it will generate a number of electrons due to secondary emission. This process repeats itself a number of times before the electrons leave the MCP and are accelerated further by the strong electric field between the MCP and the anode. The phosphor on

Fig. 23.3. Second generation image intensifier with power supply and gate unit. The different components are discussed in more detail in the text

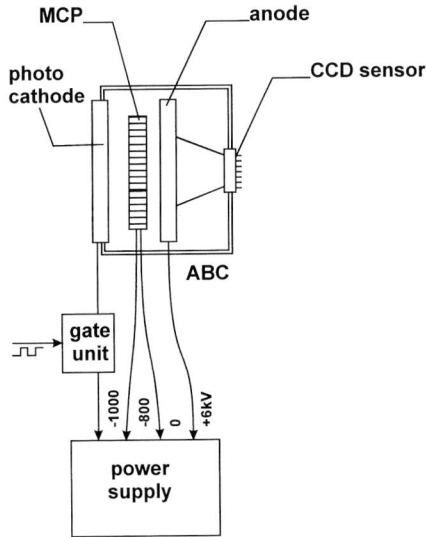

the anode transforms the "electron image" to an image composed of photons at the output window. A tapered optic fiber couples the CCD to the output window of the image intensifier.

The effective gain of this type of intensifier can be varied by changing the MCP voltage (up to a factor of 100). A much higher range in gain can be accomplished by gating of the intensifier: the photo cathode can be switched from "on" to "off" by changing the voltage between the photo cathode and the MCP input. Depending on the intensifier and the gate unit, integration times down to 100 or even 20 ns can be reached.

Apart from extending the useful light range, gating is also used for fluorescent life time imaging.

Selecting the Correct Camera for Your Application

Now that we have the necessary theoretical background, in this paragraph we will cover the most important factors to be considered during selection of a camera for your setup.

Perhaps the most important selection criterion for a camera is the light level that you plan to operate at. For most fluorescence applications a standard CCD camera will not be sensitive enough, so you will have to

Sensitivity

choose between a camera using long integration and an intensified camera. The long integration camera is superior with regard to uniformity, spatial resolution and signal-to-noise ratio. However, there is no live image on the monitor so setting up your experiment is difficult and the camera can only be used for static images. Also, to get a high sensitivity the integration times may get impractically long, and the price of the camera is high because of the advanced cooling techniques that are necessary.

Of the intensified cameras, two types may be considered: the second generation and the hybrid intensifiers. The hybrid intensifier has the highest gain of the two, but is more expensive and has a lower spatial resolution and even less uniformity than the second generation intensifier.

Spectral range   Another important criterion is the spectral range at which you plan to use the camera. This is of course dependent on the type of probes that will be used in your setup. An advantage that is valid for all image intensifiers is that the user can choose a photo cathode with a spectral sensitivity that matches the fluorescent probes. The better the quantum efficiency of the photo cathode at the wave length of the probe you are using, the lower the noise level in your image.

Spatial resolution   How important the spatial resolution is depends on your application. If you are mainly interested in measuring intensities and the objects are relatively large, the resolution is not of much concern for you. If, however, you want to be able to discern small details in your image you have to keep in mind that intensified cameras have a lower resolution than nonintensified cameras. Also, the resolution of an intensifier is worse at a high intensifier gain (low light level) than at a low gain.

Uniformity   The uniformity of a camera is derived from two sources: the pattern in black (that is, with no light on the camera), and the uniformity in white (with light on the camera). The pattern in black is mainly caused by the dark current of the CCD (especially noticeable with long integration times and a noncooled CCD camera). The uniformity in white is multiplicative and has several sources:

- Shading cause the sides and corners of the image to be darker than the center. Up to 20 % shading for a second generation and up to 50 % for a hybrid intensifier is normal.

- Local nonuniformity can appear as small black spots (caused by broken fibers) or larger areas that have a lower or higher sensitivity than other areas (caused by a nonuniformal sensitivity of the photo cathode or MCP). A local nonuniformity of up to 20 % is considered to be normal for an image intensifier.

You will have to take the uniformity into account if you compare the brightness of objects at different positions in the image.

Geometric distortion

The geometric distortion of a CCD sensor is extremely low and can be neglected compared to other sources. However, all image intensifiers suffer from geometric distortion in the range of 2 %–10 %. This distortion is fixed, so it can be compensated for in the software if needed. One special type of distortion is caused by the fiber optic coupling to the CCD sensor. If a number of fibers are shifted with respect to their neighbors this will appear as a break line in the image. This type of distortion can be very difficult to compensate for.

Sensor size and lens mount

For all microscopes one can buy accessories for the attachment of a video camera. From the optical point of view, it is important to choose the correct relay lens in the microscope (the image is projected on the sensor via this lens). Choose a lens that is made for the optical format of your camera (consult your manual). A wrong lens will cause an improper magnification factor (the image you see through the ocular has a different size than the image on the monitor) and possibly shading.

The lens mount on the microscope must of course match with the one on the camera. If possible, mount the camera facing down in order to prevent excessive mechanical force on the lens mount.

Analog vs digital video output

Most cameras have an analog video output, and the image is digitized by the frame grabber. There are, however, also cameras with an internal analog to digital (A/D) converter, providing a digital video output. This is often the case with cameras for long integration that have a 10 bits or more resolution. For most fluorescence applications there is no need for a digital output on the camera, whereas an analog output is more flexible and cost effective.

Interface to your computer

Realize that the camera and frame grabber that you select have to work together as a team. The frame grabber must be able to work with the video standard and spatial resolution of the camera. Also, it must be able to control any special functions of the camera such as integration time control and gain (extra hardware or software may be needed for this).

If you feel uncertain, let your dealer deliver the camera, frame grabber and software as a set. Should you experience any problems in getting the system to work properly, you have then only one person to talk to.

Comments

The discussion above should have given you some general information about selecting a camera. Here we will give you some extra tips to be kept in mind if you are setting up and using your system. We assume that you will use your system for quantitative luminance measurements.

Camera Settings

To ensure a linear relationship between the amount of light at the input and the video level at the output:

- Set your γ to 1 (some manufacturers call this γ off).
- Switch functions like auto black or flair compensation off.
- Set the camera gain control to manual (AGC off).
- Set the camera integration time at its maximum value.
- If the image intensifier is equipped with an automatic brightness control (ABC), switch it off if possible or set it to its highest setpoint. This ABC circuit will prevent you from doing absolute light level measurements.

Setting Imaging Parameters

- Keep in mind that the best way to get a good signal-to-noise ratio is to have a high light level at the camera input. So do not use a lower light level than necessary in your application. If you are afraid of bleaching, shorten the time that you switch on your light source to a minimum. (Note that the noise in the video signal at low light levels is mainly shot noise that is caused by the secondary emission process at the photo cathode. Even the best camera cannot change physics)

- If you can adjust the gain of the image intensifier, keep it low and increase the video gain instead. This generally will give a better signal-to-noise ratio. Do some tests with your configuration to find out what settings are the best.

- The best light level to work at (not taking into account that your specimen will bleach) is a level at which the image from the intensifier is almost noise-free. Further increase of the light level will have only a small influence on the signal-to-noise ratio, whereas the lifetime of the intensifier decreases rapidly. Also, the image will defocus at even higher light levels.

How much noise you can tolerate depends on your application. Again, do some tests with your configuration to find out what the best light level is.

- If the light level at the camera input is too high, lower the integration time of the image intensifier. (Remember that the sensitivity is linear with the integration time.) If your camera has no integration time control, use neutral density filters.

Increasing the Lifetime of Your Image Intensifier

There are a number of things to keep in mind when you use a second generation or hybrid image intensifier with regard to its lifetime.

- Image intensifiers are extremely sensitive to overexposure. Image intensifiers can burn-in (meaning that the photo cathode or MCP is degraded) in seconds.

- The sensitivity of an image intensifier decreases with time (and depends on the light level at the input). In practice you will notice that the amount of light that you need for a specific signal-to-noise ratio will get higher as time passes. Some manufacturers specify the lifetime of an intensifier at an impractically low light level. In real life the lifetime will be shorter. Do not expect a useful lifetime of more than 1000 operating hours.

- Because degradation is a local process, static images will burn-in. The speed at which this occurs depends on the light level.

- Switch off the light source if the system is not in use for some time, or use a different method to prevent light from falling on the image intensifier while the system is not being used for measurements.

Measurement Algorithms and Accuracy

The accuracy of your measurements is not only influenced by the camera and its settings, but also by the measurement algorithms.

- Noisy images may give inaccurate or poorly reproducible results. Averaging over a number of images will improve the signal-to-noise ratio. Also, spatial averaging in an image will improve the signal-to-noise ratio but lower the spatial resolution of the image. If the noise level is very high, averaging will not give accurate results because of clipping of the noise signal in the camera or frame grabber.

- Compensate for nonuniformity in the image. An object in a corner of the image may appear darker than one in the middle.

- Do not rely on long-term accuracy of the system. There are many factor influencing the sensitivity of the camera. Calibrate the system regularly, or do relative measurements.

Troubleshooting

- Poor reproducibility of measurement
  This may be caused by:

  - Too much noise in the image
  - Sensitivity variation in time
  - Light variation in time (bleaching of the specimen, main voltage variations, interference between the main frequency and the field frequency of the camera)
  - Nonuniformity of the image. The same object at two different positions may give different results.

- Poor luminous linearity

This may be caused by:

  - Clipping of noise peaks in the camera or frame grabber (in very noisy images)
  - AGC in the camera
  - ABC of the image intensifier
  - γ, auto black or flair compensation switched on in the camera
  - Improper setting of gain and offset of the frame grabber
  - Improper setting of gain of the image intensifier or video amplifier

- Poor image contrast

This may be caused by:

  - Bad or bleached specimen.

– Improper blocking filter in the optical path of the camera. A blocking filter at the wave length of the light source will prevent the black level in the video image from increasing. Note that the image intensifier may be sensitive for wave lengths that the human eye cannot see.

Chapter 24

Strategies for Studying Intracellular pH Regulation

JAN H. RAVESLOOT

Background

The short answer to the question as to why one wants to investigate intracellular pH (pH_i) regulation is that it is fascinating to study acid-base transporters, their diversity, regulation, how they handle cytoplasmic acid-base insults and how they work in concert in defending the pH_i. After having established a working setup, strategies for studying intracellular pH regulation are needed. This chapter aims to provide a succinct theoretical background from which experimental protocols and ways to analyze pH_i records can be developed. The experimental conditions and solutions that are described below are designed for studying mammalian cells.

24.1
Buffering

The first line of defense of a cell against cytoplasmic acid-base disturbances is the buffering power (β) of the cytoplasm. By definition, β is the amount of strong base (or acid) that has to be added to the cytoplasm to give a 1 pH unit change:

$$\beta = dOH^-/dpH(= -dH^+/dpH) \tag{1}$$

Cytoplasmic buffering power depends on the absence or presence of membrane permeant weak acids or bases in the extracellular solution. It makes a big difference whether the solutions used in pH_i experiments are buffered with laboratory buffers such as HEPES or with the much more physiologic CO_2/HCO_3^-.

Nonbicarbonate Cytoplasmic Buffering Power (β_i)

Together with subcellular organelles, cytoplasmic constituents such as proteins and phosphates collectively comprise the cytoplasmic buffers. They reversibly sequester or bind H^+, respectively. Inasmuch as these constituents and organelles cannot leave the cell this buffering system is said to be closed. The intrinsic cytoplasmic buffering power (β_i) normally is a function of pH_i. The β_i vs pH_i relationship, however, is unpredictable and has to be determined experimentally for each cell type as is outlined below.

Bicarbonate Cytoplasmic Buffering Power (β_{open})

In vivo, mammalian cells are exposed to an extracellular fluid with, by definition, a partial CO_2 pressure (pCO_2) equal to that of alveolar air: 38–40 Torr. Henry's law predicts that per liter body fluid at 37 °C, 0.03 mmol CO_2 dissolve per Torr. Dissolved CO_2 forms carbonic acid, H_2CO_3, which dissociates virtually completely into HCO_3^- and H^+. The pH of the overall equilibrium

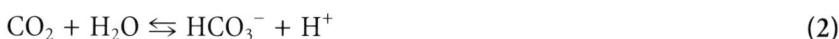

$$CO_2 + H_2O \leftrightarrows HCO_3^- + H^+ \tag{2}$$

is calculated with the well-known Henderson-Hasselbach equation

$$pH = pK_a + \log \{[HCO_3^-]/[CO_2]\} \tag{3}$$

from the pK_a, which is equal to 6.1, the dissolved CO_2 concentration ($[CO_2]$) and $[HCO_3^-]$. At a normal extracellular pH of 7.4, approximately 24 mmol HCO_3^- and 1.2 mmol CO_2 are dissolved per liter body fluid.

In the laboratory, the composition of the body fluids is mimicked by vigorously bubbling a HCO_3^--containing physiologic solution at 37 °C with a gas mixture consisting of 5 % CO_2 balanced with either 95 % O_2 or 95 % air. A pCO_2 of 5 % yields 38 Torr if the barometric pressure is assumed 760 mm Hg. To these salines, 22 mM of a HCO_3^- salt has to be added to achieve a pH of 7.4 (at 37 °C). Apparently the ionic conditions in a simple laboratory solution change the pK_a and/or the solubility of CO_2 such that 22 mmol of HCO_3^- per liter solution suffices to achieve a pH of 7.4 (37 °C).

Dissolved CO_2 molecules readily traverse plasma membranes by means of nonionic diffusion. Thus intracellular $[CO_2]$ ($[CO_2]_i$) and extracellular $[CO_2]$ ($[CO_2]_o$) are equal. $[HCO_3^-]_i$ is then calculated from the observed pH_i, pK_a and $[CO_2]_i$ ($= [CO_2]_o$). At a normal pH_i of 7.2, about 15 mmol HCO_3^- and 1.2 mmol CO_2 are dissolved per liter cell water.

The physiologic CO_2/HCO_3^- buffer pair constitutes an extremely important buffer system to the cell. Cytoplasmic HCO_3^- can bind excess H^+ and be converted to CO_2 which immediately leaves the cell. Conversely, an alkaline load can be alleviated by converting more CO_2 into HCO_3^- and H^+. The lost CO_2 is rapidly replenished. Inasmuch as in both cases CO_2 readily traverses the plasma membrane, this buffer system is said to be open. It has a theoretical buffering power of $\ln10^*[HCO_3^-]_i \approx 2.3^*[HCO_3^-]_i$. Obviously, β_{open} is a function of pH_i: the higher pH_i, the higher $[HCO_3^-]_i$, the higher β_{open}. When working with CO_2/HCO_3^- buffered solutions the total cytoplasmic buffering power (β_{tot}) is computed as follows:

$$\beta_{tot} = \beta_i + \beta_{open} \tag{4}$$

24.2
The Acid-Base Transporters

The second and very active line of defense of a cell against cytoplasmic acid-base disturbances is the controlled entry or exit of acid or base from the cell mediated by acid-base transporters. These are well-defined, highly regulated, plasma membrane-bound transport molecules which are capable of correcting pH_i.

The activity of most acid-base transporters is primarily regulated by pH_i itself, and in a predictable manner. Acid loaders usually become active when pH_i rises, thereby fighting the cytoplasmic alkaline load. Conversely, acid extruders become active when pH_i falls, thereby fighting the cytoplasmic acid load. An overview of some important acid-base transporters is given below, although the list is certainly not complete.

Acid Loaders

Cl^-/HCO_3^- exchangers and two variants of Na^+-HCO_3^- cotransporters are typical acid loaders (Fig. 24.1). With normal ion gradients, Cl^-/HCO_3^- exchangers exchange per cycle one extracellular Cl^- for one intracellular HCO_3^-. Na^+-HCO_3^- cotransporters transport per cycle one Na^+ ion in the same direction as one, two or three HCO_3^- ions. The number of cotransported HCO_3^- ions depends on the isoform of the transporter. The direction of Na^+-HCO_3^- cotransport, in turn, depends on the Na^+ and HCO_3^- electrochemical gradients and on the stoichiometry of trans-

Fig. 24.1. Cl^-/HCO_3^- exchangers and 1:3 Na^+-HCO_3^- cotransporters load the cell with acid. They transport HCO_3^- from the cell interior (*i*) to the outside (*o*)

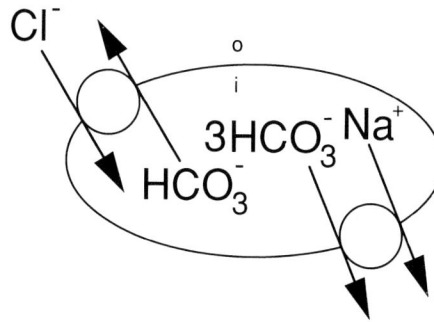

port. Perhaps the 1:2 isoforms but certainly the 1:3 isoforms transport all four ions in the outward direction and thus act as acid loaders.

Acid Extruders

The acid extruders can be subdivided into HCO_3^--independent and HCO_3^--dependent transporters.

HCO_3^--independent acid extruders. These include Na^+/H^+ exchangers, H^+ pumps (Fig. 24.2) and H^+ channels. Na^+/H^+ exchangers are found in virtually all mammalian cell types and exchange per cycle one extracellular Na^+ for one intracellular H^+. H^+ pumps, including the K^+/H^+ pumps, hydrolyze adenosine triphosphate to drive proton transport. H^+ channels behave as ionic channels and are opened by membrane potential depolarization. When open they allow protons to leave the cell.

Fig. 24.2. Na^+/H^+ exchangers, H^+ pumps and K^+/H^+ pumps are typical HCO_3^- independent acid extruders; inside (*i*); outside (*o*)

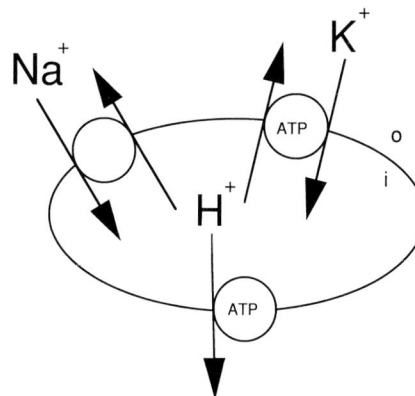

Fig. 24.3. Na^+-dependent Cl^-/HCO_3^- exchangers and 1:1 Na^+-HCO_3^- cotransporters are well-known HCO_3^--dependent acid extruders; inside (*i*); outside (*o*)

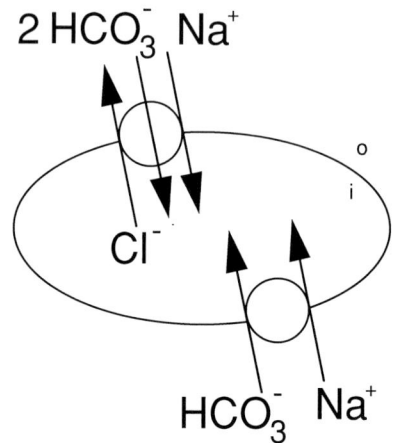

$$2\,HCO_3^- \quad Na^+$$

$$Cl^-$$

$$HCO_3^- \quad Na^+$$

HCO_3^--dependent acid extruders. Na^+-dependent Cl^-/HCO_3^--exchangers and Na^+-HCO_3^- cotransporters belong to this group (Fig. 24.3). The former exchange extracellular Na^+ and HCO_3^- for intracellular Cl^- in a process which is electroneutral. A possible stoichiometry is the exchange of one extracellular Na^+ and two HCO_3^- for one intracellular Cl^-. The 1:1 Na^+-HCO_3^- cotransporters move per cycle one extracellular Na^+ and one extracellular HCO_3^- into the cell.

(It is difficult to exclude the possibility that HCO_3^--dependent acid loaders and extruders also transport OH^- ions or carbonate CO_3^{2-} ions instead of HCO_3^- or, alternatively, in the same direction as HCO_3^-.)

24.3
HCO_3^- Transport-Mediated pH$_i$ Changes

By transporting HCO_3^- rather than H^+, pH$_i$ is indirectly altered. So the question often arises as to how much HCO_3^- needs to be transported to cause a certain pH$_i$ change. The answer depends on the initial and final pH$_i$ values and on β_i.

HCO_3^- Efflux and Cytoplasmic Acid Loading

Suppose 1 liter cell water has been given an alkaline load and acid loaders have to correct pH$_i$ by bringing it back from, say, 7.7 to 7.2. Assume further that the (mean) β_i in this pH trajectory equals 10 mM. It can be

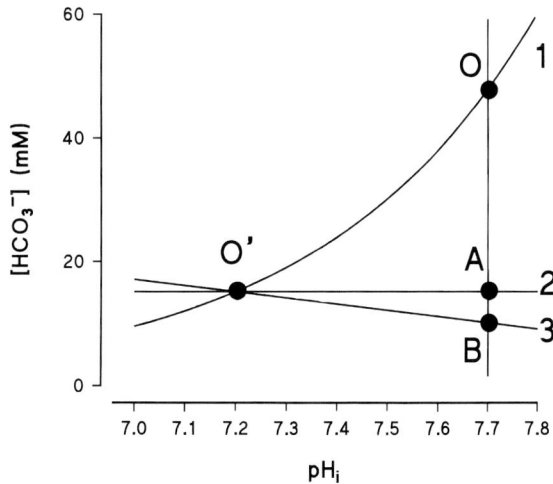

Fig. 24.4. Davenport diagram of recovery of pH_i from a metabolic alkalosis. Curve labeled with *1* represents the 40 Torr CO_2 isopleth ($[HCO_3^-] = 1.2*10^{(pH - 6.1)}$. Points *O* and *O'* on the isopleth indicate the graphical coordinates corresponding to the initial and final pH_i values. Line *2* represents the $\beta_i = 0$ nonbicarbonate buffer line. Distance *OA* represents the fall in $[HCO_3^-]_i$ yielding the H^+ that remained free to acidify the cytoplasm. The true nonbicarbonate buffer line is indicated with *3* and has a slope of -10 mM. The y-intercept is chosen such that the nonbicarbonate buffer line passes through point *O'*. In that case the vertical distance *AB* is the amount of HCO_3^- removed to generate the hydrogen ions which are all used to protonate the nonbicarbonate buffers

computed that at pH_i 7.7 $[HCO_3^-]_i$ equals ~ 48 mM whereas $[HCO_3^-]_i$ is ~ 15 mM at pH_i 7.2 ($pCO_2 = 40$ Torr). During the recovery, 43 nmol HCO_3^- are formed together with the same amount of H^+. The H^+ remain free and are added to the ~ 20 nmol H^+ already present. Thus, at the very least, 33.000043 mmol of HCO_3^- have to be removed. (Eight significant figures are of course ludicrous and serve only to show that you can safely ignore the HCO_3^- formed together with the H^+.) But many more H^+ are needed to protonate the nonbicarbonate buffers. According to the definition of β, a 0.5 pH unit change requires addition of 5 mM H^+. This is generated by converting 5 mmol CO_2 to 5 mmol H_2CO_3 which dissociate into 5 mmol H^+ and 5 mmol HCO_3^-. All of the H^+ formed binds to the nonbicarbonate buffers, none remains free. However, the HCO_3^- thus generated may not be added to the cell water as it would bind free H^+ causing a rise of the pH_i. The solution is to transport these 5 mmol HCO_3^- out of

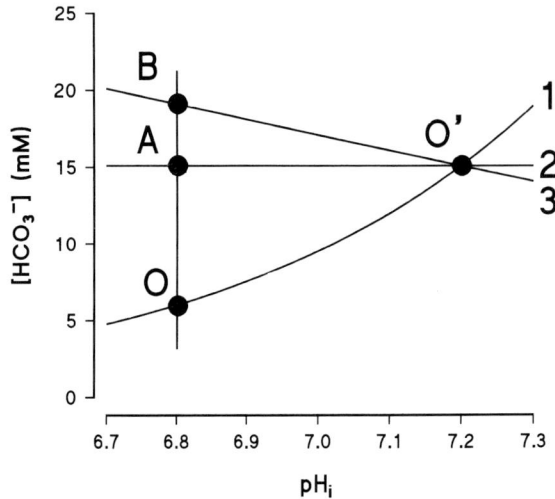

Fig. 24.5. Davenport diagram of recovery of pH_i from a metabolic acidosis.Curve labeled with *1* represents the 40 Torr CO_2 isopleth ($[HCO_3^-] = 1.2*10^{(pH - 6.1)}$). Points *O* and *O'* on the isopleth indicate the graphical coordinates corresponding to initial and final pH_i values. Line *2* represents the $\beta_i = 0$ nonbicarbonate buffer line. Distance *OA* is the increase in $[HCO_3^-]_i$ necessary to arrive at the the final pH_i value. The true nonbicarbonate buffer line is labeled *3* and has a slope of -10 mM. The y-intercept is chosen such that the line passes through point *O'*. In that case the vertical distance *AB* is the amount of HCO_3^- that has to be added to bind all hydrogen ions that dissociate from the nonbicarbonate buffers upon the pH_i increase

the cell. In this example a total of 38.000043 mmol of HCO_3^- have to exit 1 liter cell water to achieve the desired pH_i change. Furthermore, $\sim 5.000\,043$ mmol of CO_2 were consumed but were also instantaneously replenished. Figure 24.4 shows the graphical analysis of this problem.

HCO_3^- Influx and Acid Extrusion

Now suppose acid extruders have to raise the pH_i from 6.8 to 7.2, and assume that β_i in this pH trajectory again equals 10 mM. Using the same line of reasoning it can be shown that ~ 13 mmol of HCO_3^- have to be transported into 1 liter cell water to achieve the desired pH_i change. Incidentally, transporting HCO_3^- into the cell gives the same result as moving the same amount of H^+ out of the cell. In total, ~ 4 mmol of CO_2 were formed but immediately exited the cell. Figure 24.5 shows the graphical analysis of this problem.

24.4
Setpoint pH$_i$

Acid-base transporters defend a "setpoint" pH$_i$ value. When the actual pH$_i$ is equal to setpoint pH$_i$, the acid-base transporters are minimally active or have shut themselves down. If pH$_i$ is suddenly perturbed in either an acidic or alkaline direction, then the acid-base transporters return the pH$_i$ to the setpoint pH$_i$. The speed with which they accomplish this task is a measure of how well pH$_i$ is regulated. However, not only the H$^+$ (or HCO$_3^-$) fluxes mediated by the acid-base transporters govern behavior of pH$_i$, as will be discussed below.

24.5
Fluxes Determining the Behavior of pH$_i$

At all times pH$_i$, or the rate of change per unit time (dpH$_i$/dt) thereof, is the resultant of two fluxes. The rate at which H$^+$ leaks into the cytoplasm (or HCO$_3^-$ out) is called the acid loading flux. Conversely the rate at which H$^+$ is transported out of the cytoplasm (or HCO$_3^-$ in) is termed acid extrusion flux.

The H$^+$ fluxes can be active or passive. Active H$^+$ fluxes are mediated by the aforementioned acid-base transporters. Passive H$^+$ fluxes occur via ill-defined unregulated pathways collectively called background acid loading/extrusion processes.

Active and passive H$^+$ fluxes can antagonize each other or work together. Under resting conditions, most cells experience a background acid loading process, because the electrochemical driving forces favor H$^+$ influx. The normal outside pH is 7.4 whereas pH$_i$ is normally around 7.2. Most cells have membrane potentials more negative than -12 mV, the H$^+$ Nernst equilibrium potential under these conditions. Thus H$^+$ tends to leak into the cell. In addition, cellular metabolism also produces H$^+$. To obtain a steady state pH$_i$, passive inward H$^+$ flux has to be balanced by (active) outward H$^+$ flux. Here the fluxes cancel each other out.

If the same cell now experiences an acute acid load with a pH$_i$ lower than, say, 7.0, and if the acid extruders have been blocked, then a slow increase in pH$_i$ will occur. This indicates background H$^+$ extrusion. This passive H$^+$ efflux may be expected to support active acid extrusion.

It follows from the foregoing that passive H$^+$ fluxes show pH$_i$ dependency. Depending on pH$_i$ the H$^+$ fluxes may change from inward to out-

Fig. 24.6A, B. A Commonly observed pH_i transient upon exposure of a cell to a NH_3/NH_4^+ prepulse and its withdrawal. **B** Commonly observed pH_i transient upon exposure of a cell to a CO_2/HCO_3^- prepulse and its withdrawal

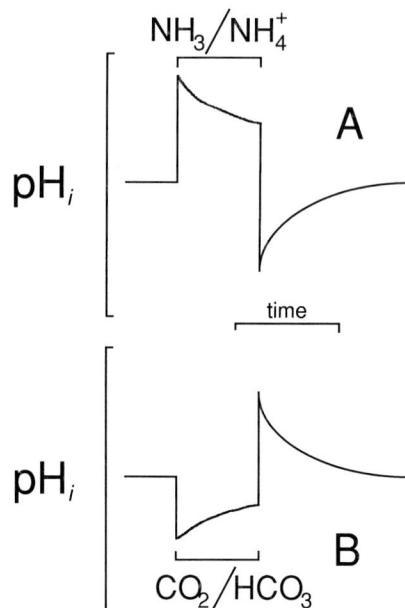

ward. The passive H^+ flux vs pH_i profile defines the magnitude of the flux in terms of moles H^+ transported into or out of the cell per liter cell water per unit time as a function of pH_i.

Acid-base transporters also have their own unique flux vs pH_i profile. The profile shows at which pH_i H^+ flux is minimal and with what sensitivity the transporter responds to acid-base disturbances. Sensitive transporters show a steep pH_i dependence and respond to small deviations of pH_i from setpoint with an enormous increase in transport activity. Insensitive transporters hardly show any change at all, even for large discrepancies between actual pH_i and setpoint pH_i.

24.6
Studying pH_i vs Studying pH_i Regulation

The aforementioned considerations have great bearing on how to design pH_i experiments. Consider the following protocol: steady state pH_i is monitored and the cell is treated with some compound with the aim of evaluating its effect on acid-base transporters. Either pH_i goes up and reaches a new steady state, goes down and reaches a new steady-state, or remains the same. This approach may not be very informative. For if pH_i went up, did the treatment stimulate outward H^+ fluxes or did it inhibit inward H^+ fluxes? If nothing happened, then both inward and outward H^+ fluxes may have been stimulated but to the same extent, or may not have been influenced at all. A better way to design pH_i experiments is to characterize in detail both the active and passive components of outward and inward H^+ fluxes. In other words, aim at constructing the active and passive flux pH_i profiles under control conditions and test conditions. The first step is to perturb pH_i and analyze the recovery of pH_i from the perturbation.

24.7
Perturbing pH_i

Cells can be delivered either a cytoplasmic acid load or alkaline load.

Acid Loading Cells

Acute acid loads are imposed upon the cell by briefly (1.5–3 min) exposing the cells to weak bases. The most popular weak base is NH_3/NH_4^+ (pK_a 8.9). Replacing 20 mM NaCl by 20 mM NH_4Cl yields 0.6 mM NH_3/19.4 mM NH_4^+ in the acid loading solution (pH 7.4).

NH_3/NH_4^+ **prepulse.** The first thing that happens during the so-called NH_3/NH_4^+ prepulse is the rapid nonionic diffusion of highly membrane permeant NH_3 into the cell. NH_3 subsequently associates with cytoplasmic H^+ to form NH_4^+ in order to establish a new NH_3/NH_4^+ equilibrium, but now inside the cell. As H^+ combines with NH_3, the pH_i shoots up within a few seconds (Fig. 24.6A). Quickly $[NH_3]_i$ and $[NH_3]_o$ become equal and remain so during the prepulse. After the rapid initial pH_i rising phase, the plateau phase follows. Now NH_4^+ enters the cell presumably via K^+ ionic channels or other K^+ transporters. Passive ionic NH_4^+ diffusion into the cell is greatly facilitated by the negative membrane potential and a favorable NH_4^+ chemical gradient. A portion of the incoming NH_4^+ is converted into NH_3 and H^+. As a result, pH_i falls during the plateau phase, though slowly. The incoming NH_4^+ ions carry the surplus protons that are going to acidify the cytoplasm upon washout of the NH_3/NH_4^+. When all extracellular NH_3/NH_4^+ is removed, NH_3 rapidly leaves the cell and within a few seconds all intracellular NH_4^+ is converted into NH_3, which also leaves the cell. The H^+ acidifies the cytoplasm such that pH_i falls rapidly. Thus, at the end of the NH_3/NH_4^+ prepulse, the cell is free of NH_3/NH_4^+ and is left with a cytoplasmic acid load. If, however, during the plateau phase pH_i neither increases nor decreases or continues to increase, then pH_i is likely to return to the pre-existing steady state pH_i value or a more alkaline pH_i value, respectively.

Alkaline Loading Cells I

Acute cytoplasmic alkaline loads are imposed by exposing cells to weak acids. The result is the mirror image of the acute acid load. One method is to use a solution that contains a higher than normal CO_2/HCO_3^- concentration: the CO_2/HCO_3^- prepulse.

CO_2/HCO_3^- **prepulse.** Before the prepulse the cell is bathed in a normal 5% CO_2/95% O_2 bubbled solution with $[HCO_3^-] = 22$ mM. Then the cell is switched to a 44 mM HCO_3^- containing solution vigorously bubbled with a 10% CO_2/90% O_2 gas mixture (pH 7.4, 37 °C). CO_2 will rapidly enter the cell and be converted to HCO_3^- and H^+ with a consequent drop in pH_i (Fig. 24.6B). After the rapid initial acidification, the plateau phase follows. In this phase HCO_3^- may enter the cell by means of ionic diffusion if the membrane potential is not extremely negative. However, depending on the cell type, alternative, nonelectrogenic ways may exist for HCO_3^- to access cells. If HCO_3^- indeed enters, a portion dissociates into CO_2 and water. In the process H^+ is consumed which causes pH_i to rise slowly during the plateau phase. The H^+ thus consumed causes the alkalinization of the cytoplasm upon return to the normal 5% CO_2/22 mM HCO_3^--buffered solution. During the washout, $[CO_2]_i$ rapidly halves and within a few seconds all redundant intracellular HCO_3^- is converted into CO_2 (which also leaves the cell) consuming H^+ in the process. Consequently pH_i rises rapidly. At the end of the maneuver the cell has a normal $[CO_2]_i$ and is left with a cytoplasmic alkaline load. Activation of acid extruders during the plateau phase is expected to increase or cause the ensuing alkaline load. If during the plateau phase pH_i remains stable or continues to decrease then pH_i is likely to return to the pre-existing steady state pH_i value or to a more acidic value, respectively.

Alkaline Loading Cells II

An alternative way of alkaline loading cells is to take advantage of Cl^-/HCO_3^- exchange. Most cell preparations are endowed with these ubiquitous transporters. They run backwards if all extracellular Cl^- is replaced

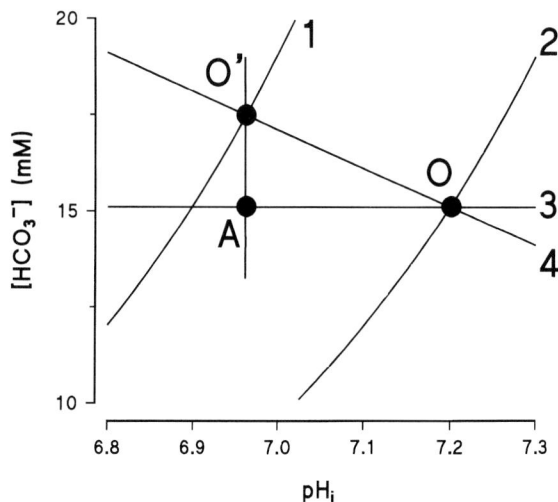

Fig. 24.7. Davenport diagram of a cellular respiratory acidosis. Curve labeled *1* represents the 80 Torr CO_2 isopleth ($[HCO_3^-] = 2.4*10^{(pH-6.1)}$, whereas curve labeled *2* is the 40 Torr CO_2 isopleth. Line *3* represents the $\beta_i = 0$ nonbicarbonate buffer line. The true nonbicarbonate buffer line is indicated with *4* and has a slope of -10 mM. The y-intercept is chosen such that the nonbicarbonate buffer line passes through point *O* on the 40 Torr CO_2 isopleth. Point *O* marks the graphical coordinates corresponding to the initial pH_i value. The intersection of the nonbicarbonate buffer line with the 80 Torr CO_2 isopleth defines the final pH_i value, indicated by *O'*. The vertical distance *O'A* represents the amount of HCO_3^- generated together with an equal amount of hydrogen ions. All of the H^+ protonates the nonbicarbonate buffers

by gluconate anions or some other anions that are not transported by the Cl^-/HCO_3^- exchangers. When running backwards, Cl^-/HCO_3^- exchangers move HCO_3^- into the cell, with a resultant alkaline load.

24.8
pCO₂-Mediated pHᵢ Changes

Raising the pCO_2 of the extracellular solution causes an intracellular acidosis the magnitude of which depends on the initial pH_i value and β_i. We will analyze one example based on the CO_2/HCO_3^- prepulse.

Suppose pH_i was 7.2 before the CO_2/HCO_3^- prepulse and assume that the (mean) β_i in the pH trajectory 7.2 to, say, 6.8 equals 10 mM. When $[CO_2]_o$ doubles $[CO_2]_i$ also doubles. In the complete absence of nonbicarbonate buffers this would yield an acidification of 0.3 units and a negligible 63 nmol increase in intracellular HCO_3^- and the same consumption of CO_2. But additional H^+ are needed to protonate the nonbicarbonate buffers. For this purpose more carbonic acid has to be formed. The H^+ ultimately produced will all bind to the nonbicarbonate buffers whereas the simultaneously generated HCO_3^- remains free in solution. To find the new $[HCO_3^-]_i$ it should be realized that for every millimole of HCO_3^- (and H^+) formed on top of the ~ 15 mmol already present at pH_i 7.2, the pH_i drops 0.1 unit (because $\beta_i = 10$ mM). Thus the following equation has to be solved:

$$15 + x = 0.03*80*10^{(7.2-0.1x)-6.1} \tag{5}$$

in which x represents the unknown amount of HCO_3^- formed. This equation cannot be solved analytically. A few "trial and error" iterations teach us that $x \approx 2.4$ mmol and the pH_i falls from 7.2 to 6.96. Figure 24.7 shows the graphical analysis of this problem. In this example $[HCO_3^-]_i$ settles at ~ 17 mM whereas $[HCO_3^-]_o$ was 44 mM. The HCO_3^- Nernst equilibrium potential equals -25 mV. The membrane potential should be less negative than -25 mV to allow for passive ionic diffusion of HCO_3^- into the cell during the plateau phase of the CO_2/HCO_3^- prepulse.

24.9
Typical Data Records and Their First Analysis

Upon withdrawal of NH_3/NH_4^+ prepulse, the pH_i may fall within a few seconds. Immediately after this drop, pH_i recovers towards, ideally, the steady state pH_i that was observed before application of the prepulse (Fig. 24.6A).

In such experiments you will notice that during the acid load recovery the rate of pH_i change per unit time (dpH_i/dt) changes from its highest value when pH_i is at its lowest, most acidic point to 0 when pH_i reaches its steady state value again (Fig. 24.6A). To quantify this behavior the data have to be analyzed such that we end up with dpH_i/dt as a function of pH_i.

Tip. The easiest way is to fit a 2nd, 3rd or 4th order polynomial function (whatever gives the best result) to the recovering pH_i vs time trace. Subsequently, the polynomial function is differentiated with respect to time, which, for polynomials, is not a difficult thing to do. Next, successive time points are plugged into the polynomial function and its time derivative. The time points are chosen such that they yield a series of pH_i values that are ~ 0.05 units apart. When the exact same time points are entered in the time derivative of the polynomial function, dpH_i/dt as a function of pH_i is found. Of course the same procedure can be followed for recoveries from acute alkaline loads.

24.10
From dpH_i/dt to Actual H^+ Fluxes

H^+ fluxes into or out of the cytoplasm cause pH_i to change. The quantity that links H^+ fluxes to pH_i changes is the cytoplasmic buffering power. Therefore we have to know the nonbicarbonate buffering power.

How to Experimentally Determine β_i

An elegant experimental approach to measure β_i is beautifully described by Boyarsky et al. (1988). It is called the "stepwise reduction in extracellular NH_3/NH_4^+ approach."

For this approach to work, all acid-base transporters of the cell have to be made nonfunctional. This is accomplished by working with CO_2/HCO_3^--free solutions, so as to incapacitate all HCO_3^--dependent transporters. Because in most mammalian cells under these conditions only the Na^+/H^+ exchangers remain active, all extracellular Na^+ has to be removed too. Na^+-free solutions do not affect β_i and render the Na^+/H^+ exchangers ineffective. In fact, Na^+/H^+ exchangers will start to operate in the reverse direction for a while, because of the reversal of the Na^+ gradient across the membrane. Let pH_i reach its new, (much) more acidic steady state value. Then apply a Na^+-free HEPES-buffered solution containing 20 mM NH_4Cl and watch pH_i shoot up reaching a new, alkaline plateau phase. When the plateau pH_i value is reached, switch to a 10 mM NH_4Cl containing Na^+-free solution and see pH_i settle at a less alkaline, more acidic plateau. When this pH_i value is reached switch to a 5 mM NH_4Cl containing Na^+-free solution and so on and so forth. In a typical experiment the cell is exposed to 20, 10, 5, 2, 1 and 0.5 and 0 mM NH_4Cl, and a series of six corresponding pH_i plateau values are observed. It is a prerequisite that there may not be too much NH_4^+ entry during the whole maneuver, ideally none at all. One way to find out is to compare pH_i before and after the maneuver. Both values should be equal or almost so.

For each of the six NH_4Cl containing solutions the dissolved $[NH_3]_o$ can be calculated because all solutions had a pH of 7.4. At each of the six observed plateau pH_i values we know $[NH_3]_i$ inasmuch as it equates with $[NH_3]_o$. If it is assumed that the pK_a governing the NH_3/NH_4^+ equilibrium is also 8.9 in the ionic environment of the cytoplasm, then $[NH_4^+]_i$ can be calculated at the six observed plateau pH_i values. The amount of H^+ delivered to the cytoplasm, dH^+, upon decrease of $[NH_3]_i$ $(= [NH_3]_o)$ is calculated by subtracting the two corresponding $[NH_4^+]_i$. The difference must have left the cell as NH_3, thus liberating an equal amount of H^+. The resultant change in pH, dpH, is obtained by simply subtracting the two observed plateau pH_i values. The ratio dH^+/dpH (more accurate: $d[NH_4^+]_i/dpH_i$) is β_i and is assigned to the mean of the two successive plateau pH_i values. One successful experiment yields six pH_i, β_i data points. From more experiments a pH_i vs β_i relationship can be constructed. It may be found, for example, that β_i decreases curvilinearly from 25 mM at pH_i 6.4 to 6.0 mM at pH_i 7.4. A polynomial function is also fit to these data points in order to obtain an empirical mathematical description of the data.

Conversion of dpH_i/dt to Actual H^+ Fluxes

The final step is to convert dpH_i/dt values to actual H^+ fluxes. This is done by multiplying dpH_i/dt with β ($= dH^+/dpH$) as to yield dH^+/dt. When the recovery from the acid-base disturbance was obtained in a CO_2/HCO_3^- free solution then the pH_i vs β_i relationship is used to obtain the β_i values corresponding to the pH_i values for which the dpH_i/dt values are known. When the recovery from the acid-base disturbance was recorded in CO_2/HCO_3^- buffered solutions, then the pH_i vs β_{tot} relationship has to be used.

dH^+/dt represents the H^+ flux per unit time for example in moles H^+ per liter cell water per second. If the surface to volume relationship of the cell is known, then the flux in moles per second per square cm of membrane can be calculated.

24.11
Investigation of Active and Passive H^+ Fluxes

When pH_i is perturbed and recovery from the cytoplasmic acid or alkaline load is recorded, two questions have to be answered: (1) which acid-base transporters contribute to the recovery and (2) what is the contribution of the passive H^+ fluxes to the recovery. The first question can be addressed by a clever choice of solutions. The first and the second problem can be solved by the use of drugs that block all active acid-base transport.

Tip. In the initial stage of your research things should be kept simple. Make sure that your solutions are nominally free of CO_2/HCO_3^- to render the HCO_3^--dependent transporters incapable of changing pH_i. Among these are all types of acid loaders. Now the HCO_3^--independent acid extrusion can be studied in your cell, most notably Na^+/H^+ exchange. Typically HEPES-buffered solutions are used with a pH of 7.4. When both the HCO_3^--independent acid extruders and β_i have been extensively defined, it is time to introduce CO_2/HCO_3^- into your solutions.

24.12
Working with HEPES-Buffered Solutions

Procedure

Assay for HCO_3^--Independent Acid Extruders

Amiloride. A convenient way to assay for HCO_3^--independent acid extruders is to analyze the cellular response to a NH_3/NH_4^+ prepulse. If the cell recovers from the acid load in a normal, 140 mM Na^+-containing solution, then a Na^+-free solution is used to wash out the prepulse. If only Na^+/H^+ exchangers are present then the acid load recovery should be completely absent and commence only once the extracellular Na^+ is given back. Final proof for the presence of Na^+/H^+ exchangers comes when you wash out the NH_3/NH_4^+ prepulse in a 1 mM amiloride-containing, normal, 140 mM Na^+-containing solution and observe a substantially slowed down or absent acid load recovery.

Tip. It takes a while for amiloride powder to go into solution. Stirring the solution while mildly heating it to $\sim 40\,°C$ or so should work.

Tip. In addition, 50 µM of the potent amiloride analogs such as ethyl isopropyl amiloride (EIPA) or hexamethylene amiloride (HMA) completely block all isoforms of the Na^+/H^+ exchangers. These isoforms differ with respect to the 50 % inhibitory concentration (IC_{50}) of amiloride and its analogs. The analogs are prepared as 50 mM stocks in DMSO and go readily into solution.

Bafilomycin A1 and NBD-Cl. When you observe an acid load recovery in Na^+-free or amiloride-containing solutions, your cell may have H^+ pumps in its plasma membrane. Vacuolar type H^+ pumps, the most common type, are blocked by $\sim 10\,\mu M$ bafilomycin A1 or the much cheaper drug 7-chloro-4-nitrobenz-2-oxa-1,3-diazol (NBD-Cl) at a concentration of $\sim 100\,\mu M$. NBD-Cl is prepared as a 100 mM stock in DMSO. You also might try energy depleting the cells with cyanide or iodoacetic acid.

Omeprazole. K^+/H^+ pumps are blocked by omeprazole. H^+ pumps and Na^+/H^+ exchangers may exist simultaneously.

Caution. There is an outside chance that even with HEPES-buffered solutions, Na^+-dependent Cl^-/HCO_3^- exchange may contribute to the acid load recovery. In osteoclasts we observed alkalinizations with HEPES-buffered, Cl^--free solutions. This indicates that the Cl^-/HCO_3^- exchangers of these cells are able to operate with trivial amounts of CO_2/HCO_3^- ($\sim 6.5\,\mu M/\sim 145\,\mu M$, pH 7.4) present in solutions equilibrated with room air. Room air contains 0.03 % CO_2. Indeed, when HEPES-buffered 0-Cl^- (chloride-free) solutions were vigorously bubbled (>100 min) with 100 % N_2 or 100 % O_2 in order to remove all dissolved CO_2/HCO_3^-, the 0-Cl^--induced alkalinizations were largely absent.

24.13
Working with CO_2/HCO_3^--Buffered Solutions

Assay for Cl^-/HCO_3^- Exchange

DIDS. Recovery from an alkaline load should proceed normally in the absence of extracellular Na^+ but should not occur in the absence of extracellular Cl^- and be blocked by 4,4'-diisothiocyanato-stilbene-2,2'-disulfonic acid (DIDS). At $500\,\mu M$ DIDS knocks out most of the Cl^-/HCO_3^- exchange isoforms known. These isoforms differ with respect to the IC_{50} of DIDS. DIDS powder goes readily into solution.

Caution. DIDS should be dissolved a few minutes before applying it to the cell, as it degrades rapidly. A DIDS containing solution should be protected from bright light. After 15 min or so, DIDS binding to the transporters becomes irreversible. As the DIDS reacts covalently with free amines, one should expect trouble when it is used together with NMDG, a popular organic substitute for Na^+ (see below).

SITS. SITS (4-acetamido-4'-isothiocyanato-stilbene-2,2'-disulfonic acid), another inhibitor of anion exchange, fluoresces which makes its use limited in quantitative fluorescence experiments.

Alternatively, removal of all extracellular Cl^-, leads to a reversible alkalinization if Cl^-/HCO_3^- exchangers are present. The 0-Cl^--induced alkalinizations must also occur in Na^+-free solutions and be blocked by DIDS. Giving Cl^- back allows the cell to recover from the alkaline insult.

Assay for Na^+-Dependent Cl^-/HCO_3^- Exchange

This potent acid extruder must be considered if upon switching from HEPES to $CO_2/$ HCO_3^--buffered, normal solutions a substantial alkalinization is observed after the initial acidification. Gathering proof for the presence of Na^+-dependent Cl^-/HCO_3^- exchange could include the following solution changes. Wash out a HEPES-buffered NH_3/NH_4^+ pre-pulse solution into a Na^+-free HEPES-buffered solution. The cell attains an acidic pH_i from which it cannot recover if H^+ pumps are absent. Then switch to a Na^+-free CO_2/HCO_3^--buffered solution and observe a (small) further acidification. Next, switch to a Na^+-free CO_2/HCO_3^--buffered solution to which 1 mM amiloride has been added. This dose of amiloride should block all of the Na^+/H^+ exchange. Finally, switch to a full Na^+-containing, CO_2/HCO_3^--buffered solution also containing 1 mM amiloride. The pH_i recovery observed is most likely due to Na^+-dependent Cl^-/HCO_3^- exchange. The recovery should be blocked by $10-100\,\mu M$ DIDS and not proceed in the absence of extracellular Na^+. Furthermore, if the cell is depleted of intracellular Cl^- then Na^+-dependent Cl^-/HCO_3^- exchange should also be inhibited. Depleting cells of intracellular Cl^- is difficult and may be approached by pretreating the cells with Cl^--free solutions.

Assay for Na^+-HCO_3^- Cotransport

Na^+-HCO_3^- cotransport may turn out to be elusive. All of the isoforms act as acid loaders when extracellular Na^+ is removed. So the first step in the assay is to remove all extracellular Na^+ using, of course, a CO_2/HCO_3^--buffered solution. An acidification that is blocked by $\sim 500\,\mu M$ DIDS is a good argument in favor of the presence of Na^+-HCO_3^- cotransport. The 1:2 and 1:3 isoforms are electrogenic. A DIDS blockable,

0-Na$^+$-induced depolarization would be further proof for Na$^+$-HCO$_3^-$ co-transport. Furthermore, recovery from an alkali load that proceeds in the absence of extracellular Cl$^-$ and is accelerated by the removal of extracellular Na$^+$ also indicates 1:3(1:2?) Na$^+$-HCO$_3^-$ cotransport. Finally, Na$^+$-HCO$_3^-$-dependent recovery from acid loads that proceeds in Cl$^-$-depleted cells (Na$^+$/H$^+$ exchange blocked by amiloride) would argue in favor of 1:1 Na$^+$-HCO$_3^-$ cotransport.

24.14
Quantifying the Passive H$^+$ Fluxes That Contribute to pH$_i$ Recovery

Passive fluxes also contribute to recovery from pH$_i$ perturbations. Thus to arrive at the active H$^+$ fluxes the passive components have to be measured. This is achieved simply by taking advantage of drugs that block the transporters such as amiloride (analogs), DIDS and NBD-Cl.

Passive H$^+$ Fluxes in the Absence of CO$_2$/HCO$_3^-$

To establish the pH$_i$ dependence of the passive fluxes the cell is acid loaded and the pH$_i$ recovery in the presence of 1 mM amiloride is observed. This is done by washing out the NH$_3$/NH$_4^+$ prepulse in a normal, 140 mM Na$^+$ containing, solution to which amiloride is added. After pH$_i$ has fallen to acidic values, it may start to recover albeit at a very slow rate. By varying the amount of NH$_3$/NH$_4^+$ of the prepulse solutions, or the duration of the plateau phase, larger or smaller acid loads are produced and data points of these amiloride-insensitive H$^+$ fluxes can be collected over a wide range of pH$_i$ values. Finally, the cell is treated with 1 mM amiloride without having pH$_i$ perturbed first. In that case, it may be noticed that pH$_i$ starts falling at a slow rate. This indicates that, at steady state pH$_i$, Na$^+$/H$^+$ exchange exactly balances a background acid loading process that has just been unmasked. In general, at acidic pH$_i$ values background acid extrusion prevails, whereas at resting pH$_i$ values background acid loading is predominant. With some luck there is a linear relationship between the background H$^+$ fluxes and pH$_i$. After characterization of the background H$^+$ fluxes, or better the amiloride-insensitive H$^+$ fluxes, these can be subtracted from the raw outward H$^+$ flux vs pH$_i$ relationship to finally arrive at the amiloride-sensitive H$^+$ flux. This yields the closest approximation of Na$^+$/H$^+$exchange-mediated H$^+$ flux vs pH$_i$ profile.

Caution. Nigericin is an antibiotic often used to calibrate fluorescent pH_i sensitive dyes after conclusion of an experiment. The drug acts as an artificial K^+/H^+ exchanger that positions itself in the plasma membrane. Its usefulness lies in the fact that when the $[K^+]_o$ and $[K^+]_i$ are equal, $[H^+]_i$ and $[H^+]_o$ also equalize. With normal K^+ and H^+ gradients, however, nigericin will cause an acidification of the cytoplasm. Nigericin is prone to stick to the tubing that delivers the solutions to the perfusion chamber. It is commendable to flush that tubing with ethanol so as to remove all nigericin before it is used for the next experiment. This is to avoid unwanted exposure of the fresh cells to nigericin, which would cause or augment substantial background acid loadingprocesses.

Passive H^+ Fluxes in the Presence of CO_2/HCO_3^-

The passive H^+ fluxes that contribute to pH_i recovery in the presence of CO_2/HCO_3^- can be assessed by treating the cells with $500-1000\,\mu M$ DIDS after alkali loading. By varying the duration of the CO_2/HCO_3^- prepulse, DIDS-resistant flux at a wide range of pH_i values can be determined. After constructing the passive H^+ flux vs pH_i profile this is used to correct the raw recovery fluxes.

24.15
How To Prepare Na^+-Free and Cl^--Free Solutions

Standard HEPES-Buffered Solution

Clean experiments hinge on well-prepared solutions. pH_i experiments on mammalian cells should always be carried out at $37\,°C$. Therefore, the solutions should be titrated to 7.4 also at $37\,°C$. A standard HEPES-buffered mammalian solution may contain 10 mM glucose, 32 mM HEPES, 130 mM NaCl, 5 mM KCl, 1 mM $CaCl_2$, 1.4 mM $MgSO_4$ and 2 mM H_3PO_2 titrated to 7.4 with ~ 12.8 mM NaOH. In the CO_2/HCO_3^--buffered variant of this solution, the 32 mM HEPES is replaced by 22 mM $NaHCO_3$. Like HEPES-buffered solutions CO_2/HCO_3^--buffered solutions can be stored in a refrigerator for well over 5 days without being bubbled or any other special precautions. Before use they just have to be warmed to $37\,°C$ and vigorously bubbled with a 5 % CO_2 gas mixture before they are delivered to the cell.

How To Prepare a Na$^+$-Free HEPES-Buffered Solution

Na$^+$ is best replaced by N-methyl-d-glucamine (NMDG) or choline. K$^+$ depolarizes the membrane potential of the cell and may activate H$^+$ channels. Some cell types may have choline transporters, so NMDG is the safest choice. In the HEPES-containing solution, 145 mM of the free base NMDG has to be titrated with HCl to achieve pH 7.4.

How To Prepare a Na$^+$-Free CO$_2$/HCO$_3^-$-Buffered Solution

The NMDG-HCO$_3$ salt is not for sale (whereas choline-HCO$_3$ is) but it can be prepared in the laboratory without too much trouble. Replace all the NaCl of the standard solution with 123 mM NMDG and titrate with HCl to 7.4 (37 °C). In the absence of HEPES you run the risk of over-shooting this value so be careful. After setting the pH to 7.4 add enough water to achieve the final volume. Next, add 22 mM NMDG powder and watch the pH of the solution shoot up again. Then bubble this alkaline solution with a 20 %CO$_2$/80 % O$_2$ mixture while at the same time monitoring its pH. Bubbling with 20 % CO$_2$ causes the pH to go down in a couple of minutes. (It takes forever if you bubble with only 5 %CO$_2$). When the pH reaches 7.4 bubble with 5 % CO$_2$; now the right amount of HCO$_3^-$ (22 mM) is present. Note: the whole procedure has to be carried out at 37 °C!

How To Prepare a Cl$^-$-Free HEPES-Buffered Solution

Cl$^-$ is best replaced by gluconate, the conjugated weak base of gluconic acid. Buy the various gluconate salts and d-gluconic acid lactone. Dissolved in water the lactone is hydrolyzed within 2–3 h to gluconic acid. In the HEPES-containing solution, $\sim$ 132 mM gluconic acid has to be titrated with $\sim$ 142.8 mM NaOH, 5 mM KOH and 1.0 mM Ca(OH)$_2$ to achieve a pH of 7.4 (37 °C). Alternatively K-gluconate, Ca-gluconate (hemisalt!) and Na-gluconate may be powdered in. Gluconate ions complex Ca^{2+} ions. To compensate for this, it is commendable to quadruple the Ca^{2+} content of the solution.

How To Prepare a Cl⁻-Free CO₂/HCO₃⁻-Buffered Solution

The 132 mM gluconic acid-containing solution, with K-gluconate, Ca-gluconate powdered in, is titrated to 7.4 with NaOH. The solution is brought to a final volume. Then 22 mM NaHCO₃ is added whereupon pH shoots up. If this solution is bubbled with 5 % CO_2 the pH comes down to 7.4 and remains there. The whole procedure has to be carried out at 37 °C.

How To Prepare a Na⁺-Free Cl⁻-Free HEPES-Buffered Solution

One way is to add 130 mM NMDG powder to 130 mM *d*-gluconic acid lactone powder and stir it for 3 h. Add K-gluconate, Ca-gluconate, HEPES and the remaining components and titrate with a 1 M NMDG-OH solution to 7.4 (37 °C).

How To Prepare a Na⁺-Free Cl⁻-Free CO₂/HCO₃⁻-Buffered Solution

Add 130 mM NMDG powder to 130 mM *d*-gluconic acid lactone powder and stir it for 3 h. Add K-gluconate, Ca-gluconate and the remaining components but not HEPES and titrate with a NMDG-OH solution to 7.4. Bring to final volume. Next, add 22 mM NMDG powder and watch the pH of the solution shoot up. Then bubble this alkaline solution with 20 % CO_2 while at the same time monitoring its pH. The pH will go down. When it reaches 7.4 bubble with 5 % CO_2; now the right amount of HCO_3^- (22 mM) is present. The whole procedure has to be carried out at 37 °C.

References

In the following papers the principals outlined in this chapter were used to design the experiments and also guided the analysis thereof.

Boyarsky G, Ganz MB, Sterzel RB, Boron WF (1988) pH regulation in single glomerular mesangial cells. I. Acid extrusion in the absence and presence of HCO_3^-. Am J Physiol 255:C844-C856

Ravesloot JH, Eisen T, Baron R, Boron WF (1995) Role of Na-H exchange and vacuolar H pumps in intracellular pH regulation in neonatal rat osteoclasts. J Gen Physiol 105:177–208

Wilding TJ, Cheng B, Roos A (1992) pH regulation in adult rat carotid body glomus cells. Importance of extracellular pH, sodium and potassium. J Gen Physiol 100:593–608

Fluorescence To Measure Intracellular Ions

Flow Cytometer Measurements

Chapter 25

An Introduction to the Working Principles of the Flow Cytometer

Sandor Damjanovich, Carlo Pieri, Laszlo Bene, Attila Jenei, and Rezsö Gáspár, Jr.

Background

Flow cytometry is a very fast optical (spectroscopic) analysis of cells on a cell-by-cell basis. The measurement, data acquisition and analysis were made possible by an innovative application of optical spectroscopy and high tech electronics, used originally in nuclear physics. A unique characteristic of the method is the analysis of individual cells at such a high speed that biological variations and their distribution over a conveniently large cell population are accessible to statistically sound analysis within a time scale of minutes among practically physiological conditions of the cells (Shapiro 1985; Ormerod 1994).

Figure 25.1A shows the scheme of the operation of the instrument. The cell suspension is forced into a fluid stream which passes an orifice. A sheath (buffer) fluid – having a slightly higher pressure than that of the sample fluid – directs the latter into the orifice, or sample head, in a concentric way. This hydrodynamic trick of Dr.Fulwyler prevents mechanical injury of the cells when they pass the nozzle (orifice). The cells can thus be aligned into a single file and studied separately (Fig. 25.1B). Nevertheless, the speed of the cells which pass the spot of the optical analysis is so high that several thousands of cells can be analyzed each minute. The actual staying time of a cell in the volume that is illuminated by a strong (in many cases, but not necessarily laser-) light-beam is not more than a few microseconds. The cell passing through the line(s) of the analysis can be discarded or collected after an individual analysis. Due to the fast multichannel analyzers, which can recognize fluorescence, or light scatter signals derived from native, or pretreated, cells, a particularly valuable way of collecting the cells is cell sorting, which is based upon the analyzed individual optical characteristics. This fact gave a name and an acronym to the method: fluorescence activated cell sorting, or briefly, FACS.

Working principle

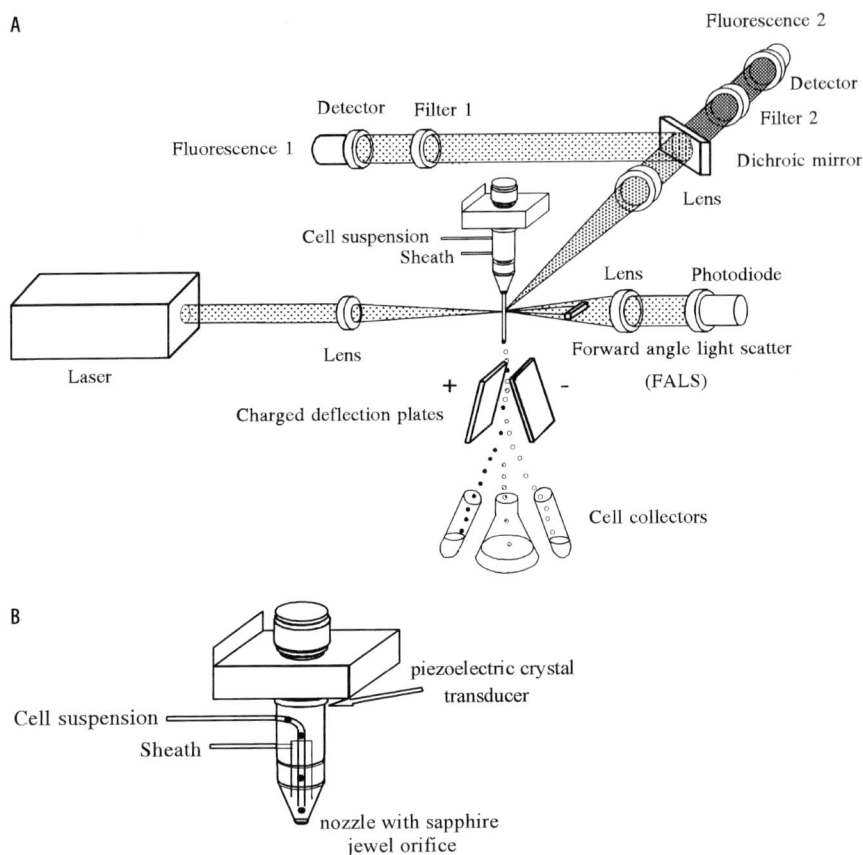

Fig. 25.1A, B. A Scheme of operation of flow cytometer and cell sorter. **B** Sample head of flow cytometer

Basic Elements of the Cell Sorter and Analyzer

The basic elements of the cell sorter and analyzer as shown in Fig. 25.1 are discussed briefly below:

- Sample holder (sample tube or chamber). This unit accommodates cell suspensions in physiological (buffer) solutions. The suspension is moved into the "flow chamber" by an indifferent gas (air, nitrogen) pressure.
- Flow chamber. In this chamber the sample line and the sheath fluid line meet. They behave according to the Fulwyler principle, i.e., there is a pressure difference in favor of the sheath fluid.

– Sheath fluid container: This unit contains the properly selected sheath fluid, generally a properly chosen buffer, which is introduced to the flow chamber by gas pressure.
– Ultrasound generator, cell sorting devices. An ultrasound generator is connected to and can "shake" the nozzle of the instrument where the cells come out in single file. This procedure (at a vibration frequency around 40 kHz) generates droplets of the sheath fluid, which can contain generally not more than a single cell. An electric unit can charge the droplets to positive or negative charges, based upon simultaneous analysis data (e.g., red or green, little or large, or combinations of such parameters) which are available by the time the droplet is separated from the sheath fluid line. These droplets can be deflected individually by an electrode pair (in an electric field of 5000 V) according to the sign of their charge into separate test tubes.

Cell Sorting

This method provides an excellent tool to separate cells based upon their size, structure or a distinct fluorescence property, labeling of chemical components or physiological parameters of the cell membrane or the cell interior.

The easiest way to handle this unit briefly is to consider it as a black box which has the unique capacity to differentiate between optical signals. **Electronic units**

For those who are more interested, a brief summary can be given as follows: The optical signals evoked by the interaction of the cells and the light beam trigger photomultipliers, or photodiodes. The original signals are amplified by the sensors and processed by a multichannel analyzer. The latter device can store size-related data in different channels. This procedure generates a quantitative scale of the acquired data. Since data are taken on a cell-by-cell basis, the optical signals generated during the interaction between the (laser-) light-beam and the cells are stored by the electronics for each cell separately. However, identical quantitative data of distinct cells are collected into the same channel of the multi-channel analyzer. Taken together, the "histogram" will indicate a numerical value of the optical signal (the serial number of a channel which collected a particular size of signals) and the number of the cells having an identical optical signal in a two- , or in some cases when two optical signals are taken in correlatedly, in a three-dimensional fashion. The numerical values of the signals from a single cell can also be represented by a dot in a two-dimensional coordinate system. Here the x-axis indicates the size of one signal. The y-axis is the size of another signal obtained simultaneously from the same cell (e.g., forward-angle light-scatter (FALS) vs a fluorescence signal, the former representing the size and the latter a particular parameter, like DNA content, or the number of a receptor at the cell surface). Since, the dots can frequently be found in distinct accumulation geometries due to different cell populations, these representations are called dot plots.

Dot Plots

By analyzing a given cell number at a time and subsequently repeating the analysis several times, a time dependence of the development of a particular cell population (e.g., the number of activated cells) can also be followed in a three-dimensional geometric formation (for histograms and dot plots see Fig. 25.2A, B).

As we have seen above, signals can come from the scattered light, the native (auto-) fluorescence of the cells, or artificially attached fluorescence dyes. An interesting new possibility is the generation of the fluorescence compound by the illumination itself. The possibility arises from the so-called green fluorescence protein (GFP), when a photo-effect itself generates a transition of the amino acids of a particular peptide to form a fluorescence compound.

The FALS indicates mostly cell size and viability. The scattered light is collected at $5°$ to the direction of the light-beam.

The $90°$ scatter signal indicates structuredness of the intracellular domain (nuclei, compartments, etc.).

Fluorescence signals can indicate the presence or absence of cell surface and intracytosol components and receptors. Their mobility and proximity can be further studied by applying anisotropy measurements or fluorescence energy transfer (for more details see Damjanovich et al. 1994).

Qualitative and quantitative information can be gained for any parameter that can specifically be labeled by fluorescence dyes: DNA content, ploidy of cells, enzyme activity, lipid content, the ratio of resting and dividing cells, Ca^{2+} and other ions, pH, membrane potential, etc.

The rather typical form of the histograms in Fig. 25.2A demonstrate membrane potential-induced distributions of an anionic (bis-oxonol) and a cationic (carbocyanine) dye. The dot plot shown in Fig. 25.2B represents the distribution of two optical parameters relative to each other.

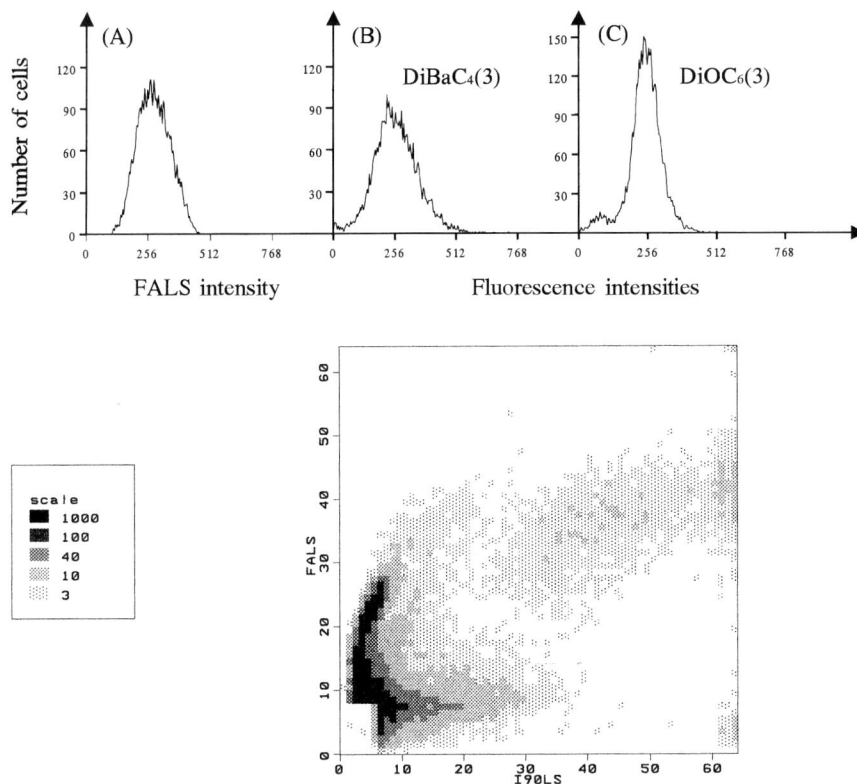

Fig. 25.2A, B. **A** Histograms showing a way to present data obtained with flow cytometric measurements. Forward angle light scatter (FALS) intensity, as well as fluorescence intensities of two membrane potential sensitive fluorescent dyes (DiBaC$_4$(3) and DiOC$_6$(3) are shown). **B** Dot plot showing FALS as a function of the 90° scatter signal

These drawings demonstrate some of the principles involved in visualizing flow cytometric data.

References

Damjanovich S, Edidin M, Szöllösi J, Trón L (1994) Mobility and proximity in biological membranes. CRC, Pearl River, NY
Ormerod MG (1994) Flow cytometry: A practical approach. 2nd edn. IRL, Oxford
Shapiro HM (1985) Practical flow cytometry. Alan R. Liss, New York.

Chapter 26

Flow Cytometric Membrane Potential Measurements

Sandor Damjanovich, Carlo Pieri, Laszlo Bene, Attila Jenei, and Rezsö Gáspár, Jr.

Background

Before deciding to use a certain technique to measure membrane potentials, a critical comparison of the available techniques is required. Therefore, we will first briefly describe three different ways to measure membrane potential.

Measuring Techniques of Membrane Potential

The three major approaches described below can be used to study membrane potential differences in all cells, regardless of their origin, and including even plant cells (mostly in the form of protoplasts).

- Electrophysiological Techniques
  Electrophysiological techniques such as voltage-clamp or patch-clamp involve the application of micro-electrodes or patch pipettes to single cells in order to directly measure membrane potential and ion currents. Voltage-clamp was developed by Hodgkin and Huxley in the early 1950s and was used for studying excitable nerve and muscle cells which were large enough to endure the introduction of several electrodes into the cell interior simultaneously. For developing the patch-clamp, around 1976, Neher and Sakmann were awarded the Nobel Prize in Physiology and Medicine in 1991. Their discovery led to a better understanding of a number of widespread diseases, e.g., diabetes mellitus, cystic fibrosis and several other relatively rare but lethal diseases (Neher and Sakmann 1976; Neher 1992; Sakmann 1992).
  The accuracy and resolution power of elegant electrophysiological techniques such as patch-clamp or whole cell clamp methods cannot be denied, as they enable us to investigate the functional states of even a single ion channel in plasma membranes. However, due to the

intrinsic constraints of the technique, which allow the measurement of only a few cells per day, parameters statistically distributed over a larger cell population or subpopulations can be investigated only with great difficulty, if at all. Thus, (near-) absolute values in potential or ion current changes can easily be studied by electrophysiological methods (Neher and Sakmann 1976; Neher 1992; Sakmann 1992). However, the biological variance among individual cells or between distinct subpopulations and the inability to extend the findings to a larger cell population present serious handicaps. Nevertheless, electrophysiological confirmation of data obtained with either or both of the other techniques is always recommended.

- Isotope Techniques

  These involve determination of the partition of organic/inorganic cations or anions, using radioactive isotopes, as a measure of potential and current changes across membranes. $^{42}K^+$, $^{24}Na^+$, but also radioactive isotopes of Ca^{2+}, Mg^{2+} or the organic tritiated compound tetraphenyl phosphonium are frequently used to study the distribution of these ions between the extracellular and intracellular compartments. While routine measuremente of membrane potential is far less convenient with this technique than with almost any other method, the accuracy of the measurements cannot be denied (Sagi-Eisenberg and Pecht 1983).

  The isotope method is among the most reliable and sensitive techniques for use in cell suspensions. However, in this case the accuracy of the measurements is limited due to false signals from debris, dead cells, and other inequalities originating from the ever present heterogeneities among cell populations.

  Nowadays isotope techniques are used less and less often as other, more convenient and faster, methods become available. However in some specific cases, e.g., when increased sensitivity is mandatory, the isotopes can be extremely useful.

- Fluorescence Techniques

  Positively (carbocyanine based cations) and negatively (barbituric acid derivatives, i.e., oxonol based anions) charged fluorescent dyes have been introduced to estimate membrane potential, either in cell suspensions or on a cell-by-cell basis, using flow cytometry. The partition of the fluorescent molecules allows calculation of the membrane potential and its changes, after the necessary calibration (Damjanovich et al. 1986, 1987; Mátyus et al. 1986; Pieri et al. 1992; Waggoner et al. 1977; Waggoner 1979 1986; Wilson et al. 1985; Wilson and Chused

1985). The so-called fast dyes, which change their spectral characteristics in an electrical field, will not be discussed in detail here.

Although fluorescent dyes (if we do not consider fluorescence correlation spectroscopy, which has the capacity to detect even single fluorescent molecules; Eigen and Rigler 1994; Rigler 1995) are generally less sensitive than radioactive isotopes, fluorescence measurements have several advantages. The sensitivity can be enhanced to a very high degree by using modern optical and electronic amplifiers. Moreover, in contrast to classical spectrofluorimetric methods, flow cytometry allows data acquisition and analysis at a single cell level, without losing the capacity to extend the measurements over a significantly large cell population. Artifacts from cell debris or nonviable cells can easily be eliminated and differences among subpopulations can be observed directly and immediately parallel to data collection.

Outline

Measuring the membrane potential of a particular cell sample is deceivingly simple.

Short protocol
- Prepare cells.
- Wash or transfer them into the desired buffer.
- Add the selected membrane potential sensing dye in a concentration and solvent which is the least harmful to the parameter to be measured.
- Run the cells through the flow cytometer, applying the right excitation wavelength and the proper emission filters.
- The collected data can be evaluated in a number of ways, depending on what is required. For example, the size or the structure (or any other fluorescence parameter) of the cells could serve as the basis for selecting cells in the population to be investigated. If time-dependent changes following a particular treatment of the cells are to be evaluated, then a meaningful number of cells (5000–10 000) must be collected. The time frame must be in the order of minutes, as data acquisition from the necessary cell population demands generally between 10 and 30 s.

Materials and Procedure

Avoiding Artifacts

In simple cases, the cells are prepared and the properly selected fluorescence dyes are added. The samples are incubated for a given time and then passed through a flow cytometer collecting fluorescence intensity data. The fluorescence intensity distribution of the cell population will reflect distribution of the membrane potential values as well. However, measurement of membrane potential of a cell population is rarely this simple.

Freshly prepared **live cells** are mandatory to avoid artifacts. With less tedious preparation methods, the physiological (native) condition of the cells is more likely to be maintained. The membrane potential itself is a very sensitive indicator of the viability of cells. However, this sensitivity is also a possible source of inconvenient side effects, if one does not pay attention to fine details during the measurements.

Cells

Using fluorescent dyes to measure membrane potential demands the selection of the right concentrations and the proper solvent. The dyes in a number of cases can be dissolved only in organic solvents. Testing the identical solvent concentration on the live cells is an important step. Since membrane potential is one of the most sensitive parameters monitoring cellular viability, it is necessary to test different solvent concentrations on the cells and then measure their membrane potential with a fluorescent dye relevant to the solvent. Generally the lowest possible solvent concentrations are the least perturbing.

Organic solvents

Frequently, chemical agents – drugs, ionophores, toxins, peptides, etc. – must be applied simultaneously with the potentiometric dyes. Thus a possible (and unsolicited) interaction between these components must be excluded before the actual potential measurement. If data are not available in the literature, one reasonably safe approach is to use a spectrofluorimeter to test the spectroscopic parameters of the dyes under similar experimental conditions and in the presence of the drugs to be administered. The lack of spectral shifts or quantum yield changes are good indicators that the applied agents will not falsify the data through chemical interactions.

Application of chemical agents

The range of membrane potential changes is rarely linear in the full scale between the normal potential and the total depolarization value. This

Lack of linearity

simply means that a dye may not indicate changes in the potential field close to the full polarization value and/or the full depolarization state. Thus, concentration dependence and potential changes must be individually determined and calibrated, even if data are available in the literature. Seemingly negligible changes in a protocol may distort the data.

Selection of the Cells

Any type of cells can be used that will not aggregate during the measurement or that do not have a strong propensity to adhere to a surface during their growth. In this latter case the manipulation needed to remove cells from a surface (trypsinization or scraping with a rubber policeman) may compromise the membrane potential. The cells should be manipulated under conditions close to those required for their growth and the native environment should be maintained throughout the measurement.

Generally, one needs no more than one million cells for a set of measurements, collecting about 5000–10 000 cells per histogram.

Selection of the Dyes

Potentiometric dyes can signal changes in the membrane potential by different mechanisms and at very different time scales. There are fast and slow dyes, which can serve as potential indicators in microsecond and millisecond time scales, respectively. It is obvious that these basic properties determine the limits to their application. It should also be kept in mind that the same dye may have different response times in different systems.

Fast and slow dyes Fast fluorescence dyes change their quantum efficiency, or shift their absorption and emission spectra, under the influence of a changing electric field.

Slow dyes generally do not change their spectroscopic parameters, but partition across or inside the membrane, which separates two compartments with different electric potentials.

Those who have never worked with potentiometric dyes to measure the membrane potential of a particular cell type are advised to consult the recent literature. A published sound calibration curve can be a great help.

The most frequently applied carbocyanine compounds, e.g., $DiOC_6(3)$, may respond slowly, by translocation to the negatively charged compartment, when the equilibrium of diffusion potentials is measured in erythrocytes. However, the optical properties of these dyes may be orders of magnitude faster when responding to millisecond duration voltage pulses in planar lipid bilayers.

Cationic dyes

Note. The cationic dyes may measure the mitochondrial membrane potential together with that of the plasma membrane.

Cationic dyes signal an increase in the membrane potential by increasing fluorescence in a FACS machine.

Anionic dyes, which are generally excluded from the negative interior of fully polarized cells, increase their fluorescence intensity parallel with depolarization.

Anionic dyes

Note. Pratap et al. (1990) gave a thorough analysis of optical potentiometric indicators, which can monitor the transmembrane electrical potential.

DiBaC$_4$(3)

Currently, $DiBaC_4(3)$ (bis-(1,3-dibutylbarbituric acid thrimethine oxonol), or, briefly, oxonol or bis-oxonol, is the most frequently applied fluorescent dye due to the advantages it offers.

Advantages

– Negatively charged, it stays outside of normally polarized cells, making it less toxic than other dyes that penetrate the cell interior.
– The spectral properties, namely the excitability and the emission characteristics, which are very near to those of fluorescein (excitation at 488 nm and emission around 540 nm), provide a comfortable range of measurement in many optical devices, particularly laser operated flow cytometers.
– The high sensitivity, 10–150 nM, helps to detect signals which monitor potential changes. Naturally, $DiBaC_4(3)$ is partitioned not only by electromotive forces, but also based on solubility and concentration. However, very low concentrations in a cell population will not translocate upon membrane potential changes.

A careful and correct application of this oxonol, especially in combination with ion-selective pharmacological agents which have been tested for possible chemical incompatibility of the dye, gives a relatively easy and quick orientational test for detecting membrane potential changes in most cell types in which electrophysiology (patch-clamp) alone may not give the desired information on the cell population.

Disadvantages

– The incompatibility of $DiBaC_4(3)$ with important K^+ ionophores such as valinomycin makes its calibration somewhat inconvenient. However, Wilson and Chused elegantly circumvented this problem by using the Ca^{2+} ionophore A23187 as a tool for calibration (Wilson et al. 1985; Wilson and Chused 1985).

Evaluation of FACS Analysis of Membrane Potentials

Double labeling

FACS analysis with fluorescence dyes has the unique property that data collection is carried out on a cell-by-cell basis without losing the capability of studying large cell populations in a reasonable short time period, i.e., a few minutes. Artifacts originating from debris and nonviable cells can easily be eliminated and, what is more, the differences among subpopulations can be observed directly and immediately parallel to data collection. Collection of membrane potential data in correlation with other parameters is feasible. Suitable selected second fluorescence dyes can monitor practically any parameter which can be specifically and quantitatively labeled (e.g., enzyme activity, DNA content, subclasses of cells, accessibility of cell surface elements, etc.) simultaneously with membrane potential on a real time basis.

Flow cytometer vs spectrofluorimeter

Finally, we wish to emphasize again that measuring membrane potential by flow cytometry rather than spectrofluorimetry has the advantage that extracellular dye concentrations can be neglected in most cases. The sheath fluid that surrounds the cells during the measurement is so thin that the fluorescence signals come almost exclusively from the cell-bound dyes. This phenomenon is further accentuated by the frequently increased quantum yield of fluorescent dyes dissolved in the plasma membrane, due to the latter's apolar nature. Thus, even weakly bound ligands can also be studied, because the removal of excess label is not a prerequisite for measurement. Last, but not at all least, in working with living cells, the high power laser excitation of the dyes allows the use of an optimally low dye concentration, thereby minimizing toxic effects of the dyes.

A major limitation of the fluorescent dyes lies mainly in problems with scaling the measured fluorescence changes, i.e., calibrating the obtained data on a millivolt scale. This is an inherent but not unsurmountable problem that has its origin in the indirectness of dye measurements. A full depolarization of the cells and a fully polarized state gives two boundaries of the fluorescence scale. Assuming a linear fluorescence intensity change of the applied dye with the changing electric potential, a reasonable scaling of the potential values even in millivolts can be obtained. A further condition is that the resting potential level of the cell must be determined. There are other ways to calibrate the membrane potential based on the fluorescence intensity value (Wilson and Chused 1985; Krasznai et al. 1995). Nevertheless, the best way is to use parallel patch-clamp measurements on a few cells, if the absolute value of the membrane potential is needed (Gáspár et al.1995; Varga et al. 1996).

Calibration

Adding one cationic and one anionic dye to separate aliquots of the same cellular system can also overcome the problem of unwanted interaction between chemical components of the system and the dyes. Ionophores (e.g., valinomycin) can react easily with oxonol dyes, thereby obscuring real changes in potential.

Interaction with other chemicals

An interesting way to measure membrane potential is based on changes in the diameter of the plasma membrane upon changing the transient dipole effect of the potential field on the lipid and protein components of the membrane. A possible application of energy transfer measurements was suggested by labeling the intracellular and extracellular part of the membrane with a fluorescent donor-acceptor pair. The measured fluorescent resonance energy transfer reflects changes in membrane diameter and in transmembrane potential (Damjanovich et al. 1994). Recently, Gonzalez and Tsien used time-resolved energy transfer in an optical microscope system to determine proximity of receptor structures to the surface of the membrane. By this method structural changes in the membrane could be observed and were scaled according to the changing transmembrane electric potential field (Gonzalez and Tsien 1995).

New approaches

It is interesting to compare the sensitivities of the three methods for measuring membrane potential. Both isotopes and fluorescent dyes can be monitored at levels as low as a couple of hundred (in very special cases using correlation fluorimetry a single molecule) molecules or less. By contrast, picoampere measurements by patch-clamp (sub-picoampere values are not easily determined) demand as many as

Sensitivities

10 million ions diffusing through a single channel per second in order to be detected. However, there is little doubt about the fact that patch-clamp is a direct measurement method, whereas the above-described fluorescence determination is indirect. Radioactive cations could also compete for directness of potential measurement, but the detection and handling of radioactivity renders use of these compounds more clumsy. Applying correlation fluorimetry in a very small volume, the fluctuation analysis of the fluorescence can reveal the presence of a single fluorescent entity (Eigen and Rigler 1994).

A combination of these experimental possibilities, such as fluorescence and patch-clamp (thereby measuring fluorescence and electric signals from a single cell simultaneously), may overcome problems arising from a single technical approach.

References

Damjanovich S, Aszalós A, Mulhern SA, Balázs M, Mátyus L (1986) Cytoplasmic membrane potential of lymphocytes is decreased by cyclosporins. Mol Immunol 23: 175–180

Damjanovich S, Aszalós A, Mulhern SA, Szöllösi J, Balázs M, Trón L, Fulwyler MJ (1987) Cyclosporin depolarizes human lymphocytes: earliest observed effect on cell metabolism. Eur J Immunol 17: 763–768

Damjanovich S, Edidin M, Szöllösi J, Trón L (1994) Mobility and proximity in biological membranes.Chapt.1.,2.,6. CRC, Pearl River, NY

Eigen M, Rigler R (1994) Sorting single molecules: Applications to diagnostics and evolutionary biotechnology. Proc Natl Acad Sci USA 91: 5740–5747

Gáspár R Jr, Bene L, Damjanovich S, Garay CM, Aranda ESC, Possani LD (1995) β Scorpion toxin 2 from *Centruroides Noxius* blocks voltage-gated K^+ channels in human lymphocytes. Biochem Biophys Res Comm 213: 419–423

Gonzalez JE, Tsien RY (1995) Voltage sensing by fluorescence resonance energy transfer. Biophys J 69: 1272–1280

Krasznai Z, Marian T, Balkay L, Emri M, Trón L (1995) Flow cytometric determination of absolute membrane potential of cells. J Photochem Photobiol B 28: 93–99

Mátyus L, Balázs M, Aszalós A, Mulhern SA, Damjanovich S (1986) Cyclosporins depolarizes membrane potential and interacts with Ca^{2+} ionophores. Biochim Biophys Acta 886: 353–360

Neher E, Sakmann B (1976) Single-channel currents from membrane of denervated frog muscle. Nature 260: 799–802.

Neher E. (1992) Ion channels for communication between and within cells. Science 256: 498–502.

Pieri C, Recchioni R, Moroni F, Marcheselli F, Falaca M, Krasznai Z, Gáspár R Jr, Mátyus L, Damjanovich S (1992) A sodium channel opener inhibits stimulation of human peripheral blood mononuclear cells. Mol Immunol 29: 517–524

Pratap PR, Novak ST, Fredman JC (1990) Two mechanisms by which fluorescent oxonols indicate membrane potentiaL in human red blood cells. Biophys J 57: 835–849

Rigler R (1995) Fluorescence correlations, single molecule detection and large number screening. Applications in biotechnology. J Biotechnol 41: 177–186

Sagi-Eisenberg R, Pecht I (1983) Membrane potential changes during IgE mediated histamine release from rat basophilic leukemia cells. J Membrane Biol 75: 97–104

Sakmann B (1992) Elementary steps in synaptic transmission revealed by current through single ion channels. Science 256: 503–512

Varga Z, Bene L, Pieri C, Damjanovich S, Gáspár R Jr (1996) The effect of Juglone on the membrane potential and whole cell K^+ currents of human lymphocytes. Biochem Biophys Res Comm 218: 828–832

Waggoner AS, Wang CH, Tolles RL (1977) Mechanism of potential dependent light absorption changes in lipid bilayer membranes in the presence of cyanine and oxonol dyes. J Membrane Biol 33: 109–140

Waggoner AS (1979) Dye indicators of membrane potential. Annu RevBiophys Bioeng 8: 47–68

Waggoner AS (1986) Fluorescent probes for analysis of cell structure, function and health by flow and imaging cytometry. In: Taylor DL et al. (eds) Applications of fluorescence in the Biomedical Sciences, 1st ed Alan R Liss,New York, Chap. 3

Wilson AH, Seligmann BE, Chused TM (1985) Voltage sensitive cyanine dye fluorescence signals in human lymphocytes: plasma membrane and mitochondrial components. J Cell Physiol 125: 61–71

Wilson AH, Chused TM (1985) Lymphocyte membrane potential and Ca^{2+} sensitive potassium channels described by oxonol dye fluorescence measurements. J Cell Physiol 125: 72–81

Chapter 27

Flow Cytometric Measurement of Cytosolic Calcium in Lymphocytes

Erika Baus, Jacques Urbain, Oberdan Leo and Fabienne Andris

Background

Calcium in Signal Transduction

Calcium plays a central role in the transduction of external stimuli in many cell types. In B and T lymphocytes, stimulation of surface immunoglobulins or T cell receptors, respectively, triggers the activation of members of the src family of tyrosine kinases. These enzymes phosphorylate several intracellular substrates including phosphoinositide-specific phosphodiesterase (PLCγ), GTPase-activating protein for p21ras (GAP) and phosphatidylinositol 3-kinase (PI3-kinase). PLCγ hydrolyzes phosphatidylinositol 4,5-bisphosphate (PIP$_2$) into diacylglycerol and inositol trisphosphate. These second messengers are responsible for the activation of protein kinase C (PKC) and an increase in the concentration of intracellular ionized calcium ($[Ca^{2+}]_i$). Together, these agents are capable of initiating a cascade of biochemical events that results in new gene expression leading to functional responses such as cytokine production, up-regulation of cytokine receptor expression and cellular proliferation (for a review of this subject see Weiss and Littman 1994). The role of calcium in the response of lymphocytes to extracellular stimuli is clearly demonstrated by the observation that cells cultured in medium containing low levels of extracellular calcium fail to proliferate or produce cytokines upon receptor cross-linking (Mills et al. 1985; Dennis et al. 1987).

The quantitative analysis of calcium responses permits us to examine the effects of natural or synthetic compounds on the early steps of signal transduction, providing insight into the site of action for a particular drug or leading to the development of new receptor agonists and antagonists.

Calcium Measurements

The qualitative and quantitative study of $[Ca^{2+}]_i$ changes during lymphocyte stimulation may be achieved by the use of calcium-sensitive fluorescent dyes. These compounds combine a fluorochrome and a calcium chelator derivative and are characterized by their ability to display fluorescence changes upon complexing calcium ions. They can be introduced into the cytoplasm of cells without affecting cell membrane integrity since they are incorporated as permeable esters that are hydrolyzed by cytoplasmic esterases into membrane impermeable free acids (see also Chap. 22).

Cells loaded with these calcium selective fluorescent indicators can be used for physiological studies of cell suspensions (spectrofluorometry), of cell fluxes (flow cytometry), or of single cells (microfluorimetry or digital imaging microscopy). In this chapter, we will focus on the use of flow cytometry, a technique which permits studies of large cell populations and has the advantage of including statistical analysis. A limitation of this method, however, is that it does not permit the kinetic analysis of complex responses such as cellular oscillatory responses. For these studies, digital imaging or microfluorimetric microscopy is necessary to complement flow cytometric analysis.

Calcium Sensitive Fluorescent Dyes in Flow Cytometry

A large set of calcium selective optical probes are presently commercially available.

- Quin-2 was the first $[Ca^{2+}]_i$ indicator dye to be described (Tsien et al. 1982), however, this probe has several limitations. The high cytosolic concentrations of the dye required to measure a signal interferes with the calcium buffering capacity of the cells. Moreover, it requires excitation at UV wavelengths which are near to the cutoff for transmission through glass, it tends to excite cellular autofluorescence, and can cause biological side effects.
- The second generation of fluorescent indicators includes Fura-2 and Indo-1 (Grynkiewicz et al. 1985). These dyes show spectral shifts upon calcium binding (i.e., absorption shift for Fura-2, emission shift for Indo-1) and may then be used ratiometrically. The advantage of ratiometric fluorescence measurements is that they are essentially independent of dye loading, cell thickness and dye leakage. The major

limitation of these probes is that they require UV excitation which may affect fluorescence measurements (as described above). Moreover, UV sources are not as widely available in commercial flow cytometers as visible laser lines.

– Fluo-3 is a recently developed fluorescein chromophore-derived calcium probe (Minta et al. 1989) which has several properties that are advantageous for flow cytometric measurements (Vandenberghe and Ceuppens 1990). It exhibits absorption and emission spectra that are compatible with the 488 nm argon laser excitation source and fluorescein filter settings widely available in commercial flow cytometers. Furthermore, it displays a high (40-fold) fluorescence intensity increase after calcium binding and has a relatively low affinity for calcium thus permitting kinetic analysis. The disadvantage of Fluo-3 is that it does not show a significant wavelength shift upon calcium binding and therefore is not amenable to ratiometric measurements. This may make it difficult to interpret the data when there is variation in cellular dye content between the populations to be compared. However, this problem can be partly avoided by applying the Tsien calibration procedure (Tsien et al. 1982) which converts the fluorescence data into $[Ca^{2+}]_i$ values that are independent of the intracellular dye concentration. For this purpose, the maximal and minimal fluorescence (*Fmax* and *Fmin*) must be estimated for each cell population studied. *Fmax* represents the fluorescence of cells with Ca^{2+}-bound Fluo-3 obtained by rendering cells permeable to Ca^{2+}. In theory, *Fmin* should represent the fluorescence of the metal-free dye. This value can be estimated by adding saturating amounts of Mn^{2+}, a metal ion which binds to Fluo-3 about 70-fold as strongly as Ca^{2+} does, with only slight modifications in its spectral properties (the Mn^{2+}/Fluo-3 complex has only one fifth the fluorescence of the Ca^{2+}/Fluo-3 complex).

In this chapter, we will focus on the use of the Fluo-3 fluorescent dye.

Materials

– A flow cytometer equipped with a 488 nm argon laser and acquisition software that is capable of time-dependent plot analysis. The cytometer used in our studies was a FACSort running with the Lysys II software (Becton-Dickinson, Mountain View, CA).
– Low-speed centrifuge for test tubes (type megafuge 1.0 R, Heraeus Sepatech GmbH, Osterode, Germany)

- Heated water bath
- Polystyrene conical tubes 12×120 mm (Falcon, Becton-Dickinson).
- Polystyrene round bottom tubes 12×75 mm (Falcon, Becton-Dickinson)
- Calcium/magnesium-free HBSS (Life Technologies Inc., Grand Island, NY). Store at 4 °C.
- RPMI 1640 (Seromed, Biochem KG, Berlin, Germany) supplemented with 5 % FCS, pen-strep, glutamine, nonessential amino acids and 5×10^{-5} M β-mercaptoethanol (complete RPMI medium). Store at 4 °C.
- Pluronic acid F-127 (Molecular Probes Inc., Eugene, OR), 5 mg/ml stock solution in water. Store at 4 °C.
- Fluo-3 (Molecular Probes), 500 μM stock solution in DMSO. Store at -20 °C.
- Ionomycin (Sigma Chemicals Co., St Louis, MO), 1 mg/ml stock solution in DMSO. Store at -20 °C.
- $MnCl_2$ (Merck, Darmstadt, Germany), 100 mM stock solution in water. Store at 4 °C.
- Sulfinpyrazone (Sigma Chemicals Co.), 10 mM stock solution in water. In order to facilitate dissolution of this compound, add a small amount of NaOH and verify that the pH of the solution after its dilution in RPMI or HBSS medium is still neutral. Store at 4 °C.
- Stimulatory compounds

Note. All antibody solutions used in these experiments should be azide-free (or contain less than 0.002 % azide). Commercial antibody preparations may require dialysis before use.

- Any specialized material such as reagents required for cell preparation

Procedure

Cell Preparation

Flow cytometric analysis is restricted to suspensions of single cells. This procedure is described for cellular suspensions derived from lymphoid tissues, peripheral blood lymphocytes or cell lines adapted to in vitro growth.

The number of cells required per assay depends upon the number of experimental conditions to be tested. In general, $2–5 \times 10^5$ responding cells are required for each experiment. If the response of a specific subset within a heterogenous population must be analyzed, the cell density should be increased based on the anticipated number of responding cells in the population.

Isolation of Lymphocytes from Lymphoid Organs (Small Animals)

Additional materials and reagents required for this procedure are:
- 60×15 mm Petri dishes (Falcon, Becton-Dickinson)
- ACK lysis buffer prepared as follows: mix 90 ml of 0.16 M NH₄Cl and 10 ml of 0.19 M Tris pH 7.6 and adjust to pH 7.2 with HCl.
- 200 μm mesh nylon screen

1. Place freshly removed organs (spleen, thymus or lymph nodes) in a 60×15 mm Petri dish containing 1 ml of complete RPMI medium.

2. With scissors, cut the organs in several places.

3. Using a circular motion, press the pieces against the bottom of the Petri dish with the plunger of a 5 ml syringe until mostly fibrous tissue remains.

4. Add 4 ml complete RPMI medium and filter the suspension through a 200 μm mesh nylon screen into a 15 ml Falcon tube. Wash the Petri dish with 5 ml of complete medium.

5. Centrifuge 10 min at 300 g and discard the supernatant.

6. For spleen cells: resuspend the pellet in 1 or 2 ml of ACK lysis buffer, incubate at 20 °C for 1 min and then add 10 ml complete RPMI medium. Centrifuge 10 min at 300 g and resuspend the pellet in 10 ml complete RPMI medium.

7. Wash the pellet again and resuspend it in complete RPMI medium.

Purification of Specific Cell Subsets Within a Population

If lymphocyte subpopulations require further purification (by sorting, adherence to beads or plastic surfaces), use negative selection procedures whenever possible in order to minimize the contribution of cell-bound antibodies to the signaling events to be subsequently analyzed.

Preparation of Peripheral Blood Mononuclear Cells (Large Animals or Humans)

Additional materials and reagents required for this procedure are:
- Lymphoprep (Nycomed, Oslo, Norway)

– Anticoagulant (for example: citrate-phosphate-dextrose)
– Siliconized glass tubes

This procedure is adapted from the Lymphoprep manufacturer's instructions.

1. Collect the blood into a tube containing anticoagulant.

2. Dilute the blood by addition of an equal volume of complete RPMI medium.

3. Carefully layer 6 ml of diluted blood over 3 ml of Lymphoprep in a siliconized glass tube. Cap the tube to prevent the formation of aerosols. Avoid any mixing of the blood and separation fluid.

4. Centrifuge at 1000 g for 30 min at 20 °C.

5. After centrifugation, the mononuclear cells form a distinct band at the sample/Lymphoprep interface. The cells are removed from the interface using a Pasteur pipette without removing the upper layer.

6. Wash the harvested fraction of cells three times with complete RPMI medium by centrifugation at 400 g for 10 min at 20 °C.

Note. Working with human material poses special safety problems associated with the risk of infection with human disease agents. Information concerning the precautions necessary for individual infectious agents is provided in the Centers for Disease Control/National Institutes of Health handbook (Biosafety in Microbiological and Biomedical Laboratories. Health and Human Services Publication #(NIH) 88–8395, US Government Printing Office, Washington D.C.).

Preparation of Cell Lines and Hybrids

1. Culture cell lines and hybrids in a flask in complete RPMI medium at optimal cellular concentration (usually $1-5\times10^5$ cells/ml).

Note. Perform calcium measurements on cells cultured at a constant density since differences in responses to the agonists have been observed depending on the density of the cell culture.

2. Collect the cell suspension in a 15 ml Falcon tube. Centrifuge 10 min at 300 g, discard the supernatant and resuspend the pellet in complete RPMI medium.

Dye Loading

The following loading procedure is adapted from a previously described protocol by Vandenberghe and Ceuppens 1990.

1. Centrifuge the cell suspension in a 15 ml Falcon tube at 300 g for 5 min at 20 °C and discard the supernatant.

2. Wash cells twice in calcium/magnesium-free HBSS medium.

Note. Since cell adherence is decreased in calcium-free medium, try to minimize the loss of cells during this step of the procedure by careful aspiration of the supernatant.

3. After the last wash, resuspend the cells at a density of 10^7/ml in calcium/magnesium-free HBSS medium containing 100 µg/ml pluronic acid (F-127) and 5 µM Fluo-3. Incubate in a water bath at 37 °C for 30–45 min.

4. Wash the cells twice in complete RPMI medium.

5. After the last centrifugation, resuspend the pellet at a density of 5×10^5 cells/ml in complete RPMI medium.

6. Once loaded, keep the cells at 20 °C protected from light (leakage of Fluo-3 is accelerated at higher temperatures).

Use of established cell lines

In contrast to freshly isolated lymphocytes, we have observed that some established cell lines (such as murine T cell hybrids) rapidly lose intracellular Fluo-3 upon labeling, thus limiting the precision of the calibration procedure. The use of an inhibitor of organic anion transporters prevents the exclusion of Fluo-3 from the cytoplasm of these cell hybrids without affecting their viability or functional responses (Di Virgilio et al. 1989; Baus et al. 1994). Thus sulfinpyrazone (250 µM) can be added in both the labeling and the complete RPMI medium to diminish Fluo-3 leakage.

Analysis of Specific Cell Subsets Within a Population

The calcium response of a specific subset within a heterogenous cell population can be separately analyzed by staining with specific antibodies coupled to a red or far red fluorochrome (Fluo-3 is a green fluoro-chrome) and performing simultaneous immunofluorescence and $[Ca^{2+}]_i$ analysis (Greimers et al. 1996). This additional staining step should be done following the loading procedure by incubating cells in the antibody solution for 15 min at 20 °C.

Note. The binding of monoclonal antibodies to cell surface proteins can alter $[Ca^{2+}]_i$ even when these proteins are not a recognized part of the signal transducing pathway. Therefore, whenever possible, a staining strategy using negative selection should be employed so that the cellular subpopulation of interest is not labeled while the irrelevant cell subsets are identified by the antibody. When negative selection is not possible and positive selection is used, check that binding of the colored antibodies does not alter the calcium response.

Flow Cytometry

1. Display all needed graphs on the screen: FSC (size) – SSC (granula-rity) and FL1 (green fluorescence) -FSC dot plots, FL1 histogram (statistics on line) and FL1-time 2D density plot to have a good overview of the events.

2. Pipet 500 µl of the cell suspension into a round-bottomed tube adapted for the flow cytometer.

3. Incubate the cells at 37 °C for 5 min.

Note. Since calcium responses are highly temperature-dependent, all measurements must be performed at 37 °C. Cells are incubated in a 37 °C water bath prior to their acquisition and this temperature is maintained during passage of the cells by keeping the sample tube in a small iso-thermic recipient containing water at 37 °C.

4. Pass the cells in the setup mode.

5. Adjust FSC and SSC values. If needed, adjust the threshold or draw a gate around the cells of interest to eliminate debris and unwanted cells from the data acquired.

6. Optimize the photomultiplier tube gain. The easiest way is to measure the basal fluorescence of the cells and the maximal fluorescence obtained by addition of ionomycin (300 ng/ml) and then adjust these values to achieve the best discrimination while avoiding out of range points.

Setting protocol

7. Check the flow rate: it should be between 100 and 200 cells/s (a higher rate generates huge files!). If the flow rate cannot be altered manually on the flow cytometer, then adjust cell density.

Analysis of Several Cell Subsets in a Population

When different cellular subsets (identified by red or far red staining) are to be analyzed, set up FL2 (red fluorescence) and/or FL3 (far red fluorescence) values and eventually adjust the FL1-FL2 compensations. FL2 and/or FL3 parameters should be displayed on the screen as either histograms or FSC-FL2/3 and FL1-FL2/3 dot plots.

To ensure proper comparisons between different samples, FSC-SSC gates and other settings should be constant for all subpopulations. To compare the responses of labeled and unlabeled cells, the acquisition may be performed on the whole cell population and gates set during the data analysis.

Note that a total flow rate higher than 200 cells/s may be required when analyzing only a small subset of the total cell population.

Data acquisition

1. Pipet 500 µl of the cell suspension into a tube adapted for the flow cytometer.

2. Incubate the cells at 37 °C for 5 min.

3. Pass the cells (in the acquisition mode) for a few seconds prior to addition of the agonist to record the basal fluorescence of the cells.

4. The agonist is introduced into the sample by quickly removing the tube, adding the agonist, mixing the suspension and replacing the tube on the nozzle. With practice, this procedure can be completed in less than 10 s. Cell responses are analyzed continuously for 2–5 min depending upon the nature of the stimulus.

Note. Different kinds of stimuli can be added to the cells given that they are soluble. Stimulation can be achieved by cell receptor agonists or antibodies directed against a membrane molecule (in some cases, cross-linking by a secondary antibody may be required).

5. To calibrate measurements, the minimal (F_{min}) and maximal (F_{max}) fluorescence of the cells must be recorded for each population. F_{max}, the fluorescence of cells with all Fluo-3 bound to Ca^{2+}, is measured directly after addition of ionomycin (300 ng/ml). In some cases, the

addition of metabolic poisons disabling ion pumping across the plasma membrane may be required in order to achieve saturating intracellular calcium concentrations in response to ionomycin (see Greimers et al. 1996). F_{min}, the fluorescence of cells with Mn^{2+} bound Fluo-3, is measured about 10 min after addition of $MnCl_2$ (2 mM) to ionomycin treated cells.

Note. One complete assay should not exceed 90 min to 2 h since after this time it is necessary to regularly check cell responses to ionomycin to ensure that no dye leakage occurred.

1. If needed, gate populations according to their size, red or far red stain-   **Data analysis**
   ing, etc.

2. For each experiment, divide the total time into small increments and calculate the mean fluorescence of the cells for each interval, convert these fluorescence values (F) into calcium concentrations by using the Tsien equation (Tsien et al. 1982): $[Ca^{2+}]_i = K_d \times (F - F_{min})/(F_{max} - F)$ where F_{min} and F_{max} are measured as described above and K_d, the dissociation constant for Ca^{2+} bound Fluo-3, is equal to 400 nM (Minta et al. 1989). Then draw a graph of mean $[Ca^{2+}]_i$ vs time.

3. For each experiment, divide the total time in small increments and determine the percentage of responding cells (i.e., the percentage of cells that raise their FL1 fluorescence to a higher level than that shown by the majority of the resting cells) for each interval.

This data analysis will enable you to compare the latency, the amplitude and duration of the calcium response as well as the percentage of responding cells with different types of stimuli and/or in different cell populations.

Results

The measurement of calcium levels by flow cytometry offers the opportunity to analyze the response of different cellular subpopulations within a single sample. The following example illustrates the response of murine B lymphocytes from a total splenic cell population to surface immunoglobulin cross-linking.

The total population of spleen cells from a naive mouse (including T, B and accessory cell subpopulations) was prepared and loaded with Fluo-3 as described above. Following the loading procedure, spleen cells were either stained with an antibody specific for a B cell surface marker coupled to a red fluorescent dye, phycoerythrin (PE-coupled anti-B220, PharMingen, San Diego, CA), or an antibody to a T helper lymphocytes marker coupled to phycoerythrin (PE-coupled anti-CD4, Life Technologies Inc.). Fluo-3 loaded, unstained cells were used for calibration of the instrument and adequate gate settings.

Spleen cells were analyzed for 15 s prior to addition of 4 µg/ml of an antibody to mouse immunoglobulins (F(ab')2 sheep anti-mouse Igs, Sigma Chemical Co.) in order to record basal Fluo-3 fluorescence. Stimulation-induced changes in fluorescence were then monitored for 2 min.

Figure 27.1 shows dot plots of the Fluo-3 generated signal (FL1, linear scale) vs the PE-associated signal (FL2, log scale). In Fig. 27.1A, B, the B cells are stained with a PE-coupled anti-B220 antibody; in Fig. 27.1C, D, the T helper cells are stained with a PE-coupled anti-CD4 antibody. Figure 27.1A, C shows the basal FL1 fluorescence of the cells; Fig. 27.1B, D shows the FL1 fluorescence of the cells 20 s after addition of anti-mouse Igs.

As anti-mouse Igs recognize B cells and not T cells, only B cells were expected to respond to the stimulus with elevated $[Ca^{2+}]_i$ levels. Indeed, in Fig. 27.1A, B, only cells expressing B220 marker (identified as the FL2-positive cells) increased cell associated FL1 fluorescence after stimulation. Accordingly, in Fig. 27.1C, D, only cells lacking CD4 expression (belonging to the FL2-negative subpopulation) responded to anti-Ig stimulation by an increase in their FL1 fluorescence. Not all FL2-negative cells responded to anti-Ig stimulation since this population includes not only B cells but also CD4-negative CD8-positive T lymphocytes and accessory cells which lack surface Ig expression.

Note that in this example, the calcium response of B cells was not influenced by the presence of membrane bound antibodies to B220 (compare the FL1-positive spots in Figs. 27.1B, D). Thus, both positive (B220 staining) and negative (CD4 staining) criteria can be used to identify a subpopulation of responding cells in these instrumental settings.

In Figure 27.2A, a kinetic analysis of the calcium response of B lymphocytes to anti-Ig stimulation is shown. FL2-positive cells from the B220 stained cell population were positively gated and their FL1 fluorescence plotted against time.

The FL1 fluorescence of these B220-positive cells after ionomycin and $MnCl_2$ treatment is shown in Fig. 27.2B, C, respectively. Note that the

Untreated   Anti-Ig treated

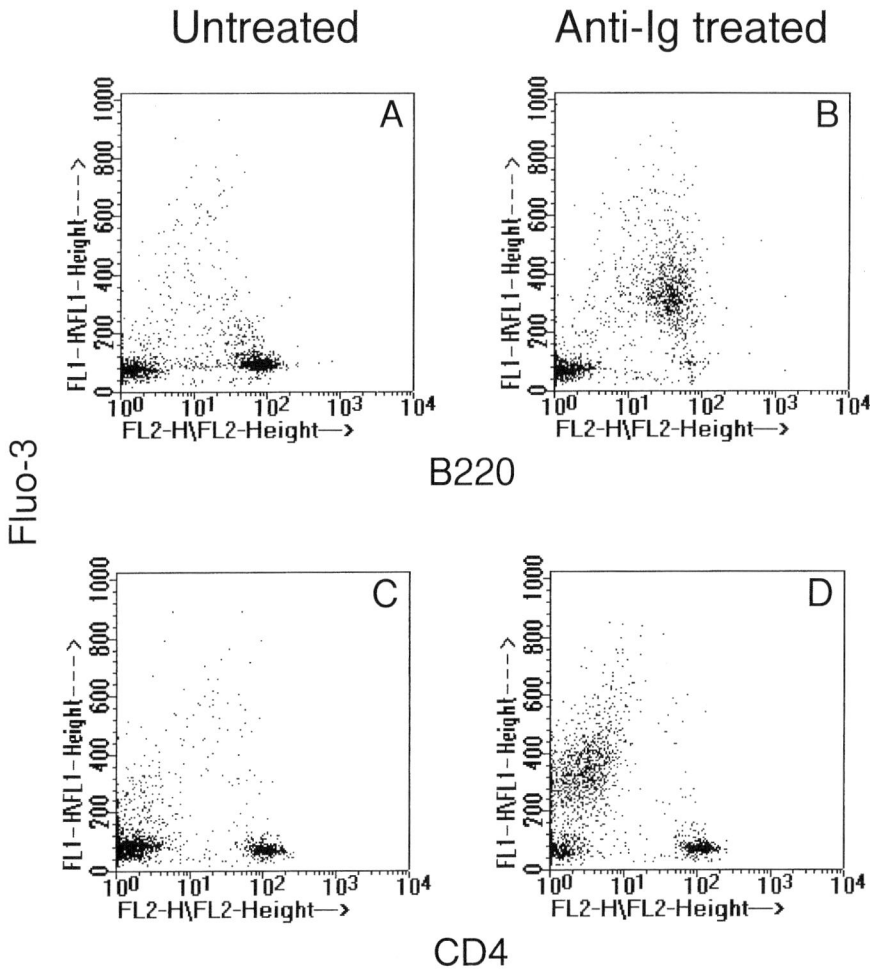

Fig. 27.1A-D. Calcium response to anti-Ig stimulation in a total spleen cell population. Cell populations stained with PE-coupled anti-B220 (**A, B**) or PE-coupled anti-CD4 (**C, D**) were treated with anti-Ig antibodies at $t = 15$ s. Fluorescence data recorded at $t = 0-10$ s (**A, C**) and $t = 30-40$ s (**B, D**) are displayed as FL1 vs FL2 dot plots

cells maintaining very low FL1 fluorescence after anti-Ig stimulation in Fig. 27.2A are most likely unloaded cells since their fluorescence does not increase even after addition of ionomycin.

In order to quantify the $[Ca^{2+}]_i$ levels during the response, the time scale of Fig. 27.2A was arbitrary divided in equal regions of about 10 s, and the

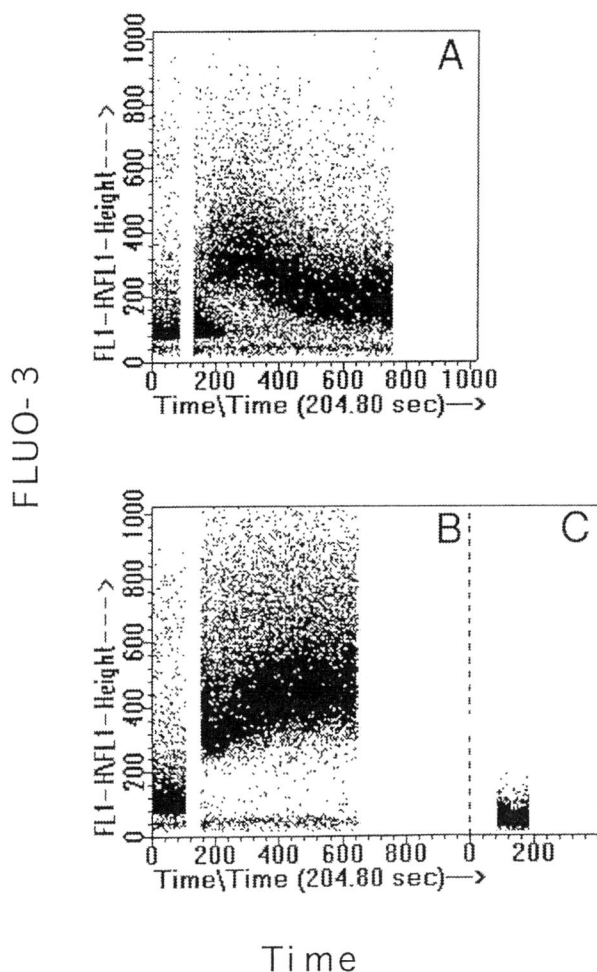

Fig. 27.2A–C. Kinetic analysis of the calcium response of B220$^+$ B lymphocytes to anti-Ig, ionomycin and MnCl$_2$ treatment. B220 stained cells were treated at $t = 15$ s with anti-Ig (**A**) or ionomycin (**B**). Ionomycin treated cells were then treated with MnCl$_2$ and analyzed after a 10 min incubation (**C**). The FL1 fluorescence of B220$^+$ gated cells is plotted against time

mean FL1 fluorescence associated with each region calculated using the Lysys II program. These data were converted into [Ca^{2+}]$_i$ using the Tsien equation (F_{max} and F_{min} were determined from Fig. 27.2B, C). These values were then plotted against time in Fig. 27.3A. Note that this graph represents the mean calcium response of a population which clearly contains a heterogenous mixture of responding cells.

Fig. 27.3A, B. Analysis of the calcium response of B220$^+$ B lymphocytes to anti-Ig stimulation. B220 stained cells were treated at $t = 15$ s with anti-Ig. **A** The FL1 fluorescence data from Fig. 27.2A are converted into [Ca^{2+}]$_i$ values according to the Tsien calibration procedure and plotted against time. **B** The percentage of responding cells calculated from Fig. 27.2A is plotted against time

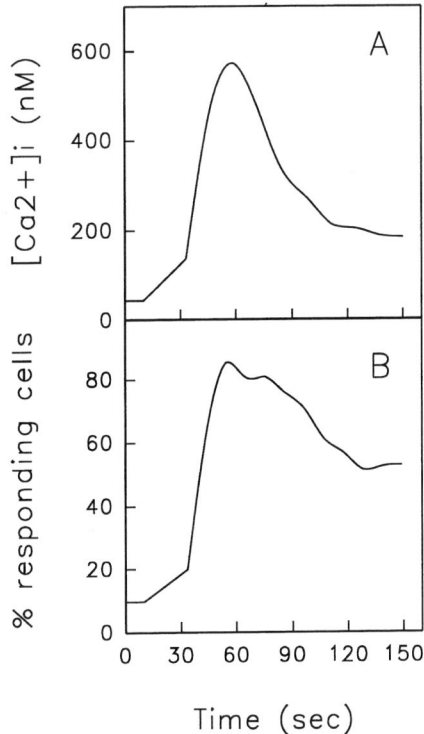

Figure 27.3B shows the percentage of responding cells – defined as the cells that raise their FL1 fluorescence to a level that is higher than the majority (at least 90 %) of resting cells – over time. However, a precise estimation of the percentage of responding cells at a certain time point within a defined population is difficult to assess by flow cytometry since cells displaying low fluorescence at any given time following stimulation may represent either nonresponding cells or cells responding in an oscillatory fashion (Choquet et al. 1994).

Acknowledgment. The authors wish to thank E. Muraille for his help in the preparation of the artwork and K. Willard for carefully revising the manuscript. This work was supported in part by the Fonds de la Recherche Fondamentale Collective and by the Belgian program on interuniversity Poles of attraction initiated by the Belgian State, Prime minister's Office Policy Programming. E.B., O.L. and F.A. are supported by the Fonds National de la Recherche Scientifique (F.N.R.S., Belgium). The scientific responsibility for this work is assumed by its authors.

References

Baus E, Urbain J, Leo O, Andris F (1994) Flow cytometric measurement of calcium influx in murine T cell hybrids using Fluo-3 and an organic-anion transport inhibitor. J Immunol Methods 173:41–47

Choquet D, Ku G, Cassard S, Malissen B, Korn H, Fridman WH, Bonnerot C (1994) Different pattern of calcium signaling triggered through two components of the B lymphocyte antigen receptor. J Biol Chem 269:6491–6497

Dennis GJ, Mizuguchi J, McMillan V, Finkelman FD, Ohara J, Mond JJ (1987) Comparison of the calcium requirement for the induction and maintenance of B cell class II molecule expression and for B cell proliferation stimulated by mitogens and purified growth factors. J Immunol 138:4307–4312

Di Virgilio F, Steinberg TH, Silverstein SC (1989) Organic-anion transport inhibitor to facilitate measurement of cytosolic free Ca^{2+} with Fura-2. Methods Cell Biol 31:453–462

Greimers R, Trebak M, Moutschen M, Jacobs N, Boniver J (1996) Improved four-color flow cytometry method using Fluo- and triple immunofluorescense for analysis of intracellular calcium ion ($[Ca^{2+}]_i$) fluxes among mouse lymph node B- and T-lymphocyte subsets. Cytometry 23:205–217

Grynkiewicz G, Poenie M, Tsien RY (1985) A new generation of Ca^{2+} indicators with greatly improved fluorescence properties. J Biol Chem 260:3440–3450

Mills GB, Cheung RK, Grinstein S, Gelfand EW (1985) Increase in cytosolic free calcium concentration in an intracellular messenger for the production of interleukin 2 but not for expression of the interleukin 2 receptor. J Immunol 134:1640–1643

Minta A, Kao JPY, Tsien RY (1989) Fluorescent indicators for cytosolic calcium based on rhodamine and fluorescein chromophores. J Biol Chem 264:8171–8178

Tsien RY, Pozzan T, Rink TJ (1982) Calcium homeostasis in intact lymphocytes: cytoplasmic free calcium monitored with a new, intracellularly trapped fluorescent indicator. J Cell Biol 94:325–334

Vandenberghe PA, Ceuppens JL (1990) Flow cytometric measurement of cytoplasmic free calcium in human peripheral blood T lymphocytes with Fluo-3, a new fluorescent calcium indicator. J Immunol Methods 127:197–205

Weiss A, Littman DR (1994) Signal transduction by lymphocyte antigen receptors. Cell 76:263–274

Part III

Fluorescence To Measure Intracellular Ions

Microfluorescence Measurements

Chapter 28

Cytosolic pH and Cell Movement Measurement in *Dictyostelium*

KEI INOUYE

Background

This chapter describes basic techniques for measuring the cytosolic pH of single motile cells by microfluorometry. A simple method of quantifying chemotactic and random cell movement is also shown.

Cytosolic pH (pH_i) plays important roles in a variety of cellular processes such as growth control, cell movement, and cell differentiation, and knowledge about its regulation is indispensable in understanding the network of signal transduction pathways controlling these cellular activities.

The cellular slime mould *Dictyostelium discoideum* is a genetically tractable eukaryotic microorganism which exhibits much higher cell motility and endocytic activity than most mammalian cells, and the mechanism whereby the external chemotactic signal controls cell movement has been extensively studied (see van Haastert 1995; Newell et al. 1995 for reviews).

There is evidence that the pH_i is involved in the control of chemotaxis in *Dictyostelium*. Upon chemotactic stimulation, the pH_i transiently rises presumably due to activation of H^+-ATPase (Aerts et al. 1987; also see Fig. 28.2) and the speed of cell movement increases (van Duijn and Inouye 1991). The pH_i has been implicated in the control of cytoskeletons via the actin-binding proteins that are sensitive to pH (Condeelis and Vahey 1982; Stoeckelhuber et al. 1996; Edmonds et al. 1995). The pH_i also plays an important role in stalk cell differentiation in *Dictyostelium* (reviewed in Inouye 1990; Gross 1994).

Various methods have been employed for pH_i measurement in *Dictyostelium*. For single-cell measurement, however, the fluorescence method has been the only practical method, since the use of intracellular microelectrodes in such small cells with high motility and a flexible plasma membrane is very difficult (van Duijn et al. 1988), and other methods, such as ^{31}P-NMR and digitonin-null point, are not applicable to single cells.

To illustrate the principles of fluorescence measurement techniques, the use of a simple setup consisting of a fluorescence microscope and a photomultiplier is assumed in the following, because more sophisticated, ready-to-use systems for fluorescence ratio imaging, though very efficient and reliable, sometimes obscure the principles and may conceal possible pitfalls in fluorescence measurements. The use of a photomultiplier is justified because pH_i is generally uniform throughout the cytoplasm (Bright et al. 1987), unlike the case of Ca^{2+} which often shows a localized distribution within the cytoplasm of a single cell. However, the arguments below also apply to systems using a video camera for the detection of fluorescence. Also the techniques and considerations described hereafter will be applicable to pH_i measurements of other eukaryotic cells.

28.1
Measurement of pH_i

Principle   The principle of pH_i measurement is very simple; to load a cell with a fluorescent pH indicator and measure its fluorescence. Since the fluorescence intensity of the probe depends not only on its ambient pH but on its amount in the cell, it is usually necessary to cancel out the effect of the amount of the probe in the cell by taking ratios of the fluorescence intensities measured at two different wavelengths (either excitation or emission or both depending on the characteristics of the probe). The two wavelengths are chosen so that the fluorescence at one wavelength is strongly pH-dependent while at the other wavelength it is relatively insensitive to pH or dependent on pH in the opposite direction.

The Choice of the pH_i Probe

Successful measurements greatly depend on the right choice of probe. Table 28.1 lists some pH_i probes that have proved to work with *Dictyostelium* cells in my hands. There are three points to consider when choosing a probe:

- pK_a of the probe
- Method of loading
- Retention of the probe in the cytoplasm

Table 28.1. A list of selected pH$_i$ probes

Probe	Permeant ester/ Dextran conjugate	Molecular mass	λex[b]	λem[b]	pK$_a$[d]	Leakage	Cost[e] (US$/ experiment)
CF	CFDA	376	490/450	520	6.4	Medium	0.004
	CFDB	431					–[f]
BCECF	BCECF-AM	688	500/450	530	7.0	Slow	1.5
	BCECF dextran	–[a]				very slow	5.4
CMF	CMFDA	465	490/450	520	6.4	Very slow	0.7
C.SNAFL-1	C.SNAFL-1-DA	510	550/510	580	7.8	Medium	1.6
			550/490	580/520[c]			
			580/490	610/520[c]			

Abbreviations: CF, 5(or 6)-carboxyfluorescein; BCECF, 2',7'-bis(carboxyethyl)-5(and 6)-carboxyfluorescein; CMF, chloromethyl fluorescein; C.SNAFL, 5(and 6)-carboxyseminaphthofluorescein; DA, diacetate; DB, dibutyrate; AM, acetoxymethyl ester.

  Positions of carboxyl substitutes are according to the Molecular Probes' catalog. 5(or 6)-: either 5-, 6-, or their mixture, 5(and 6)-: mixture of 5- and 6-. These are practically indistinguishable from each other.

  Note. Some vendors use different ways of numbering, resulting in different numbers for the same positions, e.g. Fluka's and Dojin's 4(5)- is equivalent to Molecular Probes' and Carbiochem's 5(and 6)-.

[a] BCECF attached to dextran with molecular masses of 10 000, 40 000 and 70 000 are available from Molecular Probes. For coupling of fluorescent probes with dextran, see Schlatterer et al. (1992).

[b] Approximate wavelengths for the excitation and emission of the probes. Shown on the left-hand sides and the right-hand sides of the slashes are the wavelengths for the basic and acidic forms of the probes, respectively.

[c] C.SNAFL-1 also allows dual-excitation/dual-emission measurements using filter cubes for fluorescein (ex 490 nm, em 520 nm) and rhodamine (ex 550 nm, em 580 nm), or for fluorescein and Texas Red (ex 580 nm, em 610 nm).

[d] Approximate pK$_a$ value in solution. This value may be different from the pK$_a$ in cytosol.

[e] Based on the prices listed in Molecular Probes International Price List (April 1995). Amounts of the probes (in mg) required to load 10^7 cells were calculated on the assumption that exactly the same protocol (including cell numbers) as shown in the text is used, and multiplied by the price of the probes per mg.

[f] Not available commercially. CFDB is less toxic than CFDA when loaded to high concentrations. This can easily be made by reacting carboxy fluorescein with butyric anhydride (Inouye 1985; see Thomas et al. 1979 for detailed procedure).

pKₐ of the pH probe

The fluorescence spectrum of the probe must be dependent on its ambient pH over an appropriate pH range. The pH range that can be measured with a single fluorescent pH probe is usually limited within ca. ± 1 pH unit from its pK_a value. Reported pH_i values in *Dictyostelium discoideum* range considerably, from 6.1 to 7.9, but under normal conditions the pH is slightly alkaline as in most other organisms (Inouye 1988; Furukawa et al. 1988). In addition, changes upon external stimulation rarely exceed 0.5 pH units. It follows that a fluorescent probe with a pK_a between 6.8 and 7.8 can be used for general purposes. If the phenomenon of interest involves pH changes in only one direction, either acidification or alkalinization, a pH probe with a pK_a lower than 6.8 or higher than 7.8, respectively, may be used.

Loading of the pH probe

The most commonly used method of introducing a pH probe into the cytoplasm is to use an esterified derivative which is nonfluorescent and cell-permeant. Upon entering the cytoplasm, this derivative is cleaved by the action of nonspecific hydrolase to become the original form, which is more negatively charged and hence less permeant. With amoeboid cells of *Dictyostelium* this method works well with some commonly used pH probes, although the efficiency differs considerably between probes.

Alternatively, a probe may be introduced into the cytoplasm by electroporation. With this method, cells can be loaded with a highly membrane-impermeant probe, such as dextran-coupled BCECF. In general, efficiency of loading and the viability of the electroporated cells greatly depend on the condition of electroporation. Several different protocols have been published for introducing macromolecules into *Dictyostelium* cells. I found the method shown below works well for introducing dextran-coupled probes into axenically grown *Dictyostelium* cells. It has also been shown that fluorescein-dextran can be introduced into *Dictyostelium* cells by controlled sonication (Fechheimer et al. 1986). Scrape-loading has also been successfully used for loading *Dictyostelium* cells with Fura-2 dextran (Schlatterer et al. 1992), so the same method should be readily applicable to pH_i measurement.

Microinjection has proved to be very difficult in *Dictyostelium*, mainly because of the very flexible and sticky nature of its cell membrane.

In this chapter, only the methods using esterified probes and electroporation are described.

Retention of the probe in the cytoplasm

With the method using cell-permeant derivatives, leakage of the probe does occur to some extent, even if the probe becomes more negatively charged after hydrolysis. In addition, low molecular weight probes are sequestered in vesicles regardless of their charge. This poses limitations to the length of time in which reliable pH_i measurement can be performed. This limit, however, is not always the same but can vary depending on the condition of the cells. In *Dictyostelium discoideum*, sequestration occurs quickly at the vegetative stage while probes stay longer in the cytosol as development proceeds. It is therefore necessary to check how long the cells remain uniformly fluorescent (excluding intracellular vesicles) under the condition used.

The use of high-molecular weight probes minimizes both leakage of the probe to the outside of the cells and sequestration into intracellular compartments, ensuring that the measured pH is the cytosolic pH. For instance, BCECF dextran of MW 10 000 stays in the cytoplasm of *Dictyostelium* cells for over a day without any appreciable sign of sequestration and loss of fluorescence. Newly developed chloromethyl derivatives of fluorescent dyes also stay in the cytoplasm for an extended period. Chloromethyl fluorescein has been shown to be very stable in the cytoplasm of *Dictyostelium* cells (Knecht and Shelden 1995). Although the low pK_a of this probe (6.4) limits its application to cytosolic acidification studies, this class of pH probes will extend the possibility of fluorescence pH measurement. (Chloromethyl SNARF-1, pK_a 7.8, did not give sufficient fluorescence for long-term measurement.)

Factors Affecting pH_i Measurement

A population of cells loaded with a fluorescence probe will be heterogeneous to some extent in their fluorescence intensities. When cell-permeant probes are used, heterogeneity is usually not very large, but if the brightness varies considerably from cell to cell, great care should be taken when choosing a cell for measurement, because damaged cells are likely to be more leaky and take up more probes, resulting in a highly fluorescent appearance. Fortunately, the distinction between damaged cells and healthy cells is usually clear, since uptake of excessive probes causes acidification of the cytosol (see below) and excitation of too much fluorophores within the cell apparently causes photochemical reactions which seem to damage the cell. Usually such cells become round-shaped and easily come off the substratum upon perfusion.

Heterogeneity among the cells

Choosing too dark a cell for measurement may be equally unwise because measurements with such cells will suffer from bad signal-to-noise ratios (S/Ns), and, to compensate for the limited signal intensity, one may need to increase the level of excitation light, which in turn could be a source of other artefacts (see below).

Practically the best compromise would be to choose a cell that is bright enough and at the same time actually moving by extending one or more pseudopods. Examples of such cells can be seen in Fig. 4A of Inouye (1985).

The population of cells loaded by electroporation can be very heterogeneous, and those that are visibly fluorescent are certainly damaged to various extents. It is therefore necessary to avoid irreversibly damaged cells and to allow the other cells to recover from the damage before measurement. How long it takes for the cells to fully recover will depend on the extent of damage. For electroporation using much milder conditions than necessary for loading of dextran, it still takes almost 40 min for the membrane potential to recover from complete depolarization (van Duijn et al. 1990). Considering the longevity of high-molecular weight probes such as BCECF dextran in the cytoplasm, it makes sense to load cells at an earlier stage and allow them to develop for a certain period of time and then choose actively moving cells that show uniform fluorescence in the cytoplasm. If cells at the early developmental stage are to be measured, be sure that the cells do not take up the probe by pinocytosis; vegetative cells of axenic strains (but not wild-type, see Maeda 1983), have a high pinocytic activity and it takes several hours for the fluid phase dye to be completely removed from the cells (Klein and Satre 1986).

Effects of the excitation light

Apart from bleaching of fluorescent probes, strong excitation light causes a rapid decrease of pH_i itself by an unknown mechanism. Both the intensity and duration of excitation must therefore be kept to a minimum (Bright et al. 1987; Inouye 1988). To limit the intensity of the excitation light, neutral density filters and/or a low energy light source (halogen lamp rather than xenon or mercury lamp) should be used (see Equipment). To minimize the time of exposure, the shutter for the excitation light should be opened only when necessary. It is advisable to test under the condition used how long the cells can be exposed to the excitation light (at each of the two wavelengths) without the pH_i being affected.

Effects of the probe on the pH_i

Since an esterified probe liberates acid upon hydrolysis, the amount of the probe introduced into the cells should be kept to a minimum. With minimum loading of the esterified probe, the pH_i may still be found to be affected. In such cases, it is necessary to allow a certain period of time for the cells to return to the normal pH_i after loading. As long as the concentration of the probe and the loading condition are carefully controlled, pH_i recovery will take place in no longer than a few minutes.

Materials

- Fluorescence Microscope

  An ordinary epifluorescence microscope is used. For measurements involving manipulation of cells, such as local application of chemoattractant using a micropipette, inverted fluorescence microscopes are more convenient, but upright fluorescence microscopes can also be used in combination with an appropriate perfusion chamber (see Chap. 2).

- Light Source

  If you are not going to use short-wavelength probes such as Fura-2 for Ca^{2+} measurement with the same microscope, a high-power (12 Volts, 100 Watts) halogen lamp (a lamp with a tungsten filament in a quartz casing filled with a halide gas; also called quartz-halide lamp) is quite adequate for the present purpose. It gives sufficient light for excitation of long-wavelength probes such as those listed in Table 28.1. In addition, a halogen lamp is inexpensive, safe, and energy-saving. However, exactly the same voltage must be used for all measurements because the spectrum of the tungsten filament can vary significantly if it is powered at different voltages. If the microscope is already furnished with a mercury-vapor lamp or xenon lamp, they can of course be used for pH_i measurement. These lamps, however, give too strong an illumination for living cells, so the light intensity needs to be cut down more than tenfold by neutral density filters. In addition, since the light intensities around the wavelengths required for excitation of many fluorescein-derived probes (460–500 nm) are among the lowest in the spectrum of the mercury-vapor lamp, the exciter filter will be exposed to much higher energy if a mercury lamp (and xenon lamp) is used, which would shorten the life of the filters. Regardless of the lamp used, a stable power supply is required for reproducible results (for more detailed theoretical arguments concerning fluorescence microscopes, see Taylor and Salman 1989; Wampler and Kutz 1989).

- Objective Lens

  The probes listed in Table 28.1 are all excited by light at wavelengths longer than 400 nm. For these probes, ordinary bright-field lenses of high numerical aperture (NA) can be used for pH_i measurements. If simultaneous recording by phase-contrast optics is required, phase/fluorescence lenses are convenient. Ordinary phase lenses are inefficient because of loss of the fluorescence due to their low NA.

- Filters and Mirrors
  Good filters with sharp cut-off spectral profiles are essential for successful measurement since a filter of poor quality results in low efficiency and high background. A dichroic mirror is generally used as a beam splitter for maximum yield of fluorescence. However a neutral beam splitter (mirror or prism) with small reflection may be worth considering if various probes of different spectral properties are used with the same microscope. The loss of a large fraction of the excitation light will not be a problem because the excitation light needs to be cut down any way when a strong light source is used. Omega Optical, Inc. sells high-quality interference filters and dichroic mirrors at reasonable prices. They also mount the filters and mirror on a filter holder for use with some fluorescence microscope models.

- Matching Fluorescence Intensity
  For ratiometry, it is desirable to adjust the intensities of the excitation light at the two wavelengths so that the fluorescence signals from these excitation wavelengths are about the same magnitude. This has to be empirically determined because it depends on the combination of the light source, probe, and filters used. It is therefore necessary to carry out pilot experiments: If the fluorescence obtained with one excitation wavelength is more than twice as strong as with the other, an appropriate neutral density filter should be placed on top of the exciter filter for that wavelength, so that the signals from the two excitation wavelengths become comparable. For the probes listed in Table 28.1, CF, BCECF, and CMF give much stronger fluorescence when excited at 490 nm than at 450 nm, so it is generally necessary to overlay the 490 nm filter with an ND filter (10 %–50 % depending on the light source and, if a mercury lamp is used, the exact spectral profile of the short-wavelength filter).

- Changing of the Excitation Filters
  For dual-excitation probes, quick changes of the exciter filter from one wavelength to another is required. Some recent models of fluorescence microscopes for general use are provided with cassettes for selection of excitation and emission lights of various wavelengths. On each cassette, an exciter filter, a beam splitter and a cut filter are mounted, so that change of the exciter filter alone is impossible. For microscopes of this type, it is necessary to obtain a specially designed cassette with a slider on which two exciter filters are mounted. Old models and some new microscope models (e.g., Olympus IX) have such filter holders for manual operation. They also provide motor-driven filter changers that

can be attached to the illumination unit of the fluorescence micro-scope. Some types of filter changers can operate very fast, up to a fre-quency of 1000 times/s, thus allowing measurement of fast pH changes or averaging out of any fluctuation of the light source and photomulti-plier. For most applications which do not involve very fast changes of pH$_i$, manual change of filters gives satisfactory results.

- Chamber
  To measure pH$_i$ changes in response to stimulation, use of an adequate chamber that allows quick exposure of the cells to the stimuli is very important. For nonlocalized stimulation of cells, sealed perfusion chambers are usually used. A good perfusion chamber must allow quick and complete exchange of the solution. Although any kind of perfusion chamber intended for tissue culture cells (e.g., Dvorak and Stotler 1971) may be used for *Dictyostelium* cells, simple perfusion chambers that are convenient for experiments with *Dictyostelium* cells can easily be made (see Chap. 2). Open-top chambers can be used for localized as well as nonlocalized stimulation in combination with a microcapillary.

- Photomultiplier
  Leading microscope manufacturers provide photomultipliers with variable apertures for use in combination with their microscopes. Check the spectral sensitivity because some old models do not have sufficient sensitivity at long wavelengths. If a photomultiplier of differ-ent make is available, try to use it with the microscope in hand. Per-haps they do not fit well but will be usable if the image can be brought to focus.

- Pulse Generator and Electrodes for Electroporation
  For loading of dextran-coupled pH probes by electroporation, a pulse generator capable of generating short pulses with adjustable charging voltage and a selection of capacitors and electrodes are required. Any apparatus used for introducing DNA in eukaryotic cells (such as the BioRad Gene Pulser) should be suitable for this purpose. Standard dis-posable cuvettes with electrodes can be repeatedly used without a loss of efficiency if thoroughly rinsed and dried immediately after use.

- Since esterified probes gradually decompose in aqueous solutions, great care must be taken to keep moisture from stock solutions as well as from the probe as a solid. When using, open the lid of the tube only after it reaches room temperature, otherwise it will collect moisture

Fluorescence probes

inside. Stock solutions are prepared in high-grade dry DMSO. The concentration is typically 10–20 mM since *Dictyostelium* cells require higher concentrations of esterified probes than mammalian cells to produce sufficient fluorescence for measurement.

- Stock solutions must be kept desiccated in a freezer in small aliquots, preferably under vacuum or in dry N_2 to prevent oxidation as well. For storage under vacuum, a heavy-duty tube with a side arm and a vacuum stopcock (ca. $\varnothing$ 2 cm, 15 cm) half filled with dry silica gel is convenient. A small tube containing the stock solution, with its lid **loosely** closed, is placed on the silica gel, air is pumped out, and the whole tube kept in the freezer. Use of dry N_2 is an easy yet effective alternative method for keeping the stock solution. The tube containing the stock solution is flushed with a stream of N_2 from a Pasteur pipette connected to a cylinder of N_2, and the lid is closed while flushing. When doing this, be sure that the Pasteur pipette is not pointed toward people or yourself and the valve of the cylinder must be released very carefully so that the Pasteur pipette does not shoot off.

- Stock solutions of dextran-coupled probes are made in buffer near the saturation concentration (50 mg/ml is used for BCECF dextran of MW 10 000). Stock solutions should be kept in small aliquots to avoid repeated freezing and thawing.

Solutions   Stock Solutions

- Carboxyfluorescein diacetate (10 mM): 1 mg carboxyfluorescein diacetate, 266 µl DMSO
- BCECF-AM (20 mM): 1 mg BCECF-AM, 72.7 µl DMSO
- Carboxy-SNAFL-1 diacetate (20 mM): 1 mg C.SNAFL-1 DA, 98 µl DMSO
- BCECF dextran (50 mg/ml): 10 mg BCECF dextran 10 000, 200 µl E buffer (see below)
- Loading buffer: 20 mM K/K_2 phosphate, pH 6.0

Note. An acidic pH facilitates permeation of the probe.

- Bathing solution: 20 mM K/K_2 phosphate, at desired pH
- Electroporation buffer (E buffer): 20 mM Na/K_2 phosphate pH 7.4, 50 mM sorbitol or sucrose, 2 mM $MgSO_4$, 0.2 mM $CaCl_2$

Note. Modification of the ionic composition alters the electric conductance of the buffer, which could greatly affect the setting of voltage and capacitance for electroporation.

- Loading solution for esterified probes: 1 μl stock solution, 99 μl loading buffer (exposure to up to 1 % DMSO does not affect the result).

Acidic Calibration Buffers

- 20 mM K/K$_2$ phosphate buffer pH 6.0–7.2
- 50 mM potassium propionate

Basic Calibration Buffers

- 20 mM K/K$_2$ phosphate buffer pH 7.4–8.2
- 25 mM (NH$_4$)$_2$SO$_4$

Culture and Development of *D. discoideum*

See Sussman 1987 for more detailed descriptions of culture methods.

Cell culture

Wild-type strains, and nonaxenic mutants derived therefrom, are usually grown on nutrient agar with food bacteria. Alternatively, these strains may be cultivated in suspension in buffer without nutrient in association with pregrown bacteria (Gerisch 1960).

Axenic strains (Ax2, etc., and mutants and transformants of the axenic background) are capable of growth in axenic medium without food bacteria. Although shaking culture is most commonly used for the axenic strains, stationary culture in tissue culture dishes is convenient for single cell measurements which require only a small number of cells.

Initiation of development

The most commonly used method is to harvest exponentially growing cells and wash out the nutrient by centrifugation. Maximum yield of exponentially growing cells can be obtained by harvesting cells at cell densities of ca. 5×10^6 cells/ml (for shaking culture) and 1×10^6 cells/cm^2 (for dish culture in 10–20 ml medium per ⌀ 9 cm dish).

Wash the cells free of nutrient by repeated centrifugation (five times for bacterially grown cells, once or twice for axenically grown cells) in chilled loading buffer and resuspend them in the same buffer at 10^6–10^7 cells/ml.

If cells are grown in a culture dish, remove the medium by aspiration and add fresh loading buffer to the plate. Repeat the process once or twice.

Procedure

Short protocol
1. Load cells with the probe by mixing the loading solution and the cell suspension or by electroporation.

2. Wash off the probe solution with fresh buffer.

3. Allow the cells to recover from any effect of the loading process.

4. Choose a cell and measure its fluorescence intensity at the two wavelengths.

5. Change the bathing solution as required and repeat measurement at appropriate intervals.

Esterified Probe Method

In the following, more detailed procedures are given for the use of a perfusion chamber of the types shown in Figs. 2.1 and 2.2 of Chap. 2 including discrete replacement of the solutions, as the volumes of the solutions shown are for those particular chambers (capacity 20 µl, depth 0.5 mm, width 4 mm, mean length 10 mm). If a chamber of a capacity of x µl (including the upstream capillary part and tubing) is used, the mixture of loading solution and cell suspension (step 3 in the following protocol) should be x µl and washing and perfusion solutions should at least be $2x$ µl (desirably $5x$ µl). The maximum flow rate without causing detachment of cells depends on the substratum and the developmental stage of the cells (Yabuno 1971, see Chap. 2). Since cells soon after loading with an esterified probe are less adhesive, it is safer to keep the flow rate several-fold lower than would be tolerated with nontreated cells. For a perfusion chamber of d mm depth and w mm width, $10dw$ µl/s would be the maximum that could be safely used.

Loading and Washing-Off of the Probe

Note. In the following procedures, fluorescence illumination should always be kept as short as possible.

1. Prepare the perfusion chamber.

2. Wash starved cells once by low speed centrifugation and resuspend the cells in loading buffer at ca. 10^6 cells/ml.

3. Mix 10 μl cell suspension with 10 μl loading solution in a small Eppendorf tube, and transfer the mixture onto the orifice of the perfusion chamber.

4. Draw in the cell suspension into the chamber using a syringe or pump attached to the chamber. Clean any remnant of the cell suspension around the orifice of the perfusion chamber (For an upright microscope, put the chamber upside down so that cells adhere to the coverslip).

5. At ca. 4 min after mixing, place the chamber on the microscope stage (right-side up) and fix it to the stage with a pair of stage clamps or adhesive tape or using the mechanical stage clamp.

6. Check the cells with transmitted light. Healthy cells will start to spread on the cover slip within a few minutes.

7. Place 100 μl bathing buffer on the orifice of the perfusion chamber.

8. At 5 min after mixing, perfuse the bathing buffer. Do not perfuse too quickly (allow ca. 5 s for perfusion, i.e., 20 μl/s) or the cells may detach from the coverslip.

9. Open the shutter for fluorescence illumination with its diaphragm at the open position, and check if the cells fluoresce. Note that the fluorescence increases even after having washed off the probe because hydrolysis of the esterified probe continues in the cytoplasm.

10. Choose a cell for pH determination using transmitted light and fluorescence, and bring it to the exact center of the visual field.

11. Flip the light path to the photomultiplier and adjust the size of the measurement area using the viewing telescope. If necessary, adjust the sensitivities of the photomultiplier and the recorder with fluorescence illumination.

12. With fluorescence illumination, close the diaphragm for fluorescence illumination to the minimum that covers the cell chosen for measurement. If necessary, center the diaphragm accurately.

13. Start the recorder.

14. With transmitted light, carefully move the stage for a short distance in one direction, either horizontally or vertically, so that no cell is in the area of fluorescence illumination and then switch to fluorescence and measure at the two wavelengths. This gives background fluores-

cence, which should be negligibly small. If it is negligibly small, this and the next step can be skipped afterwards.

15. Bring the cell back to the center.

16. Open the fluorescence shutter. After a few seconds of illumination with the exciter filter 1, switch to the exciter filter 2 with the shutter kept open. Go back to the exciter filter 1. If necessary, repeat the measurement with the exciter filter 2 and exciter filter 1.

17. Close the shutter.

18. Repeat this process (steps 16 and 17) at appropriate intervals.

Exchange of Bathing Solution

1. Place 100 μl of the solution to be perfused onto the orifice of the perfusion chamber.

2. Draw the solution into the chamber using a syringe or pump. Be sure not to draw the solution too far.

3. At the same time, mark the time on the chart, using a foot switch if available.

4. When the measurements is finished, remove the bathing solution using the syringe or pump, lift off the central cover slip, and thoroughly wash the chamber.

Continuous perfusion can be performed in a similar way using an appropriate configuration of the chamber (see Chap. 2).

Loading by Electroporation

This procedure is modified from Howard et al. 1988. The conditions described below give satisfactory results in our laboratory. Equally efficient loading can be obtained using different sets of parameters. If, for instance, 4 μF cannot be selected with the pulse generator, simply choose an available capacity near 4 μF and adjust the voltage so that the net effect would be the same. The efficiency of pore formation is positively correlated to the charging voltage as well as the capacity, so the use of a smaller or larger capacity needs to be compensated for by increasing or decreasing the voltage, respectively. In addition, the efficiency is inversely proportional to the conductance of the solution, which in turn is proportional to the concentration of the buffer and the volume of cell suspension (as long as the solution does not exceed the height of the electrodes). The effect of the gap distance between the electrodes is difficult to predict since it has both positive (via the resistance) and negative (via the electric field strength) components regarding efficiency of pore formation. Temperature affects not only the efficiency of pore formation (less efficient at low temperatures) but also the speed of resealing (slow at low temperatures). If everything is chilled with ice, voltage or capacitance must be increased by at least 50 % and the incubation time immediately after pulse by several-fold. It seems practical to determine optimal parameters empirically for the specific experimental conditions. The above relationships and the parameters shown below will serve as a guide.

1. Wash cells once by low speed centrifugation and resuspend the cells in E buffer at ca. 10^7 cells/ml.

2. Mix 90 µl cell suspension (equilibrated to room temperature) with 10 µl stock solution in a small Eppendorf tube, and immediately transfer the mixture into an electroporation cuvette with a 2 mm gap between the electrodes. Do not chill the cuvette.

3. Give one pulse at 0.7 kV, 4 µF, without shunt resister.

4. Transfer the cell suspension into an Eppendorf tube.

5. After 0.5 min at room temperature, add 0.9 ml E buffer and centrifuge at ca. 10 000 g for a few seconds to sediment the cells. Longer incubation at room temperature results in uptake of the probe by pinocytosis.

6. After one more wash in the same buffer, plate the cells on a coverslip attached to a holder and incubate at 21 °C in a humidified chamber for recovery.

7. At 1–2 h after electroporation, wash the plate gently with bathing buffer to remove nonadherent cells.

Calibration

In situ calibration curves are made by measuring the pH$_i$ of cells after perfusion of calibration buffers at various pHs. It takes a few to several minutes to reach equilibrium. If the amount of weak acid or base is insufficient, some recovery may occur. In such cases, increase the concentration of the acid or base. Alternatively, the use of more lipophilic acid (such as benzoate) or base (such as methylamine) may be attempted. The number of cells to be measured at each pH depends on the variance of the fluorescence ratio. At least ten cells should be measured for each pH value. Attempts to use nigericin and monensin for in situ calibration have met with unreproducible results.

Results and Comments

Effects of Weak Acids and Bases on pHi

Upon exposure to weak acid in an acidic medium (e.g., 4 mM propionate at pH 6.0), pH_i rapidly decreases over a period of 1–1.5 min, followed by gradual recovery. Removal of the acid from the bathing solution by perfusion results in rapid recovery, or often overshoot of pH_i. Weak bases at alkaline pH (e.g., 10 mM ammonium chloride at pH 7.5) affect the pH_i in opposite ways. The effects are more pronounced with increasing lipo-

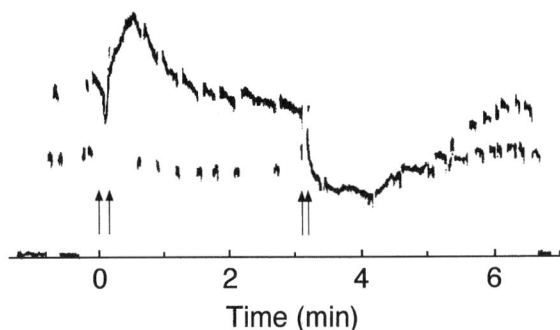

Fig. 28.1. Changes in pH_i upon exposure and removal of ammonia. Starved cells (6 h) of *Dictyostelium discoideum* Ax2 were loaded with carboxy-SNAFL 1 diacetate in a perfusion chamber of the type shown in Fig. 2.1 of Chapter 2, washed with buffer at pH 7.5, and at the time indicated with the first set of *arrows* the solution was perfused with the same buffer containing 20 mM NH_4Cl and subsequently at the second set of *arrows* the same buffer without addition. Solution exchange was completed within the short periods (5–10 s) indicated by the two successive *arrows*. The recording of the pH_i of a single cell is shown. Fluorescence was measured at an emission wavelength (580 nm) by excitation at two wavelengths (550 nm shown with the upper tracing and 490 nm the lower tracing) using a Nikon inverted fluorescence microscope equipped with Nikon P102 photomultiplier

philicity of the weak acid and base. These changes are very reproducible and can be used as a check of the system and the condition used. Figure 28.1 is an example of a chart record showing changes in the fluorescence intensity of a cell exposed to ammonia. In this recording, fluorescence was continuously recorded for an illustrative purpose mostly at the excitation wavelength that is more sensitive to pH.

Relationship Between pHi and Chemotaxis

An example of chemoattractant-induced alkalinization is shown in Fig. 28.2. Aggregation competent cells that had been loaded with carboxy SNAFL-1 DA were stimulated with 1 μM cAMP at time zero. The fluorescence intensity of a cell was measured at intervals by alternating the excitiation wavelength between 488 nm and 550 nm. The fluorescence ratios were converted to pH using an in situ calibration curve.

Fig. 28.2. An example of pH_i recording of cAMP stimulation of chemotactically active cells. Starved cells (6 h) are loaded with carboxy-SNAFL 1 diacetate, washed with buffer at pH 6.0, and at the time indicated with the set of *arrows* the solution was perfused with the same buffer containing 1 μM cAMP. Fluorescence intensity was measured at intervals as in Fig. 28.1, and pH values were calculated from the fluorescence ratios using an in situ calibration curve

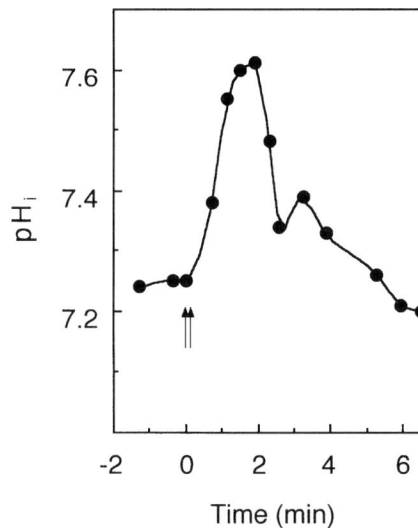

Troubleshooting

- Too weak fluorescence
- The amount of the probe may be insufficient. Try raising the concentration of the probe in the loading solution or increasing the incubation time. (Final concentration of DMSO should not exceed 1 %.) Note

that much higher concentrations are required for *Dictyostelium* cells than for mammalian cells.

- If acetoxymethyl derivatives are used, note that they are relatively impermeant compared to acetate or butyrate derivatives. If raising the concentration and longer incubation time do not give sufficient fluorescence with the system used, it may be necessary to make an acetate or butyrate derivative. They can easily be made by reacting the non-esterified probe with acetic or butyric anhydride (Thomas et al. 1979), although it is not very easy to separate the product to high purity.
- The probe may have deteriorated. Make a new stock from the solid.
- If a phase-contrast lens is used, change to a bright-field lens. If Nomarski optics are used, pull the analyzer off the light path when observing fluorescence.
- The light source may not be in the right position, in which case the excitation light can be more than tenfold lower than it should be. Check if the light beam is exactly centered according to the manufacturer's instruction.

- Fluorescence is confined to intracellular vesicles.
- Try to use a dextran-coupled probe or chloromethyl derivative of the probe.
- In the case of loading by electroporation, shorten the time of incubation in the loading buffer before and after electroporation.

- Too low pH_i values
- The cytosol is likely to be artificially acidified either during sample preparation or during measurement. Check if the cells are healthy immediately before measurement. If so, try the following in this order: (1) choose less fluorescent cells for measurement, (2) lower the excitation light by inserting a darker ND filter, (3) use a lower concentration of the probe, (4) make a new stock from the solid (the probe may have deteriorated and liberated free acid).
- In the case of loading by electroporation, the cells may not have fully recovered. Allow longer recovery time.

- No pH_i response to weak acid/base.
- Cells may be leaky, or the cytoplasm may be very acidic from the beginning. Check the points listed immediately above.

28.2
Measurement of Cell Movement

Although a number of parameters are required to fully characterize cell movement (see Portel and McKay 1979; Wessels et al. 1994), the speed and direction of cell movement alone can still give valuable information. This can be done by simply tracing the tracks of cells recorded on a time-lapse video recorder. The recorded images are either directly transferred to a computer for further analysis or drawn manually onto transparencies on the monitor screen which are subsequently digitized manually or with an x-y digitizer at regular intervals for further analysis on the computer.

Principle

Procedure

1. Plate cells in an appropriate container. A tissue culture dish or a chamber such as the one shown in Chap. 2 can be used. Note that cells move in a characteristic way depending on the substratum (Weber et al. 1995).

2. Record cell movement at a constant temperature on a time-lapse videotape recorder at a desired magnification and compression rate, preferably framewise using internal or external trigger pulses. Illumination should be kept at a minimum. A 40x phase-contrast lens and a speed equivalent to 1 frame/2–4 s may be attempted. This speed may appear very fast to the eye when played back, but it is convenient for analysis.

3. For analysis, the positions of cells at regular intervals (not necessarily every frame) are digitized and their speeds are calculated from the successive x-y coordinates and the interval length. Errors can be reduced by averaging the instantaneous speed over a certain number of intervals before and after each time point.

Simultaneous Measurement of pHi and Cell Movement

There may be cases in which simultaneous measurements of pH_i and cell movement are required. This can be achieved by using a combination of a photomultiplier and a video camera along with appropriate filters

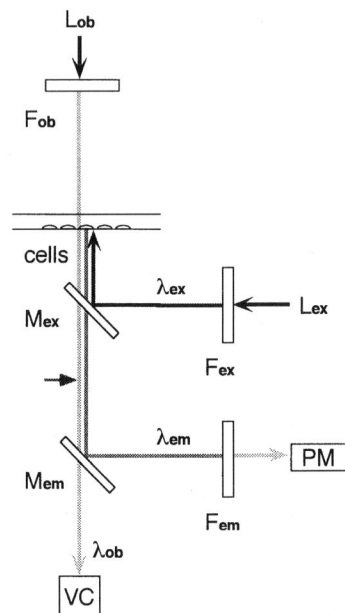

Fig. 28.3. A set-up that enables determination of pH_i while monitoring the cells by transmitted light. The *solid line* represents the path of excitation light (at wavelength λ_{ex}), *dark grey line* the path of fluorescence emission (at wavelength λ_{em}), and the *light grey line* the path of observation light (at wavelength λ_{ob}). $\lambda_{ex} < \lambda_{em} < \lambda_{ob}$. There should be no overlap between transmission of the three filters. M_{ex} is usually a dichroic mirror which reflects the excitation light but pass the fluorescence emission. If the microscope employs a system in which the filters and dichroic mirror are housed in a single block, and if the cut filter, which is in the position indicated by the *arrow*, is an interference filter which does not transmit the observation light, it needs to be moved from the block to the position indicated F_{em}

inserted at the right positions. Most inverted microscopes have two ports allowing simultaneous observation of the image through the oculars and a (video) camera. A combination of appropriate filters can then be used to separate the fluorescence and phase-contrast images of the cell, as shown in Fig. 28.3. To maximize the efficiency of fluorometry, the prism (M_{em}) should divide the light unequally so that the larger fraction of the light is directed to the photomultiplier. It would be ideal to install a dichroic mirror into the position of M_{em}. Be sure that the video camera is sufficiently sensitive at the wavelength of the observation light. An inexpensive CCD camera of relatively high sensitivity (minimum illuminance < 1 lx) can be used for λ_{ob} up to 700 nm. If a fluorescence/phase lens (e.g., CF Fluor DL 40x, Nikon) is not available, images of the contours of the cells can be obtained with ordinary lenses for fluorescence measurement so that the shapes and positions of the cells can still be monitored. The best contrast can be achieved with non-phase lenses by adjusting the aperture iris diaphragm of the condenser. For recording the movement of cells, a time-lapse video tape recorder is used.

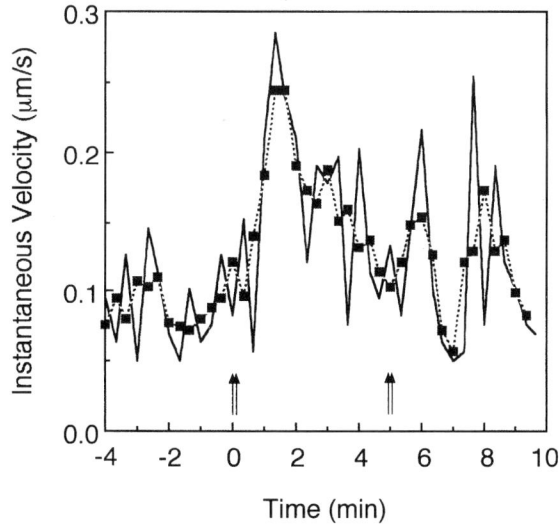

Fig. 28.4. An example of a cell movement measurement. Starved cells of *Dictyostelium discoideum* Ax2 suspended in phosphate buffer pH 7.5 were allowed to settle in a perfusion chamber of the type shown in Fig. 2.1 of Chap. 2, and their movement was recorded on a time-lapse videotape recorder (Panasonic AG 6720) at a rate of 1 frame/4 s using a Nikon ciné autotimer CFMA as an external trigger. A cell was chosen that showed continuous movement in roughly one direction and its movement was digitized manually at 20 s intervals. The *solid line* shows mean velocities determined over the 20 s periods and the *dashed line* with *symbols* represents mean velocities of the three consecutive 20 s periods around each time point. At the time indicated by the first set of *arrows*, the same buffer containing 10 mM NH_4Cl was perfused at ca. 10 µl/s, and at the second set of *arrows*, it was replaced by buffer without addition. Many other cells showed clear responses at about the same time by actively extending their pseudopods without much translocation

Results and Comments

Figure 28.4 is an example of a cell motility assay as determined by measuring the distance between successive positions of a cell's centroid at 20 s intervals. It can be seen that the speed of cell movement increased, with an appreciable delay, after perfusion of NH_4Cl. This is in clear contrast to the almost instantaneous response of pH_i (Fig. 28.1). Measurement of the area and perimeter of the cells would enable detection of subtle responses of the cells.

Acknowledgements. I thank Kyoto Kogaku and KS Olympus for the use of Nikon and Olympus microfluorometry systems, respectively.

References

Aerts RJ, de Wit RJW, van Lookeren Campagne MM (1987) Cyclic AMP induces a transient alkalinization in *Dictyostelium*. FEBS Lett 220:366–370

Bright GR, Fisher GW, Rogowska J, Taylor DL (1987) Fluorescence ratio imaging microscopy: Temporal and spatial measurements of cytoplasmic pH. J Cell Biol 104:1019–1033

Condeelis J, Vahey M (1982) A calcium-regulated and pH-regulated protein from *Dictyostelium discoideum* that cross-links actin filaments. J Cell Biol 94:466–471

Dvorak JA, Stotler WF (1971) A controlled-environment culture system for high resolution light microscopy. Exp Cell Res 68:144–148

Edmonds BT, Murray J, Condeelis J (1995) pH regulation of the F-actin binding properties of *Dictyostelium* elongation factor 1 alpha. J Biol Chem 270:15222–15230

Fechheimer M, Denny C, Murphy RF, Taylor DL (1986) Measurement of cytoplasmic pH in *Dictyostelium discoideum* by using a new method for introducing macromolecules into living cells. Eur J Cell Biol 40:242–247

Furukawa R, Wampler JE, Fechheimer M (1988) Measurement of the cytoplasmic pH of *Dictyostelium discoideum* using a low light level microspectro-fluorometer. J Cell Biol 107:2541–2549

Gerisch G (1960) Zellfunktionen und Zellfunktionswechsel in der Entwicklung von *Dictyostelium discoideum*. I. Zellagglutination und Induktion der Fruchtkörper-polaritat. Roux' Archiv für Entwicklungsmechanik 152:632–654

Gross JD (1994) Developmental decisions in *Dictyostelium discoideum*. Microbiol Rev 58:330–351

Howard PK, Ahern KG, Firtel RA (1988) Establishment of a transient expression system for *Dictyostelium* discoideum. Nucleic Acids Res 16:2613–2623

Inouye K (1985) Measurements of intracellular pH and its relevance to cell differentiation in *Dictyostelium discoideum*. J Cell Sci 76:235–245

Inouye K (1988) Differences in cytoplasmic pH and the sensitivity to acid load between prespore cells and prestalk cells of *Dictyostelium*. J Cell Sci 91:109–115

Inouye K (1990) Control mechanism of stalk formation in the cellular slime mould *Dictyostelium discoideum*. Forma 5:119–134

Klein G, Satre M (1986) Kinetics of fluid-phase pinocytosis in *Dictyostelium discoideum* amoebae. Biochem Biophys Res Commun 138:1146–1152

Knecht DA, Shelden E (1995) Three-dimensional localization of wild-type and myosin II mutant cells during morphogenesis of *Dictyostelium*. Dev Biol 170:434–444

Maeda Y (1983) Axenic growth of *Dictyostelium discoideum* wild type NC-4 cells and its relation to endocytotic ability. J Gen Microbiol 129:2467–2473

Newell PC, Malchow D, Gross JD (1995) The role of calcium in aggregation and development of *Dictyostelium*. Experientia 51:1155–1165

Potel MJ, Mackay SA (1979) Preaggregative cell motion in *Dictyostelium*. J Cell Sci 36:281–309

Schlatterer C, Knoll G, Malchow D (1992) Intracellular calcium during chemotaxis of *Dictyostelium* discoideum: a new Fura-2 derivative avoids sequestration of the indicator and allows long-term calcium measurements. Eur. J Cell Biol 58:172–181

Stoeckelhuber M, Noegel AA, Eckerskorn C, Koehler J, Rieger D, Schleicher M (1996) Structure/function-studies on the pH-dependent actin-binding protein hisactophilin in *Dictyostelium* mutants. J Cell Sci 109:1825–1835

Sussman M (1987) Cultivation and synchronous morphogenesis of *Dictyostelium* under controlled experimental conditions. Meth Cell Biol 28:9–29

Taylor DL, Salmon ED (1989) Basic fluorescence microscopy. Meth Cell Biol 29:207–237

Thomas JA, Buchsbaum RN, Zimniak A, Racker E (1979) Intracellular pH measurement in Ehrlich ascites tumor cells utilizing spectroscopic probes generated in situ. Biochemistry 18:2210–2218

Van Duijn B, Inouye K (1991) Regulation of movement speed by intracellular pH during *Dictyostelium discoideum* chemotaxis. Proc Natl Acad Sci USA 88:4951–4955

Van Duijn B, Vogelzang SA, Ypey DL, van der Molen LG, van Haastert PJM (1990) Normal chemotaxis in *Dictyostelium discoideum* cells with a depolarized plasma membrane potential. J Cell Sci 95:177–183

Van Duijn B, Ypey DL, van der Molen LG (1988) Electrophysiological properties of *Dictyostelium* derived from membrane potential measurements with microelectrodes. J Membrane Biol 106:123–134

van Haastert PJM (1995) Transduction of the chemotactic cAMP signal across the plasma membrane of *Dictyostelium* cells. Experientia 51:1144–1154

Wampler JE, Kutz K (1989) Quantitative fluorescence microscopy using photomultiplier tubes and imaging detectors. Meth Cell Biol 29:239–267

Weber I, Wallraf E, Albrecht R, Gerisch G (1995) Motility and substratum adhesion of *Dictyostelium* wild-type and cytoskeletal mutant cells: A study by RICM/brightfield double-view image analysis. J Cell Sci 108:1519–1530

Wessels D, Vawter-Hugart H, Murray J, Soll DR (1994) Three-dimensional dynamics of pseudopod formation and the regulation of turning during the motility cycle of *Dictyostelium*. Cell Motil Cytoskel 27:1–12

Yabuno K (1971) Changes in cellular adhesiveness during the development of the slime mold *Dictyostelium* discoideum. Dev Growth Differ 13:181–190

Analysis of Agonist-Induced Cell Recruitment in Terms of Intracellular Calcium Mobilization in a Population of Enzymatically Dispersed Pancreatic Acinar Cells

Rolf L.L. Smeets, Remko R. Bosch and Peter H.G.M. Willems

Background

The term cell recruitment is often used to describe the accumulation of motile cells such as neutrophils, lymphocytes, macrophages and mast cells at the site of release of some chemotactic factor. In the broadest sense of the word, however, this term applies to every situation in which a signal mobilizes cells to perform some kind of cellular activity. For instance, in a population of enzymatically dispersed pancreatic acinar cells the peptide hormone cholecystokinin (CCK) both time- and dose-dependently increases the number of cells displaying repetitive changes in free cytosolic calcium concentration (Fig. 29.1). This particular experiment clearly demonstrates that only part of the cells respond to stimulation with a submaximal concentration of 10 pM of the COOH-terminal octapeptide of CCK (CCK_8), with a delay greatly differing between individual cells. As a result, maximal recruitment is reached only at 10 min following the onset of stimulation. By contrast, the vast majority of cells respond within the first minute of stimulation with a close to maximal hormone concentration of 1 nM. Such detection of differences in sensitivity among individual cells was enabled by the development of digital imaging techniques allowing simultaneous monitoring of agonist-induced changes in $[Ca^{2+}]_i$ in large numbers of individual cells loaded with Ca^{2+}-sensitive fluorescent dye. This chapter aims to give a detailed description of the method used in our laboratory to analyze agonist-induced cell recruitment in terms of intracellular calcium mobilization.

Principle and applications

Acinar Cells. In our studies on agonist-induced cell recruitment we make use of a preparation of freshly isolated rabbit pancreatic acinar cells. The exocrine part of the pancreas produces a fluid rich in electrolytes and digestive enzymes. Approximately 90 % of the exocrine tissue consists of relatively large cells with a pyramidal shape. These cells, referred to as acinar cells, synthesize, store and release the digestive enzymes. Acinar

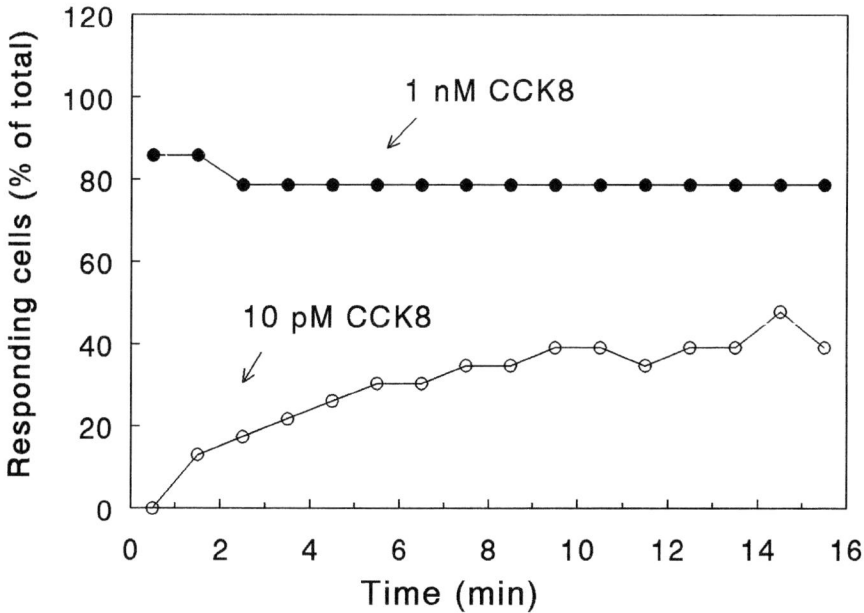

Fig. 29.1. Time-dependence of the recruitment of acinar cells displaying transient rises in $[Ca^{2+}]_i$ following submaximal and close to maximal stimulation with CCK_8. Isolated pancreatic acinar cells, loaded with Fura-2, were stimulated with the indicated concentration of CCK_8 for 16 min. Changes in fluorescence emission ratio were monitored by digital imaging microscopy of Fura-2 fluorescence from individual acinar cells. The percentage of apparently active cells was determined at 1 min intervals following the onset of stimulation with the indicated concentration of CCK_8

secretion is controlled by a variety of stimulants including the gut hormone CCK.

Signal Transduction. CCK binds to a specific receptor, the CCK-A receptor, coupled to phospholipase Cβ to promote the hydrolysis of phosphatidylinositol 4,5-bisphosphate. Triggering of acinar secretion by CCK is therefore paralleled by a rapid increase in inositol 1,4,5-trisphosphate (IP_3), releasing Ca^{2+} from intracellular stores, and 1,2-diacylglycerol (DAG), activating protein kinase C (PKC). The aim of our studies is to assess possible feedback roles for PKC in signaling through the CCK-A receptor (Willems et al. 1993, 1995a, b; Smeets et al. 1996). In order to investigate whether certain drugs such as the potent PKC activator 12-O-tetradecanoylphorbol 13-acetate (TPA) affect the dose-recruitment curve for CCK and therefore the above signaling process, video imaging microscopy of Fura-2-loaded acinar cells is used.

Video Imaging. After loading with the fluorescent Ca^{2+} indicator Fura-2 the acinar cells are transferred to a perfusion chamber placed on the stage of an inverted microscope. In our studies we routinely use an epifluorescent 40x magnification quartz oil immersion objective (NA 1.30). In combination with a video frame size of 256×256 pixels, this allows simultaneous monitoring of close to 100 individual pancreatic acinar cells. Superfusion is started and the cells are allowed to equilibrate before being challenged with the indicated concentration of the hormone. The cells are alternately excited with light of 340 and 380 nm and the corresponding images, captured at 492 nm by means of a video camera, are written into a dynamic random access memory. After ratioing of the corresponding 340 and 380 nm images, individual cells are selected and the changes in average fluorescence emission ratio, representing changes in average $[Ca^{2+}]_i$, are analyzed as a function of time. Cells displaying a rise in fluorescence emission ratio after the hormone has entered the incubation chamber are referred to as responding cells.

Response Patterns. The recordings presented in Fig. 29.2 are illustrative for the differences in response pattern that can be observed within a population of freshly isolated pancreatic acinar cells stimulated with CCK_8. Moreover, the figure shows that, in this experiment, in which the hormone concentration is stepwise increased, some cells respond to 10 pM CCK_8 (a), whereas others (b and c) do not. Dose-response curves for the recruitment of acinar cells by CCK_8 are routinely prepared by stimulating batches of cells with single hormone concentrations for 5–10 min.

Figure 29.3 shows a typical dose-recruitment curve ($EC_{50} = 16.8$ pM) for CCK_8. The number of responding cells is expressed as a percentage of the total, and in this particular experiment on average 60 acinar cells are analyzed for each data point. The figure also shows that the PKC activator TPA (0.1 µM) evokes a rightward shift ($EC_{50} = 175$ pM) of the dose-recruitment curve for CCK_8. This finding clearly demonstrates that activation of PKC reduces the sensitivity of the acinar cells to CCK_8. The next section will give a detailed description of this type of experiment.

Fig. 29.2. Different response patterns of freshly isolated rabbit pancreatic acinar cells stimulated with CCK_8. Fura-2-loaded pancreatic acinar cells were superfused with the indicated concentration of CCK_8 for the indicated period of time. Changes in $[Ca^{2+}]_i$ were measured by means of video imaging microscopy. The recordings shown are from three out of 60 individual acinar cells monitored simultaneously. (Reprinted from Willems et al. 1995 with permission)

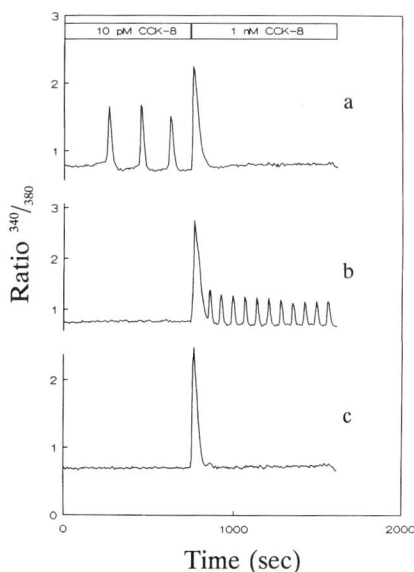

Materials

In essence, analysis of agonist-induced cell recruitment involves simultaneous measurement of receptor-mediated changes in cellular activity in large numbers of individual cells. Here we describe a method to analyze agonist-induced cell recruitment using video imaging microscopy of cells loaded with Ca^{2+}-sensitive fluorescent dye.

Imaging Setup. In our laboratory, dynamic video imaging of cells loaded with the fluorescent Ca^{2+} indicator Fura-2 is carried out using an inverted microscope (Nikon Diaphot) in combination with the MagiCal hardware and TARDIS software (provided by Joyce Loebl, at present at Applied Imaging, Hylton Park, Wessington Way, Sunderland, UK), as described in detail by Neylon et al. (1990). In this system, the light of a 100 W Xenon lamp is directed through a quartz neutral density filter (ND ≥ 1.5, Ealing Electro-Optics, Holliston, MA, USA) to avoid damage of the two excitation filters and to reduce bleaching of the intracellularly trapped fluorochrome. Fluorescence emitted at 492 nm is monitored by means of an intensified charge-coupled device (CCD) camera (Photonic Science) operating at video frame rate with the rotating filter wheel, containing the excitation filters of 340 and 380 nm (± 12 nm), in the discontinuous stepping mode.

Equipment

Fig. 29.3. Effect of TPA on the dose-recruitment curve for CCK_8 in freshly isolated rabbit pancreatic acinar cells. Fura-2-loaded pancreatic acinar cells were preincubated in either the absence (*open circles*) or presence (*closed circles*) of 0.1 µM TPA for 10 min and subsequently stimulated with the indicated concentration of CCK_8 for 5–10 min. The percentage of cells displaying a receptor-evoked increase in $[Ca^{2+}]_i$ was determined by means of video imaging microscopy

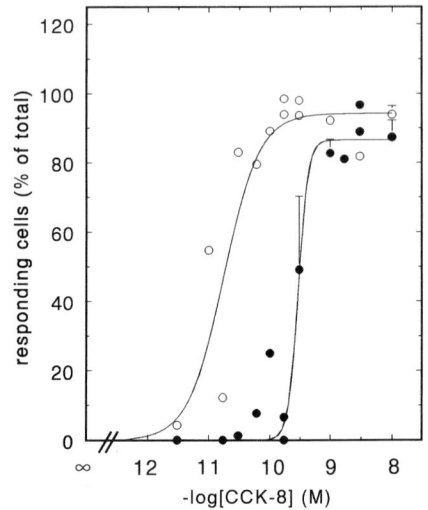

Procedure

Typical experiment

– In a typical experiment with pancreatic acinar cells, at each excitation wavelength, 16 video frames with frame capture dimensions of 256×256 pixels are collected, averaged, and written into a dynamic random access memory with a maximal capacity of 504 frames. Averaging is carried out to improve the signal-to-noise ratio. Frames to be averaged are collected at video frame rate. In the case of 16 frames this means a sampling time of 640 ms. The interframe interval between the last 340 nm sample and the first 380 nm sample is 500 ms. This is the time required for the filter wheel to change position in the discontinuous stepping mode. Sampling of one averaged 340 nm sample and a corresponding averaged 380 nm sample therefore takes 2280 ms. Obviously, in case the cells under investigation display high frequency Ca^{2+} fluctuations the sampling time has to be reduced by decreasing the number of frames to be averaged.

– At the end of the experiment, a region free of cells is selected and one averaged background frame is collected at each excitation wavelength. Before ratioing of the frames, the averaged 340 and 380 nm background frames are subtracted from the corresponding experimental frames.

– The images are ratioed pixel by pixel and color coded by a calibration table. In our experiments, we routinely use an interframe interval of

4 s between the last 380 nm and first 340 nm sample. In doing so, the interframe interval between the ratio frames is increased to 6.3 s. At the maximal storage capacity of 251 frames for each excitation wavelength this results in a total sampling time of 26 min.

– After ratioing, individual cells are selected by means of a light pen and changes in fluorescence emission ratio are plotted in a graph. A major disadvantage of the TARDIS software is that during the working out of the experiment only eight out of 40–100 cells can be analyzed at a time. Therefore, video imaging systems possessing the ability to pre-select large numbers of areas of interest for on-line ratioing are highly preferable to the system described above.

Pancreatic Acinar Cells

Studies undertaken to unravel the mechanism of action of CCK frequently make use of a preparation obtained by a procedure employing the enzymatic digestion of pancreatic tissue in combination with mechanical shearing and chelation of divalent cations to break junctional complexes (Amsterdam and Jamieson 1974).

Cell isolation

Using this procedure we routinely obtain a preparation, referred to as isolated pancreatic acinar cells, consisting of both single cells (22 %) and cells forming part of a doublet (21 %), a triplet (34 %), or a larger group of (mostly 4–5) cells (22 %) (Willems et al. 1993a).

Fluorescence Measurements

1. At the end of the isolation procedure, the acinar cells are resuspended in a HEPES/Tris medium (pH 7.4) containing 133 mM NaCl, 4.2 mM KCl, 1.0 mM $CaCl_2$, 1.0 mM $MgCl_2$, 5.8 mM glucose, 0.2 mg/ml soybean trypsin inhibitor, an amino acid mixture according to Eagle, 1 % (w/v) bovine serum albumin and 10 mM HEPES, adjusted with Tris to pH 7.4 (loading medium) and kept on ice until use.

2. For each fluorescence measurement, an aliquot of the chilled suspension is removed, centrifuged for 5 min at 100 g, and resuspended in loading medium. The temperature of the suspension is brought to 37 °C and the cells are loaded with Fura-2 by the incubation in 3.3 μM Fura-2-AM (Molecular Probes Inc., Eugene, OR, USA) for 15 min. After loading, excess Fura-2-AM is removed by washing twice with the

above HEPES/Tris medium (pH 7.4) containing 0.1 % (w/v) bovine serum abumin (incubation medium).

3. Fura-2-loaded cells are resuspended in incubation medium containing either 0.1 μM TPA (Sigma Diagnostics, St. Louis, MO, USA) or 0.1 % (v/v) DMSO. The cell density is chosen so as to form a single layer of cells when 300 μl of the suspension is transferred to the incubation chamber. The phorbol ester is added from a 100 μM stock solution in DMSO. This stock solution is stored at -20 °C.

Note. Tumor promoting phorbol esters such as TPA have to be treated with great care.

4. The cells (300 μl) are transferred to a thermostatic (32 °C) incubation chamber (Leiden Chamber; Ince et al. 1985) and allowed to attach to a glass coverslip, forming the bottom of the chamber, for 5 min. The inlet, connected to a superfusion pump, and the outlet, connected to a vacuum pump, are placed opposite to each other in such a manner that the volume of the incubation chamber is kept at 300 μl. Superfusion (1 ml/min) is started and unattached cells are automatically sucked off. The superfusion medium contains either TPA (0.1 μM) or 0.1 % (v/v) DMSO. If necessary, the cover slips can be coated with poly-L-lysine (MW $>$ 300 kDa; Sigma) to improve cell attachment.

5. The incubation chamber is placed on the stage of an inverted microscope. Using an epifluorescent 40x magnification quartz oil immersion objective in combination with video frame capture dimensions of 256$\times$256 pixels close to 100 acinar cells are monitored simultaneously for 5–10 min.

6. The acinar cells are stimulated with the indicated concentrations of CCK_8 at exactly 10 min following the onset of stimulation with TPA. In this context it is important to note that certain PKC isotypes are down-regulated when cells are incubated with TPA for prolonged periods of time (Smeets et al. 1996). Since pancreatic acinar cells do not display spontaneous changes in $[Ca^{2+}]_i$ monitoring can be started 1 min before superfusion medium containing CCK_8 enters the incubation chamber.

7. After background correction and ratioing, individual cells are selected and analyzed for changes in fluorescence emission ratio.

Results

Off-Line Analysis

• For a dose-recruitment curve the number of cells displaying an increase in fluorescence emission ratio is expressed as a percentage of the total. It should be noted that lower hormone concentrations require longer measuring times since the delay time increases with decreasing hormone concentration (Fig. 29.1).

• In case the time-dependence is studied the percentage of cells displaying repetitive changes in fluorescence emission ratio is expressed as a function of time.

Comments

Pancreatic Acinar Cells

• In our experience, freshly isolated pancreatic acinar cells retain their responsiveness to hormonal stimulation for up to 10 h following isolation when kept on ice until use.

Cell Lines

• We have extended the observation of agonist-evoked cell recruitment to a eukaryotic expression system, namely the chinese hamster ovary (CHO) cell, expressing the pancreatic acinar CCK-A receptor. In case of CHO cells, transiently expressing the CCK-A receptor, an average of 60 % of the cells display a CCK_8-induced increase in fluorescence emission ratio at 2–3 days following electroporation in the presence of an expression vector containing the full-length CCK-A receptor cDNA. The advantage of such a transient expression system is that it allows rapid screening of mutated calcium mobilizing receptors.

• In preliminary experiments, using mock-transfected CHO cells as a control, it was regularly observed that a certain percentage of the cells spontaneously displayed transient rises in $[Ca^{2+}]_i$. Careful examination revealed that the incidence of these spontaneously active cells depended on the degree of confluency of the monolayer. Thus, in confluent monolayers large numbers of cells showed spontaneous activity, whereas in subconfluent monolayers the percentage of spontaneously active cells remained below 5 % of the total. During analysis of cell recruitment the possibility and incidence of spontaneous Ca^{2+} changes should always be kept in mind.

General Conclusion

In conclusion, analysis of agonist-induced cell recruitment provides a powerful means to assess whether a particular manipulation, be it pharmacological or molecular biological, affects signaling through Ca^{2+} mobilizing receptors.

References

Amsterdam A, Jamieson JD (1974) Studies on dispersed pancreatic exocrine cells. 1. Disociation technique and morphologic characteristics of separated cells. J Cell Biol 63: 1037–1056

Ince C, van Dissel JT, Diesselhoff MC (1985) A teflon culture dish for high-magnification microscopy and measurements in single cells. Pflügers Arch 403: 240–244

Neylon CB, Hoyland J, Mason WT, Irvine RF (1990) Spatial dynamics of intracellular calcium in agonist-stimulated vascular smooth muscle cells. Am J Physiol 259: C675-C686

Smeets RLL, Garner KM, Hendriks M, van Emst-de Vries SE, Peacock MD, Hendriks W, de Pont JJHHM, Willems PHGM (1996) Recovery from TPA inhibition of receptor-mediated Ca^{2+} mobilization is paralleled by down-regulation of protein kinase $C\alpha$ in CHO cells expressing the CCK-A receptor. Cell Calcium 20: 1–9

Willems PHGM, van Emst-de Vries SE, van Os CH, de Pont JJHHM (1993a) Dose-dependent recruitment of pancreatic acinar cells during receptor-mediated calcium mobilization. Cell Calcium 14: 145–159

Willems PHGM, van Hoof HJM, van Mackelenbergh MGH, Hoenderop JGJ, van Emst-de Vries SE, de Pont JJHHM (1993b) Receptor-evoked Ca^{2+} mobilization in pancreatic acinar cells: evidence for a regulatory role of protein kinase C by a mechanism involving the transition of high-affinity receptors to a low-affinity state. Pflügers Arch. 424: 171–182

Willems PHGM, Smeets RLL, Bosch RR, Garner KM, van Mackelenbergh MGH, de Pont JJHHM (1995) Protein kinase C activation inhibits receptor-evoked inositol trisphosphate formation and induction of cytosolic calcium oscillations by decreasing the high-affinity state of the cholecystokinin receptor in pancreatic acinar cells. Cell Calcium 18: 471–483

Fluorescence To Measure Intracellular Ions

Confocal Microscopy

Chapter 30

Quantitative Confocal Fluorescence Measurements in Living Tissue

MARK FRICKER, RACHEL ERRINGTON, JULIAN WOOD,
MONIKA TLALKA, MIKE MAY AND NICK WHITE

Background

Fluorescent probes offer unparalleled opportunities to visualize and quantify dynamic events within single living cells with a minimum of perturbation. Cells or monolayers maintained in culture can be readily imaged; however, measurements are often also needed from cells within intact tissues that are operating in their correct physiological context to include the effects of cell-cell interactions and the mechanical, ionic and physiological effects of the extracellular matrix (e.g., Errington et al. 1997). A range of different measurement techniques are now available to quantify fluorescence signals from reporter molecules within biological specimens, including fluorimetry, flow cytometry, microscope photometry, camera and confocal microscopy. Each system performs well for a specific range of sampling conditions and specimens; thus, several techniques in combination may be needed to provide a sufficiently flexible balance between the spatial, temporal and spectral resolution required. Several recent volumes cover many of the technical details and practical applications of these techniques (e.g., Wang and Taylor 1989; Taylor and Wang 1990; Mason 1993; Matsumoto 1993; Nuccitelli 1994; Pawley 1995a). Here we focus on the practical advantages and disadvantages of quantitative fluorescence measurements using confocal microscopy as a tool to study living cells in intact animal, plant and fungal tissues (Errington et al. 1997; Fricker et al. 1994; White et al. 1996).

Optical Sectioning Using Confocal Microscopy

In conventional camera-based fluorescence microscopy, the excitation light is needed to illuminate a large area in the (x, y) plane to allow simultaneous measurement of the fluorescence emission for all points in the focal plane. Fluorescence is also unavoidably excited from an extended volume above and below the focal plane. This fluorescence signal is not confined to the focal plane of the image but is also spread out or blurred at the faceplate of the camera. If the specimen is very thin, in the order of a few microns, this usually presents little problem for physiological measurements; however in larger cells or cells forming part of a multicellular structure, the out-of-focus blur significantly degrades the contrast and distorts any attempts at quantitative measurements. In a confocal microscope, the excitation light is focussed to one or more discrete points in the specimen. However, fluorescence is still excited in the cones of illumination above and below the focus. This out-of-focus fluorescence is prevented from contributing to the final image by a physical barrier in an imaging plane before the detector. In-focus fluorescence from the focal point passes through a small aperture in the image plane. Thus the illumination and detection points are co-aligned or "confocal." A single optical section is generated by scanning the confocal point in the (x, y) plane to build up an image. A series of optical sections collected at different focus levels constitutes a three-dimensional (3-D) image. Repeated sampling in time forms a 4-D image. There are a wide range of commercially available confocal microscopes that achieve varying degrees of optical sectioning by scanning and detecting single points, multiple points or narrow slits (Pawley 1995a).

Quantitative Imaging Using Confocal Techniques

The main advantages of confocal microscopy in quantitative fluorescence measurements of physiological parameters arise from the well defined 3-D volume (voxel) that is sampled to form each 2-D picture element (pixel). The volume probed is always asymmetric, being at least three to four times longer in the axial (z) direction. The overall shape of the confocal probe in fluorescence is described by the point spread function (psf). Optical probe dimensions are usually given in terms of the full width at half the maximum height (FWHM) in a given direction along one of the orthogonal axes. Point scanning confocal instruments can achieve a probe size of $0.2\,\mu m \times 0.2\,\mu m \times 0.6\,\mu m$, in x, y and z, respectively, with a high numerical aperture (NA; 1.4) oil immersion lens in fluorescence. Typical probe dimensions for physiological measurements are likely to be larger, in the region of $0.4\,\mu m \times 0.4\,\mu m \times 1.2\,\mu m$, as longer working distance, lower NA lenses are often used, and the optical sectioning is relaxed by increasing the pinhole diameter.

The removal of out of focus blur allows discrimination of signals from subcellular domains, whilst contaminating signal from autofluorescence and/or stained tissues outside the focal plane is rejected. Measurements can also be made even if substantial compartmentalization has occurred, provided the organelles can be adequately resolved. In addition, the 3-D sampling available with confocal systems allows a detailed quantitative description of any subcellular or cellular morphological changes associated with physiological signals.

In many instruments there is control over the degree of confocality, the area scanned, the scan speed and the number of frames averaged that provides the user with considerable flexibility in choice of sampling speed and the volume of specimen imaged. At one extreme, repeated sampling can be made of a single point in the 1 MHz range, whilst a 3-D volume of an thick tissue specimen, such as a root tip or invertebral disc for example, can be sampled at high resolution in a few minutes. Thus different facets of a given biological question can be tackled on the same instrument.

Statistics of Fluorescence Intensity Measurements

To make absolute photometric measurements, the relationship between the image brightness and the concentration of the fluorophore has to be calibrated. Accurate measurements of low-intensity fluorescence signals are limited by the number of photons emanating from the specimen. For a homogeneous sample, the sampled photons are drawn from a population with a mean value (S) and a distribution defined by Poisson statistics in which 63 % of values occur in the range $S \pm \sqrt{S}$ (Pawley 1995b). The quantity, $\sqrt{S}$, is the intrinsic noise and provides an upper limit to the accuracy that can be achieved in any measurement. It is the task of the imaging and digitization system to record this signal as faithfully as possible under a given set of experimental conditions. First, this involves maximizing the number of photons emitted per voxel of dye that actually reach the detector. For example, increasing the NA of the lens increases the collection efficiency, whilst careful selection of wavelength filters and pinhole size may improve the contrast by optimizing the optical transfer efficiency for specified wavelengths. Second, the quantum efficiency (QE) of the detector determines the number of photons that generate a measurable signal. Photomultiplier detectors have very good linearity over a wide intensity range, but they have a low QE that is markedly wavelength sensitive. Different photocathode materials and end-window geometries can markedly increase the QE (Pawley 1995b). Third, it is important to minimize additional noise from the electronics.

In a system with true photon counting, a discriminator is used to eliminate multiplicative noise in the detector: each photoelectron arising from the photocathode contributes a single pulse that is counted. This is the optimal digitization technique for signals with a low number of photons within the linear range of the photon counting circuitry, as it provides a way of determining the true black level. The number of photons that can be reliably counted in commercially available CLSMs for a 1 µs pixel is around 10–20/scan. In practice, this is often comparable to the maximum signal that is actually present under physiological imaging conditions.

In analogue mode the current from the photomultiplier can be integrated over the pixel dwell time and converted to a voltage which is usually digitized to 8 bits. The current arising from each photon depends on the statistical variation in the number of secondary electrons produced, which typically adds Gaussian noise to the signal. Thermally produced photoelectrons contribute a constant dark current with shot noise, whilst photomultiplier and digitizer offsets will add DC voltages, and further noise to the output. Thus, although the digitized 8 bit image has 0–255 gray levels, there is no information about the actual number of photons contributing to the signal under these conditions, unless the system is calibrated.

Range of Signals

To illustrate the range of signals that might be encountered a typical example under physiological imaging conditions will be considered. The maximum signal from a fluorescein molecule operating near saturation is about 10^9 photons/s and requires a laser power of ≈ 1–2 mW at the back focal plane of the objective (Sandison et al. 1995; Tsien and Waggoner 1995), most of which ends up passing through the sampling volume. Under physiological imaging conditions the laser power is 10- to 50-fold lower, to maintain cell viability, and each voxel is effectively imaged for ≈ 1 µs in a point scanning instrument during which time each molecule would emit between 20–100 photons. A 0.4 µm $\times 0.4$ µm $\times 1.2$ µm "cubic" voxel of dye at a concentration of 10 µM contains ≈ 1200 molecules, giving 24 000–120 000 photons from the illuminated voxel. The optical transfer efficiency of a typical confocal system is 2 %–10 %, giving 480–12 000 photons at the detector. The quantum efficiency of the photomultiplier varies with wavelength from 3 % to 13 %, giving a mean of 14–1560 photons that would actually be detected per pixel. If these were both represented within a 0–255 range, the mean $\pm \sqrt{\text{variance}}$ (noise) would be about 201 ± 54 gray levels in the first case and about 249 ± 6 in the second (in the absence of additive Gaussian noise). Scanning at faster rates significantly reduces the pixel dwell time and the number of photons excited per pixel resulting in significantly increased variance. Video rate imaging at 25 Hz per frame gives pixel dwell times of usually $\approx 0.1\,\mu s$ and thus a maximum number of photons per pixel of about 1.4–156 per scan, equivalent to 138 ± 117 to 236 ± 19 gray levels for a point scanner.

In physiological measurements, there may be additional weak background signal from the autofluorescence in lens elements, sample chamber and specimen, so a practical definition of background (B) is a composite including nonspecific fluorescence in the optics and sample (B_1), and DC voltages from the photomultiplier dark current (B_2); the photomultiplier offset (B_3); and the digitizer offset (B_4), etc., such that the total signal recorded (T) is $S + B_1 + B_2 + B_3 + B_4 \ldots B_n$. An average value of B is usually measured from a region of the specimen without fluorophore and subtracted from the total to give the mean signal (i.e., $S = T-B$). However, the overall noise (N) also contains components from all these sources as well as the shot noise associated with sampling of the signal. Noise associated with sampling photons scales as the square root of the number of photons thus the total noise $N = \Sigma(\sqrt{(S+B_1)} + N_1 + N_2 + N_3 + N_4 \ldots N_n)$ for each sample (pixel). It is not possible to separate and remove the background noise components, thus noise from all sources contributes to an increase in the variance of the signal. To make an accurate estimate of the mean signal requires that both the background and the total intensity are recorded without clipping the intensity distributions (see Sect. 30.1). Integration over x, y, z or time will increase the signal-to-noise (S/N) ratio as the signal increases proportionally with the number of pixels averaged, whilst the noise only increases with the square root of the number.

Identifying and Minimizing Problems Associated with Confocal Measurements

The primary objective of quantitative physiological measurements is to maximize the S/N ratio with minimal disruption to the cell physiology. The demanding optical system required to achieve optical sectioning in a confocal microscope both accentuates many of the problems associated with accurate measurements in biological tissues using conventional microscopy and poses some unique additional difficulties. In particular, problems arise when observing cells within intact or thick microscopical specimens. A digitally stored 3-D fluorescence image represents an approximation to the intensities emitted from a corresponding volume within the specimen; however, physical interactions between the specimen and both the microscope illumination and fluorescence signal significantly affect the position and volume of the confocal probe. The most important of these interactions are refractions which occur at all non-opaque boundaries through the optical path of both microscope and specimen resulting in image blurring and distortion. As refractive index varies with wavelength in most materials (dispersion), the magnitude of the spatial errors will also vary with wavelength (chromatic aberration). Blurring of the probe results in a decrease in signal accepted by a confocal detector pinhole, leading to significant attenuation of fluorescence intensity. Thus, in confocal microscopy, spatial errors also give rise to photometric errors (Hell and Stelzer 1995; White et al 1996). Separate calibration of such spatial and photometric errors for a given specimen is required to convert the raw digital data to a true rectilinear array of corrected intensities (Errington et al. 1997; White et al. 1996).

Given the large number of variables, it is not possible or even desirable to provide a prescriptive set of rules that can be established as generalized protocols to achieve high S/N, good specimen viability and accurate measurements for each experimental system. The points below represent guidelines that we have found useful in live cell imaging projects across a range of plant, fungal and animal specimens and are grouped under the overall scheme shown in Fig. 30.1. Additional tutorial material on general aspects of practical confocal microscopy can be found in Centonze and Pawley (1995). We have concentrated on the confocal instrument calibration, data collection and analysis protocols appropriate for live cell confocal imaging (see also Terasaki and Dailey 1995). Techniques for calibration of specific dyes can be found elsewhere in this volume (see also Wang and Taylor 1989; Mason 1993; Matsumoto 1993; Nucitelli 1994).

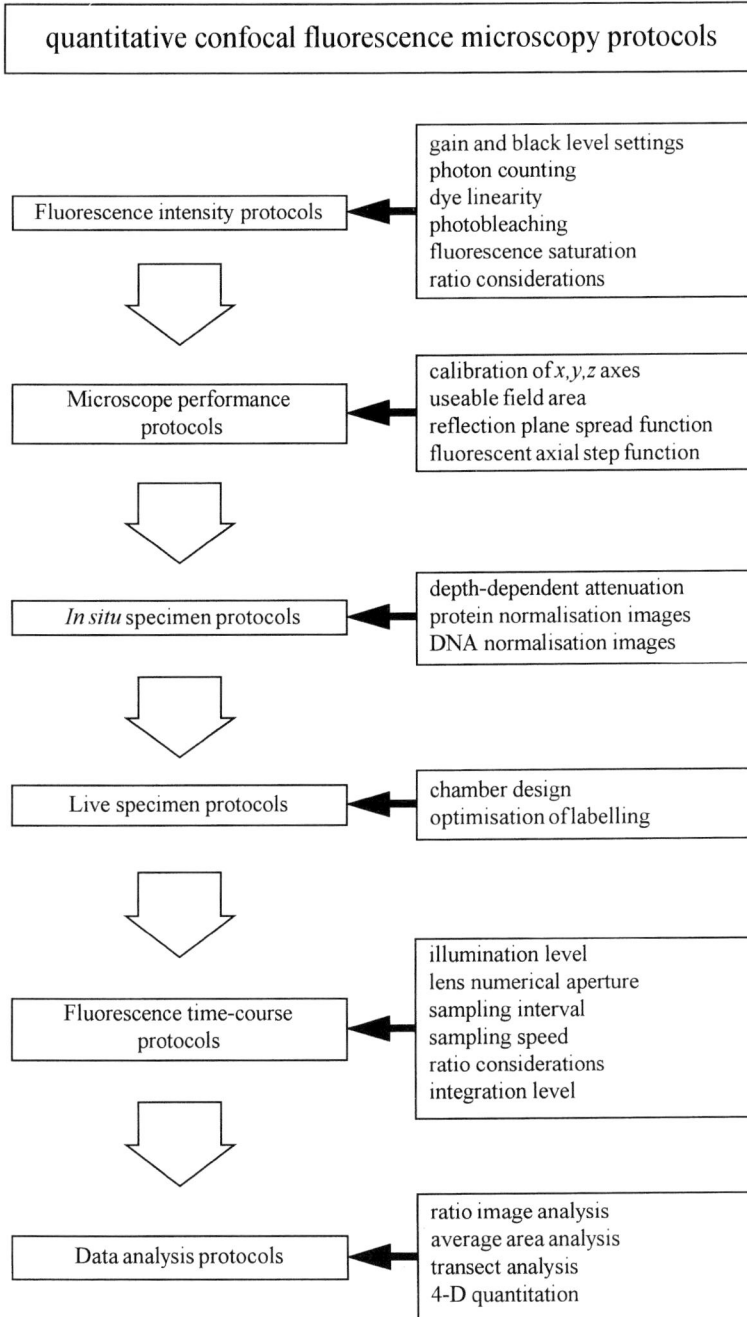

Fig. 30.1. The major stages in quantitative confocal fluorescence microscopy

Procedure

30.1
Fluorescence Intensity Protocols: Fluorescence Image Statistics

Instrument settings

The aim is to determine instrument settings for accurate measurements of fluorescence intensities.

1. Set up a slide-coverslip sandwich using spacers to make a chamber around 100 μm deep containing an appropriate buffer solution with and without 50 μM fluorochrome. Appropriate spacers include thin-coverslips, aluminium foil or dabs of nail varnish (Fig. 30.2A). All

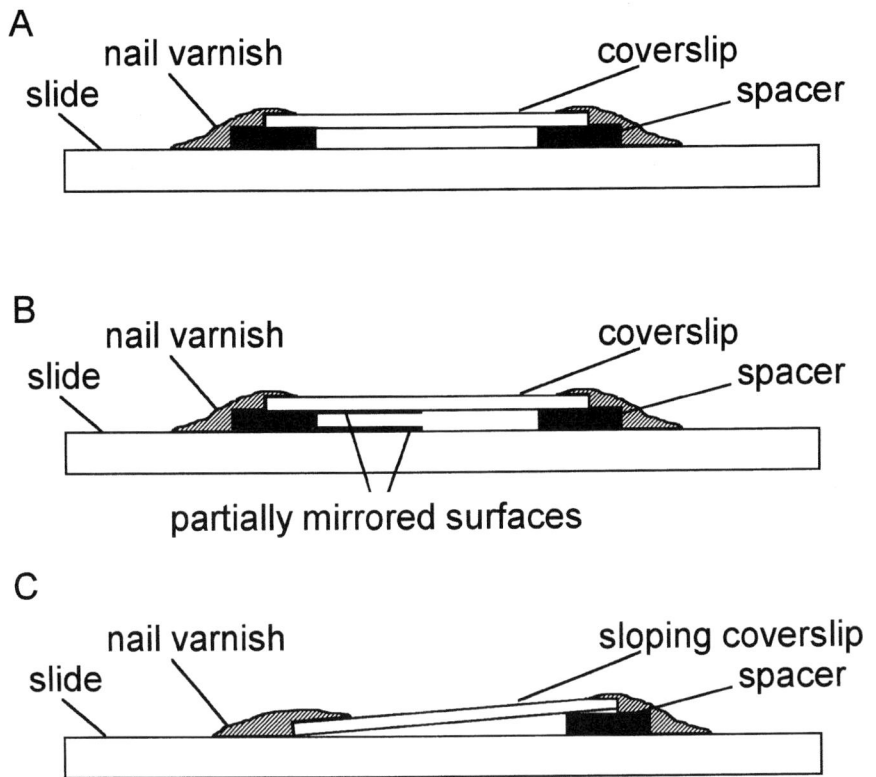

Fig. 30.2A–C. The different slide-coverslip sandwich configurations for microscope performance measurements. A General fluorescence intensity measurements; B measurement of reflection plane spread function; C rapid measurement of reflection plane-spread function and fluorescence axial step function

chambers should be made hours or days in advance and sealed with nail varnish to ensure maximum stability of the preparation.

2. Focus into the blank "sea" away from the coverslip and scan a single (x, y) plane. This provides a measure of the background including autofluorescence from the medium, etc.

3. Plot the frequency histogram of the pixel intensity distributions for the central 1/4 of the field.

4. The black level should be adjusted so that all of the fluorescence intensity histogram is recorded above a gray level of zero without clipping. It is now possible to determine the average background value.

5. Focus into a sea containing 50 μM fluorochrome and plot the frequency histogram.

6. The histogram should have an approximately symmetrical shape and should also not be clipped. If photon counting is used and the signal is low enough, the noise should be near to the square root of the mean number of photons. In analogue mode there is not necessarily a predictable relationship between the mean and variance (see "Statistics of Fluorescence Intensity Measurements").

7. The gain should be incrementally adjusted and the mean intensity plotted after inspection of the histogram distribution. If any of the histogram reaches 255, the signal will be clipped, reducing the apparent mean at higher gains. A lower dye concentration or laser setting is required and the measurement continued.

8. The linear region of the gain control can be determined as well as the effects of increasing gain on the variance of the signal.

9. A plot of the S/N at varying pinhole diameters also provides an indication of the optimal pinhole setting to maximize the S/N ratio.

10. It is often useful to generate a specific look-up table with color coding to indicate the correct black level setting and indicate values approaching saturation – avoiding the need to keep measuring the histogram distribution.

30.2
Fluorescence Intensity Calibration Protocols: Fluorochrome Properties

The aim is to determine the range of dye concentration that gives a linear response.

Dye concentration and linearity

1. With the gain and black level set correctly for 50 µM fluorochrome, the effects of varying fluorophore concentration can be examined. At low fluorophore concentrations, up to ca. 50 µM, a linear relationship is predicted between concentration and fluorescence emission. At higher concentrations, self-absorption and self-quenching gradually decrease the relative intensity of emission towards a plateau value or even a slight decrease in emission at higher concentrations.

2. **Photobleaching:** There is a finite probability that a fluorophore will undergo conversion to a nonfluorescent form with repeated cycles of excitation and emission in a process termed photobleaching. Typically the number of photons emitted before the fluorophore degrades is about 10^4–10^5 and limits the total signal that can be produced for any given concentration of fluorophore. In principle, the rate of photobleaching can be assessed from the change in mean intensity after repeated scanning of the fluorochrome sea. However, under these conditions the dye is freely diffusible and replenishment by unbleached molecules will reduce the apparent effects of bleaching in a fluorescent sea. The rate of bleaching in vivo is likely to be higher as the fluorophores are constrained to a much smaller volume and will also depend markedly on the local environment, such as oxygen level, etc.

3. **Saturation:** Fluorescent molecules also saturate at high laser intensities when electrons are depleted from the ground state, particularly in fluorophores with long decay times. The maximum rate for fluorescein, for example, is about 10^9 photons/s and this typically requires laser intensities in the order of 1–2 mW at the back focal plane of the objective (Sandison et al. 1995; Tsien and Waggoner 1995). Laser intensity settings required to keep cells viable are typically 10- to 50-fold lower than this so saturation is unlikely to be a problem during physiological measurements.

4. **Ratio considerations:** Excitation ratio measurements require a means to balance the intensity of the two laser lines using neutral density filters, for example, so that the fluorescence signal remains within the useable range for both wavelengths across the relevant range of dye

concentration and dye responses. We have found it useful to start each set of physiological measurements with an in vitro calibration response for the dye across the dynamic range anticipated in vivo, both to adjust the relative laser intensities and to define the in vitro calibration curve.

30.3
Microscope Performance Protocols: Calibration of Image (x, y) Axes

The aim is to calibrate the spatial dimensions of the image (x, y) axes.

1. Image a stage micrometer in transmission or reflection mode.

2. Measure the length of an appropriate number of graticule divisions and record the number of pixels. The pixel size (in microns)is equal to the length of the graticule units measured (in microns)/number of pixels.

3. Repeat for each excitation wavelength available.
   Lateral chromatic aberration will cause a slight change in the field size between different wavelengths which will result in misregistration of images. In fluorescence there will also be misregistration of illumination and detection probes at each wavelength leading to signal attenuation (White et al. 1996). The degree of radial falloff across the field can be measured empirically (see Sect. 30.5) or predicted in conjunction with measurements of the fluorescence axial step function (see Sect. 30.7; White et al. 1996).

Spatial dimension calibration

30.4
Microscope Performance Protocols: Calibration of the Image (z) Axis

The aim is to determine the z-axis focus correction.

Modern high numerical aperture oil immersion lenses show minimal aberration when imaging a specimen adjacent to a coverslip. However, the changes in refractive index between the immersion medium, coverslip and aqueous perfusate introduce refraction effects that are only corrected near the coverslip. Deeper into the sample the image is degraded – marginal rays are affected to the greatest extent while axial rays are not

The z-axis focus correction

affected at all. The result is a magnificational change in z and a z broadening of the imaging probe. To calibrate the distortion of axial focus position due to mismatched immersion and mountant, the distance between glass/aqueous medium/glass boundaries in a slide-coverslip sandwich can be measured and compared with the "correct" distance determined with a matched immersion system, usually water immersion into an aqueous medium (Errington et al. 1997; White et al. 1996). This problem is reduced if water immersion lenses can be used for the experiment and the specimen is largely aqueous. Appropriate high NA lenses have recently become available from several manufacturers to help address this issue.

1. Set up an aqueous sea sandwiched between a slide and coverslip approximately 50–100 μm deep using spacers (see Sect. 30.1).

2. Collect (x, z) sections at ca. 0.3 μm x, y pixel spacing in reflection mode using a water immersion lens. Sampling should start ca. 5–10 μm outside the fluorescent medium and continue at 0.4–0.5 μm z step intervals through the medium and 5–10 μm further into the slide.

3. Repeat the sampling with the "experimental lens" and the appropriate immersion medium.

4. Measure the intensity distribution along a z transect with some averaging, e.g., 16–32 pixels wide, normal to the slide-coverslip sandwich.

5. The distance between the two major reflection peaks for the water immersion lens corresponds to the matched refractive index situation (D_{water}).

6. The distance between the apparent reflectance peaks for the experimental lens/immersion medium combination (D_{expt}) will be greater than D_{water} for oil immersion and less for dry lenses.

7. The ratio of the distance with the water lens compared to the experimental lens (D_{water}/D_{expt}) gives the z correction factor that has to be applied to all measurements involving z. Typical correction factors range from 0.82–0.88 for oil into water (White et al. 1996) which is close to the refractive index ratio ($\eta_{medium}/\eta_{immersion}$).

30.5
Microscope Performance Protocols: Useable Field Area

The aim is to determine the radial fall-off in intensity across the field.

The measured intensities from a homogenous fluorescent sea are rarely uniform across the field. There is usually a central, circular peak and a radial fall-off in intensity towards the edge of the field. This arises from lateral spherical aberration, lateral chromatic aberration and field curvature, which combine to blur both the illumination and detection probes and shift their relative positions in x, y, z. The signal intensity is reduced because the blurred information is rejected at the pinhole.

Radial intensity fall-off

1. The extent of radial fall-off in signal from all sources of aberration can be measured using line transects across the (x, y) images of the fluorescent sea, preferably with an amount (e.g., 16–32 pixels) of averaging normal to the transect.

2. For ratio applications it is critical that this measurement is repeated for both excitation/emission combinations as the aberrations are wavelength-dependent.

3. The useable confocal field of view will typically be limited to 3/4 (linear dimensions) of the full field for Plan-Apo water immersion lenses and 1/2 to 1/4 of the field for high NA Plan-Apo oil immersion lenses and aqueous samples. This measurement should be repeated with all lenses.

30.6
Microscope Performance Protocols: Reflection Plane Spread Function

The aim is to assess the general performance of the microscope system in reflection.

The monochromatic reflection plane spread function is readily determined from reflection images of a mirrored surface. It provides an indication of optical section thickness, field curvature, wavelength-dependent shifts in axial focus position and the effects of lens aberrations. These functions also provide a simple assessment of the microscope performance over a period of time.

Reflection plane spread function

1. The sample used for the empirical measurement of axial distortion can be used, although greater contrast is achieved if the glass surfaces are partially mirrored (Fig. 30.2B). These can easily be generated by sputter coating in an electron microscopy unit.

2. Collect axial (x, z) sections in reflection contrast over about 20 µm centered on the coverslip/medium boundary at a reasonably high resolution, e.g., 0.1 µm z focus increments and an (x, y) pixel spacing of 0.03 µm for a 60x lens. Average at least eight lines to reduce noise. Repeat for each lens and wavelength combination.

3. The spread or extent of the mean intensity averaged normal to the axis over a central portion of the field and plotted against the corrected z distance (see Sect. 30.4) defines the overall sectioning characteristics (or plane z response) for each monochromatic wavelength in reflection.

4. The FWHM provides a measure of the optical section thickness. Values will be proportional to the wavelength and will also depend on pinhole size. Minimum values will be around 450 nm in the blue, increasing to around 600 nm in the red.
   The axial position of the maximum in the plane spread function will also alter towards the edge of the field if there is any appreciable field curvature. In addition, the magnitude and shape of the side lobes provides an indication of the lens aberrations. The axial displacement of the central peak between different wavelengths provides a measure of the axial chromatic aberration of the system (Fricker and White 1992). The effect of different pinhole sizes on optical sectioning can also be readily determined to examine the S/N ratio vs section thickness, for example.

30.7

Confocal Microscope Performance Protocols: Fluorescence Axial Step Function

The aim is to assess the general performance of the microscope system in fluorescence.

Fluorescence axial step function   Fluorescence imaging involves a minimum of two wavelengths and three are used in ratio measurements. An appropriate test sample is needed to assess the performance of the microscope under these conditions.

Although the fluorescence psf can be derived from imaging subresolution fluorescence beads, good fluorescent psf measurements are difficult to achieve as they are very noisy and prone to specimen movement. An alternative strategy relies on measurement of the change in fluorescence intensity function on entering a fluorochrome sea – the fluorescence axial step function.

1. Set up an aqueous fluorochrome sea sandwiched between a slide and coverslip approximately 50–100 μm deep using spacers (see Sect. 30.1).

2. Collect axial (x, z) sections at a reasonably high resolution, e.g., 0.1 μm z intervals and an (x, y) pixel spacing of 0.3 μm for a 60x lens. Average at least eight lines to reduce noise. Sampling should start ca. 10 μm outside the fluorescent medium and continue 10 μm further into the medium.

3. The fluorescence step function is defined by the line transect into the fluorescent sea, preferably with an amount (e.g., 16–32 pixels) of averaging normal to the transect.

4. A second measurement can also be made at the medium/slide boundary to give an indication of the fluorescence performance after traversing a reasonable depth of aqueous medium. The overall intensity will be reduced compared to the surface profile as a result of depth-dependent signal attenuation (see Sect. 30.8) and the boundary will be less steeply defined as the probe is blurred.

30.8
Microscope Performance Protocols: Rapid Performance Assessment

The aim is to assess the general performance of the microscope system rapidly in both reflection and fluorescence.

A rapid assessment of both the reflection plane spread function and the fluorescence axial step function can be made from single (x, y) images of a tilted specimen. This allows direct visualization of both functions and the effects of interactive changes in imaging parameters, such as pinhole size, alignment, etc., are immediately apparent.

Rapid performance assessment

1. Construct a wedge-shaped chamber with an inclined coverslip at an angle of ca. 1 in 50 (Fig. 30.2C) containing ca. 50 μM fluorochrome.

2. Align the slope of the coverslip parallel with the x-axis of the microscope.

3. Collect single (x, y) scans in fluorescence and reflection mode and adjust the focus until the peak intensity is roughly centered in the field of view.

4. The intensity profiles along the x-axis represent the overall shape of the two functions. To calibrate the functions, the lateral displacement of the reflection peak is measured with changing focus by collecting an (xz) series through the chamber (see Sect. 30.4). The calibration factor is equal to the change in focus/change in x position of the reflection peak.

30.9
In Situ Specimen Protocols: Depth-Dependent Signal Attenuation

The aim is to measure the signal attenuation with depth through a permeabilized specimen infiltrated with a fluorochrome sea.

Depth-dependent signal attenuation

The intensity response in the axial (z) plane is affected by the axial geometric and chromatic aberrations present along the entire optical path including the specimen. Tissues contain many additional refractive index boundaries which will all contribute to further chromatic and spherical abberation of the confocal probe geometry. The consequences of these effects will be increasing signal attenuation with increasing depth through the specimen. The complex spatial distribution of the refractive material and the overall geometry of the tissue currently prevent development of universal models for tissue-dependent attenuation. However, partial correction can be achieved by a more pragmatic approach based on determination of the axial intensity profile of a permeabilized specimen filled with a fluorescent sea' (Errington et al. 1997; White et al. 1996). The resultant response combines the effects of depth-dependent sea response and the additional contribution of the permeabilized tissue. In vivo, the effects are likely to be marginally worse than even this case, as a number of refractive boundries, particularly membranes, will be distorted or extracted during the fixation/permeabilization procedure. The overall protocol is summarized in Fig. 30.3.

1. Place the specimen in an appropriate fixative (e.g., 2 % paraformaldehyde in PBS; fresh ethanol: acetic acid (3:1); acetone or methanol at $-20\,°C$) for 30 min.

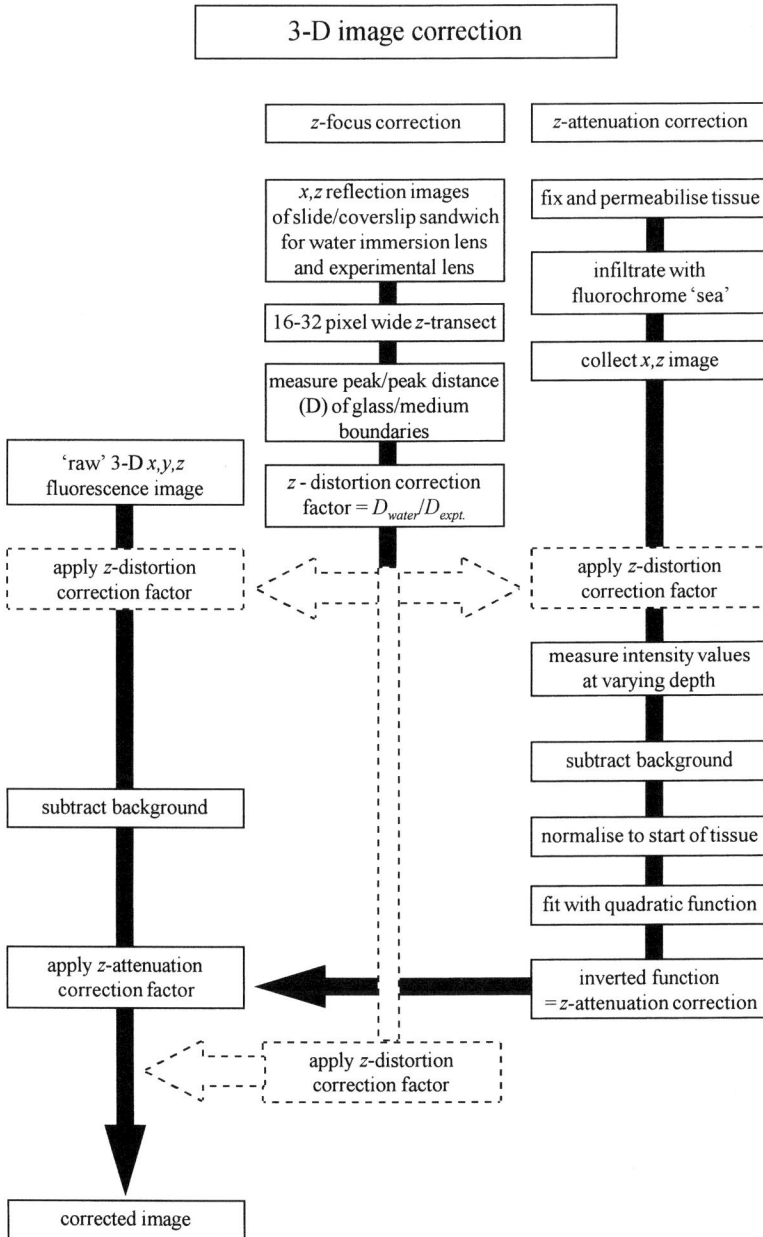

Fig. 30.3. The main steps used to correct z focus error and z attenuation in 3-D images. The z distortion factor can be applied directly to the image data to view correctly sized (x, z)images, or used to scale the numerical data extracted from the image as part of the analysis protocol

2. Specimens fixed with nonaqueous media may need to be rehydrated (e.g., through an ethanol series of 70/50/30/10/PBS) for 30 min each at room temperature).

3. Incubate in PBS containing ca. 50 µM of the appropriate fluorochrome for 24–48 h with gentle agitation to ensure good tissue penetration.

4. Collect axial (x, z) sections simultaneously in fluorescence and reflection mode into two channels through the permeabilized, infused specimen. Sampling should start ca. 10 µm outside the fluorescent medium, with a 0.3 µm x,y pixel spacing, and continue at 0.4–0.5 µm z step intervals through the medium plus specimen and ca. 10 µm further into the slide.

5. Measure average fluorescent intensities in regions of the specimen at varying depths to determine the in situ sea response. For plant material the most useful region to measure is in vacuoles where there is a large volume of homogeneous dye concentration.

6. The attenuation profiles are normalized to the start of the tissue, defined from the reflection images, and can be fitted with a variety of functions. Typically a quadratic function provides a reasonable agreement for simple flat tissues.

7. The inverse of the parametized equation is used to generate a correction factor to apply to successive z planes in the experimental data.
The correction for z axis focus error can be applied directly to the images to view correctly scaled (x, z) sections, but for measurements it is more efficient to apply the correction to the numerical values extracted during the analysis (White 1995; White et al. 1996).
In structures that have a relatively constant organization in the x, y plane a single axial correction equation may be sufficient. For tissues that show a more complex and variable organization, a series of correction equations may be required related to each zone of the tissue. The magnitude of the attenuation will depend on the depth, lens (particularly NA), immersion medium, bathing medium and wavelength (see Errington et al. 1997; White et al. 1996). Certain lens/immersion combinations may actually give an increase in intensity initially, depending on the precise conditions that the lens has been corrected for.

30.10
In Situ Specimen Protocols: Normalisation of Fluorescence Intensity Against Protein or DNA Content

It is often useful to be able to normalize the fluorescence intensity in a range of different units other than on a volume basis, such as µmole/mg protein). Conversion of fluorescence intensity to concentration can be readily achieved from in vitro calibration curves measured using identical microscope settings. If particular tissue zones, cells or compartments are segmented from the image, results can be expressed on an appropriate database, such as pmole/cell. At the moment it is rather more complex to relate fluorescence levels to other parameters such as protein levels or DNA content. The basic approach involves collection of 3-D images of protein or DNA distribution from corresponding regions of the same or an equivalent specimen and measurement of the total amount of protein/DNA. The following two protocols are given as basic approaches to labeling protein and DNA. The steps required to analyze such data are described in Sect. 30.17.

Normalization of fluorescence intensity

The aim is to label and image protein distribution to normalize quantitative fluorescence measurements for protein content.

Normalization for Protein Content

At the end of the live experiment, material can be fixed, stained for protein and re-imaged to normalize fluorescence intensity measurements for protein content. We have used FITC labeling of amine residues as a general protein labeling reagent. A number of other reactive groups with different amino acid specificity are available, linked to a wide range of different fluorochromes.

1. Fix specimen and rehydrate if necessary (see Sect. 30.9)

2. Stain protein with 50 µM FITC in PBS for 30 min.

3. Wash in PBS three times.

4. Images are collected under the same conditions as the experimental imaging.

The aim is to label and image DNA content to normalize quantitative fluorescence measurements for DNA content.

Normalization for DNA content

A wide variety of DNA fluorochromes are now available with varying spectra, sensitivity, discrimination between DNA and RNA and specificity for GC vs AT base pairs. We have used chromomycin A$_3$ staining to label nuclei in *Arabidopsis* roots as this method gives very little autofluorescence when excited with a 442 nm He-Cd laser. Other DNA probes should be selected and optimized for the tissue autofluorescence and excitation wavelengths available.

1. Fix specimen and rehydrate if necessary (see Sect. 30.9)

2. Stain in chromomycin A$_3$ for 1 h in the dark: 100 µg/ml CA$_3$ (Sigma) in 10 mM Na$_2$HPO$_4$, 10 mM NaH$_2$PO$_4$, 140 mM NaCl, 70 mM MgCl$_2$

3. Mount in fresh antifade: 0.1 % (w/v) phenylenediamine in 90 % (v/v) glycerol, 10 % (v/v) 1x PBS.

4. Store slides in dark at 4 °C for up to 1 week.

Note. Separate z distortion and z attenuation correction factors have to be determined for material mounted in glycerol-based media as these have a different refractive index compared to aqueous media alone.

30.11
Live Specimen Protocols: Chamber Design

Measurement chamber

Working with living cells can be difficult, however, there are specific (and even objective) decisions that can help establish the protocols needed to carry out successful experiments. Keeping cells alive, healthy and fully functional on the microscope is really the crux of live cell imaging (see Errington et al. 1997; Matsumoto 1993; Terasaki and Dailey 1995). Choice or design of a chamber necessitates a compromise depending on the length of observation, the nature of the experiment and the properties of the cell.

- A robust and reliable chamber is required to give reproducible imaging conditions and withstand continual z focusing without leaking and moving. Dimensions are critical, a large viewing area enables accessibility for wide, high NA, oil immersion lenses. However, large coverslips are more prone to warping resulting in drifts in focus.
- Temperature control is most easily achieved in a thermostatically controlled laboratory with air conditioning when the chamber, media and microscope are all at equilibrium. If additional temperature regulation is required, the controller must be a proportional type as this will maintain the heater at a steady state. A simple on/off type alternates between fully on or fully off and fluctuations in temperature may be

large. Temperature regulation is most precise if the sensor sits in the medium. Chambers with large viewing areas tend to cause more problems with uneven heat dispersion and increased rate of heat loss. Objectives can also be heated with coils, etc., to eliminate this component as a heat dump. Apparently, the optical properties of some lenses are not compromised up to 50 °C provided the temperature is not cycled repeatedly.

- Cells may require perfusion in order to maintain appropriate levels of dissolved gases, such as O_2 and CO_2, nutrients and pH. It is important to make sure that the mechanics of liquid moving in and out do not cause too much disturbance. It is also crucial to ensure that the liquid is preheated before it enters the bath.
- The chamber may also have to incorporate features which enable the manipulation of the cells or tissue on the microscope such as microelectrodes or a rapid or pulsed delivery system for pharmacological agents.
- Vibration isolation tables are often essential components of the imaging system to prevent vibrations reaching the microscope optics and scan system. Such effects are often apparent as minor periodic or aperiodic signal intensity fluctuations or, in severe cases, zig-zag distortion of structures. In physiological imaging with large, sometimes open, perfusion chambers, air currents also provide a major source of specimen movement, particularly in rooms with circulating air conditioning. Robust chamber design and shielding the stage area help to minimize this effect.
- Autofocus routines may also be used to overcome some problems with specimen movement.

30.12
Live Specimen Protocols: Optimizing the Labeling Environment

The fluorescent dye may interfere with the normal function of the cells as the concentration is increased. In addition, the interaction of illumination with the fluorescent dye may cause phototoxic damage, particularly if the excited dye reacts with oxygen to give free radicals. Thus, the concentration of fluorochrome introduced should be kept low to minimize buffering of the ion involved and any potential nonspecific chemical or photochemical side effects. It is also possible to incorporate free radical scavengers such as ascorbic acid at (0.1–1 mg/ml), Trolox (vitamin E), or carotenoids (see Tsien and Waggoner 1995) or deplete oxygen

Labeling environment

levels by modification of the chamber atmosphere or adding Oxyrase (Oxyrase Inc., P.O. Box 1345, Mansfield, OH 44901). However, these treatments can dramatically affect the physiological status of the tissue, particulary the redox equilibrium of the cells.

Usually the best way forward is to ascertain the upper limit of dye loading consistent with minimal disruption of cell physiology and then manipulate the image collection conditions to maximize cell viability. Typically for calcium or pH dyes this is about 50 μM. It is important to have good accessible markers of cell function to compare in loaded and unloaded cells. These may include parameters such as membrane potential, cytoplasmic streaming, growth rates, elongation rates, and cell division rates. Alternative strategies involve other "vital" or "mortal" staining techniques which probe different aspects of membrane integrity and metabolic activity.

30.13
Fluorescence Time Course Protocols: Optimizing Settings

Optimizing settings
Once the cells are loaded with the appropriate dye, the collection protocol can be optimized. In most point scanning instruments several variables contribute to the sampling achieved within the specimen. The volume sampled depends on the lens and level of optical sectioning. The period each pixel is illuminated depends on the scan speed; however, signal will only be usefully collected for a fraction of the dwell time depending on the speed of the detection electronics. The pixel spacing depends on the lens magnification and zoom. There is usually control over the degree of confocality, the area scanned, the scan speed and the number of frames averaged. However, it is also important to minimize light exposure to the sample, hence even when hunting around for the cells this should be done as fast as possible.

- The amount of laser illumination presented at the sample is probably the most critical parameter. This can be adjusted by altering the output power of the laser, inserting neutral density filters or by using an acoustic-optic modulator. If the output is measured with a power meter at the objective lens then this ensures that the optimal conditions can be achieved independent of the CLSM system. Values between 76 μW (Tsien and Waggoner 1995) and 20 μW (Errington et al. 1997) are appropriate for high NA lenses giving acceptable *S/N* ratios and cell viability.

– The NA of the objective lens dictates the initial light gathering efficiency – the higher the NA the better the collecting efficiency. It should be noted that the illumination spot is smaller with higher NA lenses and this may cause increased phototoxic damage. Therefore, when changing between lenses the laser intensity may need adjustment or the integration time, scan speed, etc., altered.

– Keeping cells alive may also be at odds with optimal sampling. Measurements of ion concentrations rarely require spatial resolution greater than $\approx 0.5\,\mu m$ in x, y, thus it is important to optimize the instrument to maximize S/N and minimize phototoxicity under these conditions. This relatively low spatial resolution may permit some flexibility in choice of objectives and pinhole settings. Pawley (1995b) also suggests using a high NA lens to maximize fluorescence collection efficiency, whilst underfilling the back focal plane to reduce the effective NA of the illumination beam giving a larger spot. The optimal pixel spacing will also be much greater than that required to give the maximum spatial resolution possible using the Nyquist criterion.

– The frame rate is determined by the frame size and scan speed, each of these parameters can be altered separately. CLSM offers many different timelapse imaging modes: xt(time), xyt, xzt, xyzt, the choice of mode depends how much imaging the specimen can handle and again the temporal and spatial resolution required. In combination, the frame rate and frame size dictate the pixel dwell time, or how long a single voxel is illuminated. In experiments in which temporal resolution is critical, the area scanned can be reduced to a single line on some instruments or even a single point.

– Scanning instruments are ideally suited to simultaneous emission ratio measurements as most are equipped with two photo-multiplier detectors. Excitation ratioing requires rapid switching between different wavelengths and adds a time delay between collection of a pair of wavelength images for ratioing. With multiple lasers this requires rapid synchronized shuttering of the two beams. For multiple line lasers, wavelength selection requires filter wheels, acousto-optic wavelength selection or separation and shuttering of the individual lines.

– Integrating multiple frames or slowing the scan speed reduces noise, but in live cell imaging this increases the overall excitation exposure and may not be appropriate for rapidly changing phenomena or dynamic systems with rapid rates of streaming. At the moment it is also still not clear whether the best S/N regime is to have a single slow scan with no integration or several fast scans incorporating integration. The debate hinges on the recovery time of some reversibly excited states that deplete the fluorochrome without generating fluorescence.

- Intermittent sampling with a time delay between each time point can also be used for slowly changing phenomena.

- It is very useful, if not essential, to have some form of information about the progress of the experiment on-line. Specific time course software often allows the user to specify a number of regions of interest (ROI) and report the average fluorescence intensities (with or without dark current and/or background subtraction) or ratios from these regions during the experiment. Graphical presentation of the intensities/ratios from these regions allow dynamic changes to be followed during the experiment and synchronized via on-screen annotation of the traces. Detailed analysis is better undertaken post-capture but requires additional time to save the image data.

- Some instruments have 24 bit display capacity during image collection and allow on-line "merging" of three signal channels as the individual red, green and blue channels of the 24 bit display. In ratio measurements, the color from a two-channel merge of each wavelength image can be used to provide a visual pseudocolor indication of the instantaneous fluorescence ratio, although there is no on-screen record of the history of the experiment.

30.14
Fluorescence Time Course Protocols: Data Collection

The aim is to establish optimal collection conditions whilst maximizing cell viability.

Data collection
1. Find the cell(s) of interest with the pinhole aperture fully open, the laser intensity at a low setting and the scan speed to fast. (This ensures you can move the sample around and find the cells of interest as fast as possible, while exposing them to the minimal amount of laser illumination.).

2. Select the appropriate objective lens and adjust frame size and zoom to encompass the region of interest.

3. Adjust the black level settings to ensure the background is recorded

4. Close the pinhole down.

5. Select the scan speed

6. Adjust the gain to ensure the signals are not saturated.

7. Set the number of frames to be integrated.

8. Set the sampling interval between time points.

9. Four set of controls should be run: (1) sample alone with no dye or laser to test the effects of the microscope perfusion regime compared to an equivalent sample under normal incubation conditions; (2) sample plus laser to test the biological effects of intrinsic absorbance and to measure the levels of autofluorescence; (3) sample plus dye (but without laser) to test the effects of dye loading on physiological function; (4) sample plus dye plus laser to test the potential phototoxic effects of illumination levels and dye concentrations. All of these steps require good markers of cell function and physiological response for comparison with unloaded cells.

30.15
Data Analysis Protocols: Ratio Image Analysis

The aim is to visualize spatial and temporal fluorescence dynamics

Visualization of fluorescence dynamics

Once the images have been collected correctly there are a wide range of different analysis techniques that can be applied to extract useful quantitative information. Images collected at two different wavelengths can be ratioed pixel-by-pixel to generate a ratio image that compensates in principle for varying dye levels, dye leakage and bleaching. This method has found wide application in both conventional and confocal imaging (see Kurtz and Emmons 1993; Mason 1993; Diliberto et al. 1994; Girard and Clapham 1994) and provides a good visual indicator of the magnitude of the response and the level of spatial heterogeneity within or between cells. The major steps in calculating a ratio image are given in the left column of Fig. 30.4.

1. The *S/N* ratio can be increased at the expense of spatial resolution by an averaging filter (e.g., a 3×3 box reduces noise by $\sqrt{9}$).

2. Images taken at each wavelength can be manually aligned in x, y to correct for any minor misregistration between the two wavelength images. Typically the misalignment is less than $1-2$ pixels for a dual excitation system, depending on the (x, y) pixel spacing, and should be less than 1 pixel for a dual-emission system.

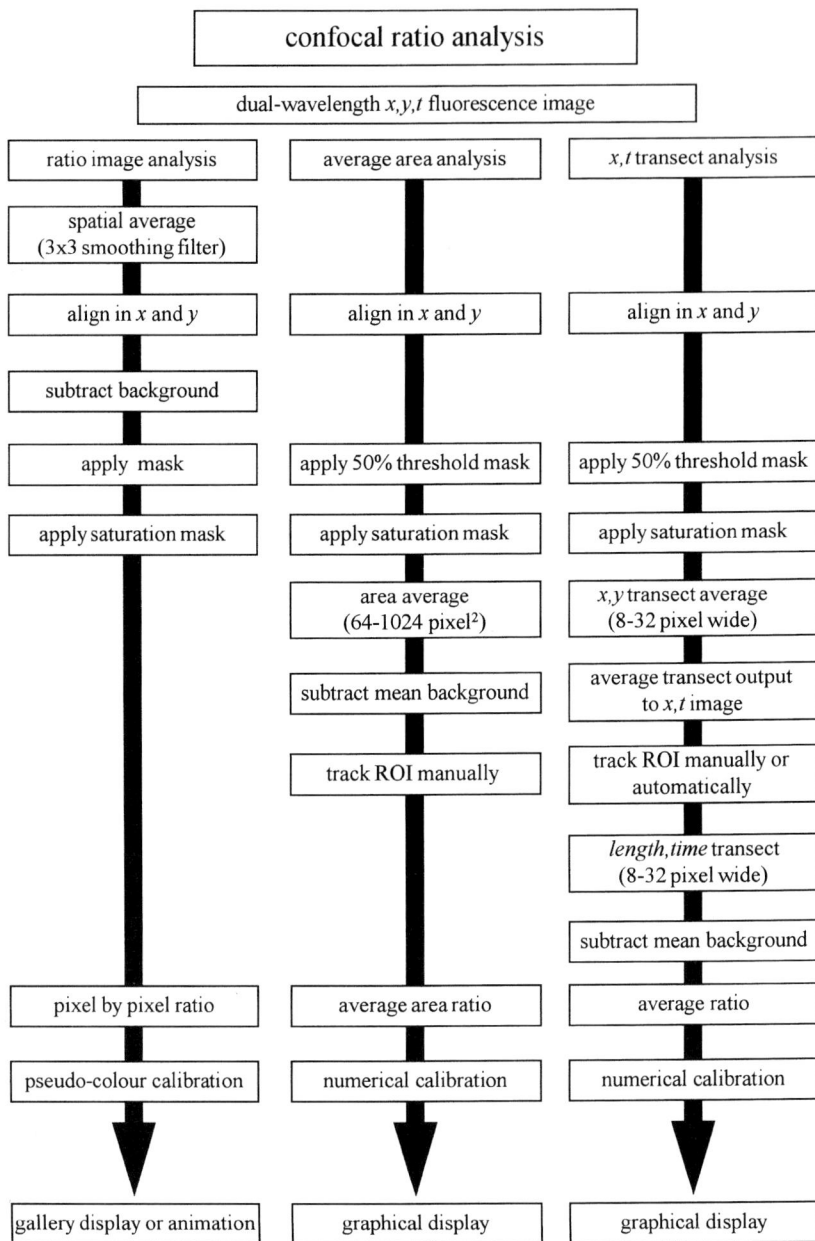

Fig. 30.4. The main steps in confocal ratio analysis. Visualization of the spatial distribution of ratio dynamics is best achieved using ratio imaging (*left column*). Quantitative analysis can be achieved from a region of interest (ROI) using an area average (*center column*) or a transect analysis (*right column*)

3. Measurement of the background signal is best achieved from an adjacent region of tissue that is unloaded. Background corrections should be applied for each wavelength image and time point in the series.

4. Pixels with low values or those outside the object are normally masked to exclude regions of corresponding high variance from the ratio image by setting the intensity to zero below a threshold intensity value or outside a spatially defined mask. Three protocols may be used to define the mask: (1) An intensity value at a fixed number of standard deviation units above the mean background intensity, typically 2 SD units; (2) at a 50 % threshold between the fluorescence intensity within the object and the background (e.g., Errington et al. 1997); (3) A morphological boundary, such as the edge of the cell, defined from a separate image, such as a bright-field view.

5. Masking is also required to exclude values approaching saturation of the 8 bit range. Saturation is related to an assessment of the number of photons contributing to the signal and requires knowledge of the conversion from photons to gray levels (see "Statistics of Fluorescence Intensity Measurements"). A pragmatic approach is to measure the distribution of intensities in a fluorescent sea at about the concentration of fluorochrome encountered in vivo and determine the highest mean value where the distribution is not clipped.

6. Background subtraction, threshold masking and saturation masking can all be accomplished in a single rapid operation by manipulation of the look-up table (Fig. 30.5) and application to the image.

7. The ratio image is calculated pixel-by-pixel.

8. Pseudocolor look-up tables are often used to enhance the viewer's perception of changes, particularly in publications where gray scale images are not reproduced well.

9. A variant on this overall scheme that avoids selection of a masking threshold is to code pixel ratios by hue and pixel intensities by brightness. In 8 bit displays this can be achieved using 6 bits for hue and 2 bits for brightness. On 24 bit displays, a wide range of hue and brightness values are available.

10. Graphical presentation of data derived directly from a region of the ratio image should be avoided. Ratio images are notoriously noisy and it is difficult to interpret the statistics from spatial averaging of

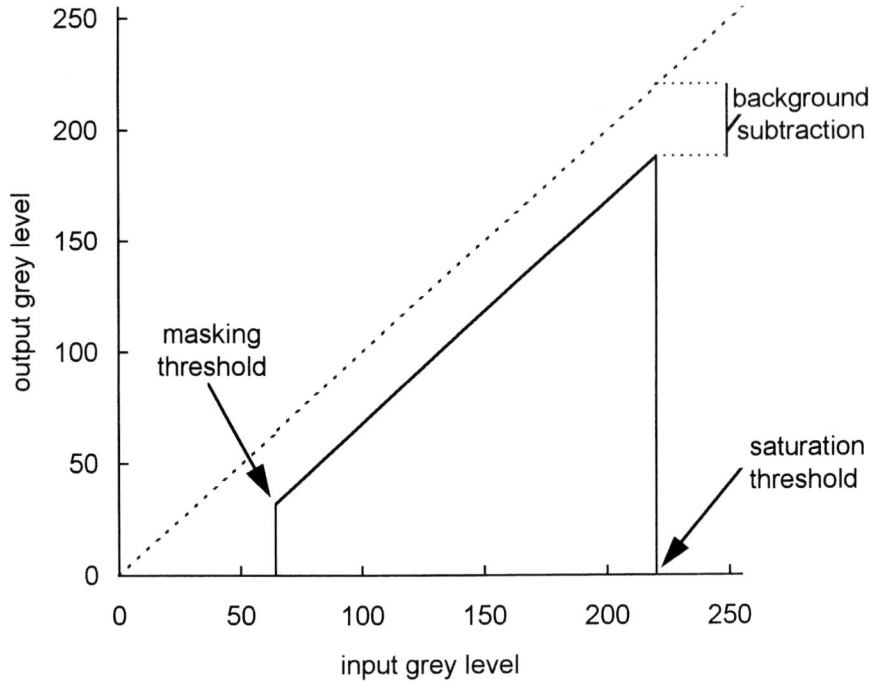

Fig. 30.5. Manipulation of the look-up table for rapid image processing. The *dotted line* represents a 1:1 correspondence between input and output values. The *solid line* is a function that combines subtraction of a mean background value, threshold masking and saturation masking in a single operation

the ratio values, as the distribution is not symmetrical. Thus it is usually more appropriate to visualize changes using ratio images, but to perform quantitative analysis on the original intensity data directly with an average area analysis or a transect analysis as described below.

30.16
Data Analysis Protocols: Average Area Analysis

The aim is to quantify fluorescence dynamics.

Quantification of fluorescence dynamics

The basic protocol for average area measurements is shown in the middle column of Fig. 30.4. In some cases it is useful to segment the object using a 50 % threshold mask prior to area measurements to facilitate measurements from regions encompassing irregular structures such as cytoplasmic strands, filopodia and subcellular compartments, without recourse to detailed manual delimitation of the area.

1. A series of ROI are defined on one wavelength image and the average intensity and standard deviation measured from that region. The corresponding region is measured on the other wavelength image with appropriate image or area alignment if necessary.

2. The average background values are subtracted independently for each ROI at each wavelength.

3. The average ratio is calculated and displayed graphically.

4. Error bars can be calculated from the SD of the individual wavelength populations. If the 90 % confidence limits of the individual wavelengths are used, the ratio values will have 81 % confidence limits.

30.17
Data Analysis Protocols: Transect Analysis

The aim is to extract and measure changes in fluorescence along a user-defined transect.

A number of biological experiments involve systems that show well-defined growth patterns or physiological movements. We have developed a modification of the area analysis protocol that facilitates measurements of ROI that are defined with respect to a particular structure in the specimen, such as the tip of a pollen tube or leading edge of an epithelial cell (Fricker et al. 1994). It is usually appropriate to segment the object using a 50 % threshold mask prior to the transect analysis. The major steps are shown in Fig. 30.4, right column.

Transect analysis

1. A transect is drawn along the structure in the direction of movement and the average intensity 8–32 pixels normal to the transect measured for each wavelength image with appropriate image or transect alignment if necessary.

2. The averaged transects are used to construct successive lines in intermediate length,time (l, t) images.

3. If the region of interest has moved, the intensity profiles can be aligned vertically for the ROI and values extracted using a second transect along the vertical time axis.

4. The average background values are subtracted independently for each wavelength.

5. The average ratio for each time point is calculated and displayed graphically.

30.18
Data Analysis Protocols: 4-D (x, y, z,Time) Single Wavelength Fluorescence Quantitation

x, y, z, t quantitation

Quantitative measurements from 3-D images over time require correction of the data for axial distortion, measured using the protocol described in Sect. 30.4, and axial attenuation, measured using the protocol described in Sect. 30.9, to relate the measured fluorescence intensity to the "real" level of fluorescent probe. A schematic representation of these stages is shown in Fig. 30.6 (see Errington et al 1997). After such correction, the fluorescence is calibrated per unit volume sampled throughout the specimen. The object(s) of interest are typically segmented at the 50 % intensity between an average "internal" value and the background in each x, y section in turn. Alternatively, objects can be segmented at the 50 % threshold from the 3-D image using a "seed-fill" function. In many cases, additional manual or semi-automatic editing of the x, y sections is necessary to extract the object completely. We have not elaborated on these methods as the range of functions available and implementation depend to a large extent on the software available.

If particular zones or cells are segmented, the results can immediately be expressed on a per cell basis. It may also be appropriate to normalize the fluorescence levels to other parameters such as protein or DNA content. At the moment this is a rather complex operation and involves collection, correction and segmentation of an equivalent 3-D image of protein or DNA distribution and measurement (usually) of the total amount of protein/DNA in a corresponding region; however, the results give an incredibly detailed quantitative cellular map of fluorescence kinetics in units that can be directly compared with results from other techniques, such as flow cytometry or conventional biochemical and molecular biological analysis. This approach should have wide application in single cell biochemical measurements with the increasing range of fluorescent enzyme substrates. With the advent of green fluorsescent protein (GFP) and its derivatives, it is also now possible to construct fluorescent chimeric fusion proteins to follow protein dynamics in vivo. 4-D confocal

4-D fluorescence quantitation

live 4-D x,y,z
fluorescence image

fix tissue

correct z-distortion and
z-attenuation (see Fig. 3)

8-16 pixel transect across
object boundary

define object boundary as
threshold at 50% intensity
between internal value and
background

seed fill

manual editing of
objects

Measure parameters:
e.g. total intensity, mean,
variance, frequency
histogram, volume

Calibrate fluorescence
intensity with standard
curve *in vitro*

express results in relative
fluorescence per:
unit volume
cell

express results in moles per:
unit volume
cell

label for protein or
DNA

collect 3-D protein /
DNA reference image

correct z-distortion and
z-attenuation (see Fig. 3)

3-D x,y,z protein / DNA
reference image

segment region
corresponding to object

measure total fluorescence

Calibrate fluorescence
intensity with standard
curve *in vitro*

express results in moles per:
unit protein
unit DNA

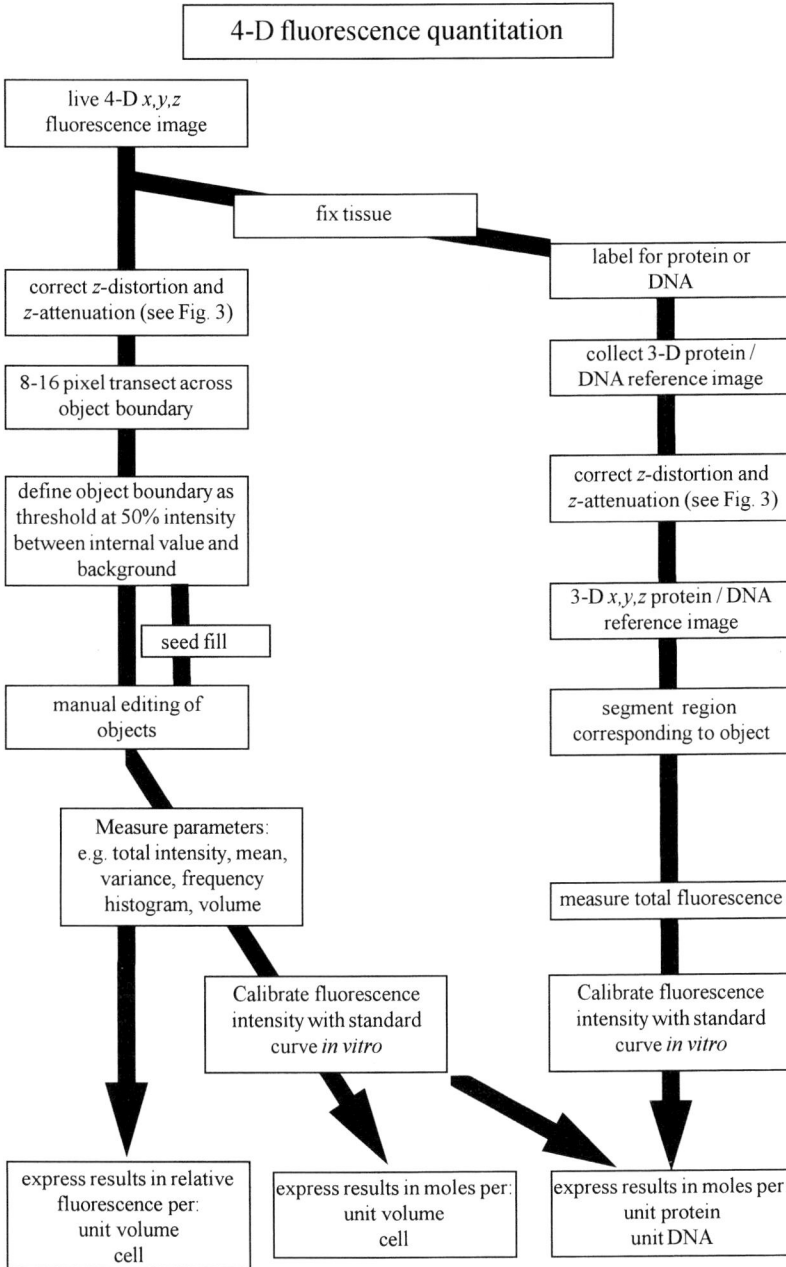

Fig. 30.6. The main steps to extract quantitative information from a four-dimensional (x, y, z, time) image and normalize the data to cell volume, protein content or DNA content

imaging allows quantitation of these events in the correct physiological and morphological context and also allows measurements on rare, highly specialized cells.

References

Centonze V, Pawley J (1995) Tutorial on practical confocal microscopy and the use of the confocal test specimen. In: Pawley JB (ed) Handbook of confocal microscopy. 2nd edition. Plenum, New York, pp 549–570

Errington RJ, Fricker MD, Wood JL, Hall AC, White NS (1997) Four-dimensional volume imaging of living chondrocytes in porcine articular cartilage using confocal laser scanning microscopy: a pragmatic approach. Am J Phys 272: 1040–1051

Diliberto PA, Wang XF, Herman B (1994) Confocal imaging of Ca^{2+} in cells. Meth. Cell Biol. 40: 243–262

Fricker MD, Tlalka M, Ermantraut J, Obermeyer G, Dewey M, Gurr S, Patrick J, White NS (1994) Confocal fluorescence ratio imaging of ion activities in plant cells. Scanning Microscopy Supplement 8: 391–405

Fricker MD, White NS (1992) Wavelength considerations in confocal microscopy of botanical specimens. J Microscopy 166: 29–42

Girard S, Clapham DE (1994) Simultaneous near ultraviolet and visible excitation confocal microscopy of calcium transients in *Xenopus* oocytes. Meth Cell Biol 40: 263–286

Hell SW, Stelzer EHK (1995) Lens aberrations in confocal fluorescence microscopy. In: Pawley JB (ed) Handbook of confocal microscopy. 2nd edition. Plenum, New York, pp 347–354

Kurtz I, Emmons C (1993) Measurement of intracellular pH with a laser scanning confocal microscope. Meth Cell Biol 38: 183–194

Mason WT (1993) Fluorescent and luminescent probes for biological activity. Academic Press, London

Matsumoto B (1993) Cell biological applications of confocal microscopy. Meth Cell Biol, vol. 38

Nuccitelli R (1994) A practical guide to the study of calcium in living cells. Meth Cell Biol, vol. 40

Pawley J (1995a) Handbook of confocal microscopy. Pawley JB (ed) 2nd edition. Plenum, New York

Pawley J (1995b) Fundamental limits in confocal microscopy. In: Pawley JB (ed) Handbook of confocal microscopy. 2nd edition. Plenum, New York, pp 19–38

Sandison DR, Williams RM, Wells KS, Strickler J, Webb WW (1995) Quantitative fluorescence confocal laser scanning microscopy (CLSM). In: Pawley JB (ed) Handbook of confocal microscopy. 2nd edition. Plenum, New York, pp 39–54

Taylor LD, Wang Y-L (1990) Fluorescence microscopy of living cells in culture. Part B. Meth Cell Biol, vol. 30

Terasaki M, Dailey ME (1995) Confocal microscopy of living cells. In: Pawley JB (ed) Handbook of confocal microscopy. 2nd edition. Plenum, New York, pp 327–346

Tsien RY, Waggoner A (1995) Fluorophores for confocal microscopy. In: Pawley JB (ed) Handbook of confocal microscopy. 2nd edition. Plenum, New York, pp 267–280

Wang Y-L, Taylor LD (1989) Fluorescence microscopy of living cells in culture. Part A. Meth Cell Biol, vol. 29

White NS (1995) Visualisation systems for multidimensional CLSM images. In: Pawley JB (ed) Handbook of confocal microscopy. 2nd edition. Plenum, New York, pp 211–254

White NS, Errington RJ, Fricker MD, Wood JL (1996) Aberration control in quantitative imaging of botanical specimens by multidimensional fluorescence microscopy. J Microscopy 181: 99–116

Chapter 31

Cytosolic Calcium Measurements with Confocal Microscopy

Wim J.J.M. Scheenen

Background

In the last decade, measurement of intracellular calcium concentrations ($[Ca^{2+}]_i$) using intracellularly trapped fluorescent dyes has become a popular way to determine changes in $[Ca^{2+}]_i$ in individual cells. More recent advances in confocal laser scanning microscopy have made it possible to perform these measurements with even higher spatial resolution. A detailed description of quantitative analysis of confocal microscopy was described in the previous chapter. This chapter will provide an overview of the basics of confocal microscopy and will subsequently describe several ways of using confocal microscopy in determining $[Ca^{2+}]_i$.

Principle of Confocal Miscroscopy

The idea of confocal microscopy is based on the convergent and divergent pathways of light through a microscope (Inoue 1995). In normal microscopy all light coming from the specimen is passed through to the oculars. This leads to an image that, although bright, does not contain much spatial resolution. Confocal microscopy, by contrast, is based on focusing of the laser beam into a spot, or Airy disk, on the specimen. Therefore, only the spot in the plane of focus will be illuminated optimally; any other plane is illuminated by a more diffuse disc of light. Subsequently, when studying fluorescence only the fluorescence coming from the plane of focus is let through to the detector. This is generally achieved by placing an iris diaphragm, or pinhole, in front of the detector.

Pinhole. Figure 31.1 shows the pathways of excitation light focusing on the specimen and fluorescence light coming from a specimen going through the objective and finally reaching the detector. In general, a confocal microscope will have a dichroic mirror in the fluorescence path and

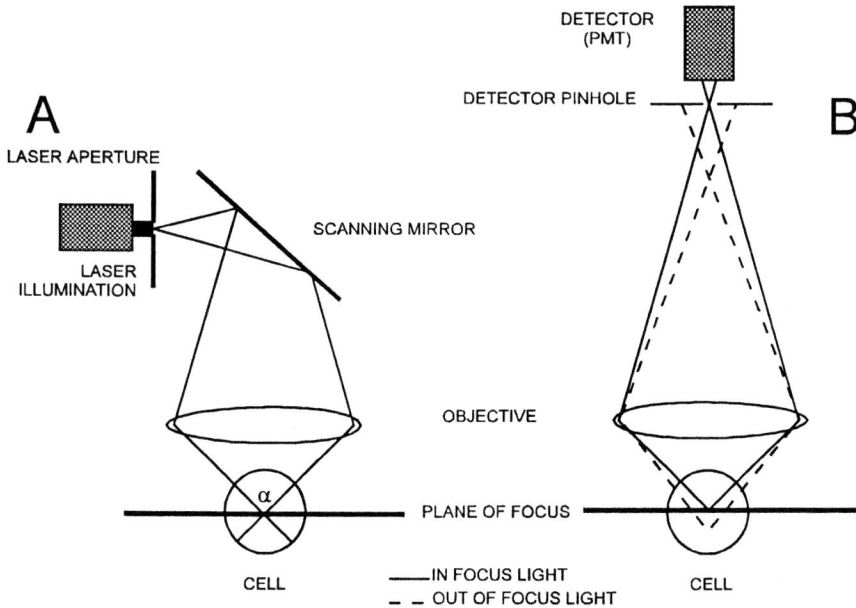

Fig. 31.1A, B. Simplified explanation of the light paths of the laser and the fluorescence in a confocal microscope. **A** The laser beam will be focused onto one point in the specimen. By tilting the mirror, scanning of the entire cell is achieved. **B** Fluorescence coming from the plane of focus will converge onto the pinhole, thereby passing it and reaching the detector. Light coming from an out of focus plane will not converge onto the pinhole and therefore will be largely blocked from reaching the detector

two detectors, giving the possibility of capturing two different fluorescence signals at the same time. Note the pinhole in front of the detector that prevents the passage of light coming from planes other than the confocal plane. Only fluorescence light from the confocal plane converges onto the pinhole and is thereby able to pass it. Light coming from a plane different from the confocal plane will not converge onto the pinhole. In this way the pinhole serves as a filter by largely blocking out of focus fluorescence. In other words, only a small fraction of that light will be able to pass and reach the detector.

Numerical Aperture. Apart from the pinhole being important in achieving a sharp confocal image, the numerical aperture (NA) of the objective lens is also important, since the latter determines the focusing of the laser beam onto the specimen. Increasing the NA of the objective will lead to a better spatial resolution on the xy and the xz directions, since the angle α (Fig. 31.1A) at which the laser beam focuses onto the specimen will be higher, leading to a smaller Airy disk. Therefore, a higher NA will lead to a sharper confocal image.

By scanning the laser beam spot across the specimen an image of the specimen is generated. There are three types of laser scanning mechanisms:

- Specimen scanning. This scanning technique moves the microscope stage in the x-y-z direction to obtain an image.

- Lens scanning. In this scanning technique not the microscope stage, but the objective lens is moved in the x-y-z directions.

- Mirror scanning. This technique uses a mirror to scan the laser beam through the objective in the x-y direction. Positioning of the specimen in the z-direction is achieved by moving the specimen stage with a stepping motor.

Note. Although specimen scanning is the ideal technique for obtaining high-resolution confocal images, the vast majority of the commercially available confocal microscopes uses the mirror scanning technique. A problem with the latter technique is that the laser beam will be off axis with respect to the microscope objective. Therefore, any chromatic aberration of the objective lens can lead to problems. This shows up as a decrease in the apparent light intensity of the specimen when a laser beam is directed through the periphery of the objective (Sandison et al. 1995).

Fig. 31.2. Explanation of the different steps of the experiment (see text)

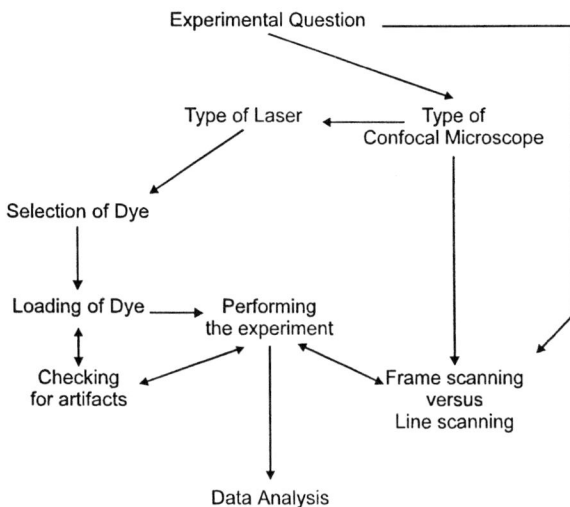

The image generated by scanning the laser is transformed into a computer image containing both spatial (distance between pixels) and intensity information (gray levels of individual pixels). The minimum detectable distance between pixels is described by the Nyquist criterium, which states that the sampling frequency should be at least twice the spatial frequency present in the specimen (Webb and Dorey 1995). Apart from the spatial information, imaging of dynamic processes like $[Ca^{2+}]_i$ requires that a series of images have to be taken over time. Depending on the speed of the $[Ca^{2+}]_i$ changes of interest, the sampling interval should be changed accordingly. In the following sections examples will be discussed of how to set up and perform an experiment in which a time course of $[Ca^{2+}]_i$ will be followed (Fig. 31.2). Finally, a few words will be spent on the data analysis.

Materials and Procedure

31.1
Determination of the Pinhole Size and Appropriate Acquisition Speed

In generating an image, spatial resolution, image brightness, scanning speed and probe bleaching are interconnected to each other. In other words, if one wants a very bright and sharp confocal image, one should use a slow scanning speed and should allow for bleaching to occur. However, performing an experiment on living cells requires that bleaching and bleaching-induced toxicity should be kept to an absolute minimum.

Minimizing bleaching

Therefore, the laser power used should be set as low as possible. This can be done by:

- Reducing the laser output power

- Introducing neutral density filters in the optical path
  As a consequence the obtained fluorescence signal becomes very low. In order to increase this signal, the experimenter can choose to increase the size of the pinhole (Sheppard et al. 1991). By increasing the size of the pinhole, light from a thicker plane is allowed to pass to the detector, leading to a reduced optical sectioning.

Note. For the experiments performed on pituitary cells I found that in general setting the pinhole size to 60 % or 70 % of the maximum led to sufficient fluorescence intensities and sufficient optical sectioning (Scheenen et al. 1996).

Another option in obtaining better signals is to change the acquisition mode of the confocal microscope. For example, the BioRad MRC-600 has the possibility to perform fast photon counting. Using this acquisition mode, low fluorescent signals can be detected more easily, while maintaining a good signal-to-noise ratio. Similar sensitive sampling techniques most likely can be found in other commercial confocal microscopes and it is therefore wise to consult the technical specifications for the confocal microscope you are using.

Setting acquisition speed

The theory described above applies to a point scanning microscope. The scanning of a point scanning microscope is a slow process. In general, it takes between 2 and 5 s to scan one image with most types of microscopes commercially available today. It will be clear that for most Ca^{2+} measurements a time resolution of 2–5 s is an absolute minimum, and often experiments with faster acquisition speeds are required. Therefore, the experimenter should try to anticipate both the length of the experiment and the speed of the Ca^{2+} changes to be studied. The speed of a Ca^{2+} wave can vary considerably, depending on the type of cell. As an indication, observed speeds for several cell systems are in the range of 2 μm/s to more than 80 μm/s (Jaffe 1993).

A useful protocol to follow in order to determine the optimal acquisition speed is the following:

1. Study a few cells, using low laser power settings in the standard acquisition mode.

2. If the expected Ca^{2+} processes cannot be followed both in time and space using these settings, the scanbox should be subsequently reduced and, if possible, an optical zoom performed.

3. If, under these conditions, the Ca^{2+} processes are still too fast to follow, then a line scanning experiment should be set up as described in Sect. 31.3.

Note. Reducing the scanbox in practice means that the spatial resolution is decreased. A disadvantage of reducing the screen size is that the spatial resolution can get too low (i.e., only a few pixels per cell), and thus one of the advantages of confocal microscopy is lost. In most commercially available confocal microscopes this can be overcome by performing an optical zoom, i.e., reducing the area which is scanned by the laser. However, a smaller area gets excited more intensely, and thus bleaching and bleaching-induced toxicity become serious problems.

Recent developments in confocal microscopy have led to faster scanning types. This faster scanning is achieved by scanning the specimen with the laser, which is transformed into a line, or by using fast scanning mirrors. A full description of all confocal microscopes is beyond the aim of this chapter and therefore I have limited the discussion to slow point scanning systems. For the interested reader, an extensive overview of the types of confocal microscopes commercially available can be found in Pawley (1995).

31.2
Choosing the Ca^{2+} Sensitive Dye in Accordance with the Type of Laser

Type of Laser

Apart from considering the scanning speed, another major consideration needs to be made, namely, the type of laser that the confocal microscope is equipped with. This has important consequences with respect to the proper choice of the calcium sensitive dye. In the more traditional Ca^{2+} imaging techniques the UV excitable probe Fura-2 is used to visualize changes in $[Ca^{2+}]_i$. Although recently a UV confocal microscope has been developed for the use of Fura-2, most experimenters will have to perform their experiments using a microscope equipped with a visible wavelength laser. A visible wavelength laser usually emits light in the blue or green region. Typically, for a confocal microscope equipped with a low power argon laser, the two excitation wavelength that can be used are 488 nm and 514 nm. The currently more popular krypton-argon laser has three major excitation wavelengths, at 488 nm (blue light), 568 nm (green) and 620 nm (red).

Selecting the appropriate dye

Intrinsically coupled to the type of laser that is available is the selection of the Ca^{2+} sensitive dye (Lipp and Niggli 1993; Yao et al. 1995). Of the Ca^{2+} dyes currently available that can be excited at 488 nm, Fluo-3 and calcium-green are the most widely used. These dyes are excited at 488 nm and emit at 520–530 nm. Upon binding of Ca^{2+}, the fluorescence intensities of both Fluo-3 and calcium-green are increased. Another probe that increases its fluorescence intensity upon binding of Ca^{2+} is calcium-orange. This dye is excited at 540 nm and has an emission maximum at 580 nm. Apart from these dyes, Fura-red can be used for confocal experiments (Lipp and Niggli 1993; Scheenen et al. 1996). After excitation at 488 nm, the emission spectrum of Fura-red has its maximum at 620 nm. In contrast to the previously mentioned dyes, Fura-red emission intensity decreases upon binding of Ca^{2+}. A disadvantage of all these dyes is that they are single wavelength emission dyes, and therefore dye distribution, bleaching and organellar uptake can lead to serious errors in estimating real $[Ca^{2+}]_i$. The experimenter should always be aware of the existence of these artifacts and should check for them by performing a few simple tests, like those described in the section "Troubleshooting." However, if one is interested in dynamic properties, such as the spreading of a Ca^{2+} signal through the cell and the speed of a Ca^{2+} wave, then single wavelength emission dyes are very useful.

Lately, the use of high power argon laser's has been implemented in commercially available confocal microscopes. The emission lines from a high power argon laser differ from that of a low power argon laser in that the former has a strong UV emission line at 351 nm. The addition of this UV emission line to the spectrum of available lines makes confocal microscopy suitable for the use of the dual-emission ratioing probe indo-1 (Millot et al. 1995). After being excited at 351 nm, indo-1 emits at 405 nm in its Ca^{2+}-bound form, whereas the Ca^{2+}-free form emits at 485 nm. Therefore, the use of a UV laser combined with indo-1 as the Ca^{2+} sensitive dye is useful when absolute $[Ca^{2+}]_i$ is required.

All dyes described above are available from Molecular Probes.

31.3
Loading of the Dyes and Setting the Filters

Dye Loading   After the experimental conditions have been determined an experiment can be started:

1. Loading of the dye. In general, the same loading conditions as for video imaging apply. For loading of the dyes, using the acetoxymethyl (AM) ester form, the concentrations vary considerably between cell types. For all dyes concentrations of 5–10 µM are generally used. Loading times vary between 30 min and 1 h (Millot et al. 1995; Scheenen et al 1996; Yao et al. 1995).

   Note. For Fura-red, higher concentrations (up to 20 µM) have been reported although concentrations below 10 µM might prove sufficient (Scheenen et al. 1996). Therefore, before a series of experiments is started, it is best to test a few loading concentrations of the dye used and a few loading times in the range indicated. Longer loading conditions in general are not preferred, since intracellular sequestering becomes a more serious problem.

Tip. When intracellular dye loading is too low, 0.002 %–0.2 % pluronic F127 (Molecular Probes) can be included in the loading medium (Poenie et al. 1986).

2. After loading, cells should be washed to allow complete de-esterification of the AM form. Subsequently the experiment can be started. For experiments with indo-1 this is a relatively straightforward process. The excitation wavelength should be set at 351 nm by selecting a proper excitation filter, and the fluorescent light should be passed onto a dichroic mirror that splits the light at 445 nm. This means that all fluorescence below 445 nm (i.e., indo-1 in its Ca^{2+}-bound form) is reflected onto the first photon multiplier tube (PMT), whereas all fluorescence above 445 nm passes through to PMT 2 (i.e., Ca^{2+}-free indo-1). By setting the gains of the PMTs independently, ratio values can be obtained which can be converted into Ca^{2+} signals after proper calibration (Millot et al. 1995).

3. The experimental setup for use of a single wavelength dye is in principle equally straightforward. For example, for Fluo-3 and calcium-green, an excitation filter of 488 nm is needed. Fluorescence emission is generally passed through a bandpass filter of 525±15 nm, meaning that all light below 510 nm (primarily being reflected excitation light from the coverslip) and above 540 nm is blocked from reaching the PMT.

Note. Whenever the loading conditions are poor for the studied cell type, it might be a good idea to change this bandpass filter for a long-pass filter that allows all light above 510 nm to reach the PMT. In this way a larger amount of fluorescence light will be captured.

4. Performing the initial scans. Now that everything is set correctly, the scanning process can be started. During the first few scans the user should try to focus through the sample either using a z direction stepping motor or by manually turning the micromovement knob of the microscope. By doing this, the user can select the appropriate plane in the z direction. It is best to perform these initial scans with a low laser power to reduce bleaching of the intracellularly trapped dye.

Settings of Excitation and Emission Filters

31.4
XY vs XT Scanning

XY scanning   As described in the previous section, the user will have to experiment a bit with the setting of the screen size, the optical zooming factor and eventually setting up a line scanning experiment. The way in which a screen is filled on a BioRad MRC-600 during the laser scanning process is shown in Fig. 31.3. Although the actual number given might differ

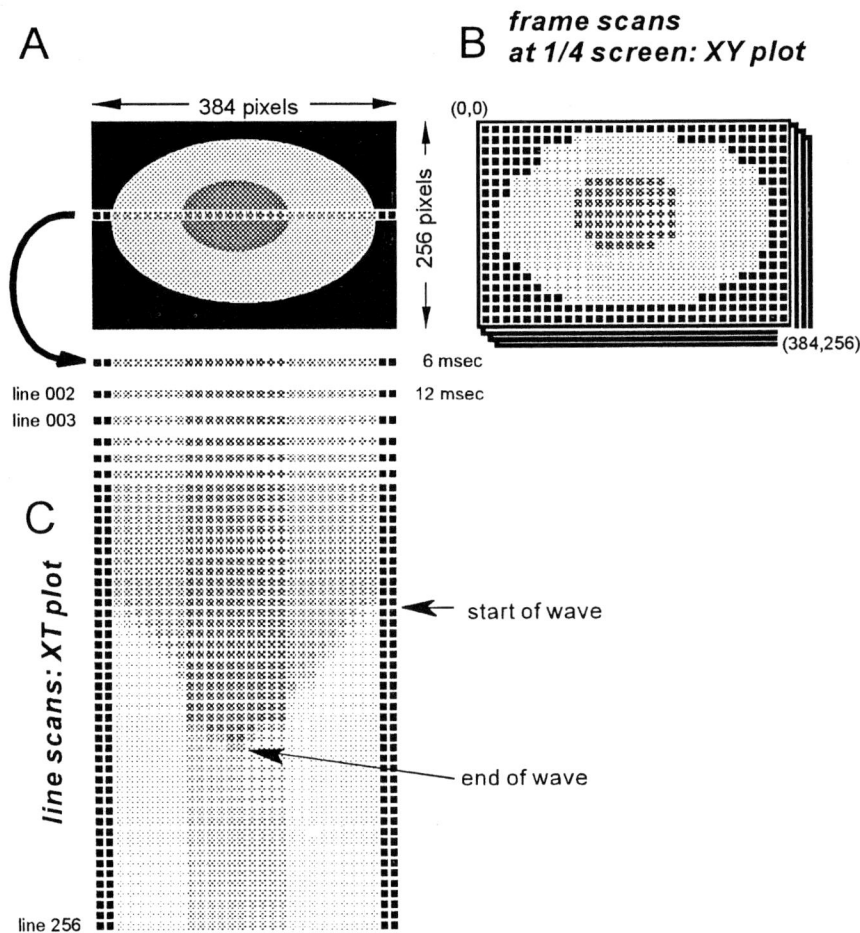

Fig. 31.3A–C. Schematic representation of image acquisition in the xy scanning mode (**A, B**) or xt scanning mode (**A, C**). **B** Scanning of the specimen starts in the upper left corner (0,0) and ends in the lower right corner (384,256)

between different confocal microscopes, the principle is the same. When the scanning of the laser is started, the scanning will start at position $x = 0$, $y = 0$ (0,0). First the laser will scan the $y = 0$ scanning line until the final x position is reached ($x = 648$). This process takes 2 ms in the normal scanning mode and 6 ms in the slow scanning mode on the Bio-Rad MRC-600. After completing the first line, the line $y = 1$ will be scanned in the same way. This process will be repeated until all y lines have been filled, hence the screen is filled. After this process is completed, the laser scanning will start again from point 0,0. This scanning mode is generally called xy scanning.

Reducing the scanning box size has the following effects:

1. The time interval between successive frames will also be decreased.

2. The number of pixels per cell will be reduced. Consider for example the case in which in whole screen scanning a given cell is $256 (16 \times 16)$ pixels. By reducing both the x and y points by half, the scanned frame will become one fourth of the original, leading to $64 (8 \times 8)$ pixels/cell.

   For a BioRad MRC-600 the maximal scan rate at full screen is one frame every 2 s, whereas the maximal scan rate is three frames per second at a frame of one eighth screen.

It can happen that a scan rate of three frames per second is not sufficiently rapid for the Ca^{2+} signaling process to be studied. For Ca^{2+} waves with speeds of 30–40 µm/s and a cell diameter of 10–20 µm, scanning speeds in the millisecond range are required. Although the traditional slow point scanning devices do not allow for such scanning rates in an xy screen, they offer the possibility of performing a so-called line scanning. This means that the laser point is not scanned through the entire frame, but is fixed to one y position.

XT scanning

The experimenter has only to select the desired scanning line. Take for example the cell shown in a frame in Fig. 31.3A. If one wants to perform a line scanning experiment the scanned line is set at a certain y position (arrow). The laser will be subsequently scanned over this line. After the scanning of this line is finished, instead of positioning the laser line to the next y position, the same line is scanned again. This line will be placed on the screen in the next y position in the frame. This might be confusing at first, but after a little practice, the user will get used to this type of frame filling. It is important to remember that subsequent y positions in the frame actually represent different times. Since the y information is replaced by time information, this type of experiments is also called xt scanning.

Note. It will be clear that a line scanning experiment leads to a major loss in spatial information. A user is therefore well advised to obtain as much information as possible about a given cell type by performing xy experiments and only go to xt experiments when speeds of Ca^{2+} waves are being considered. The positioning of the laser scanning line in these experiments should be performed with great care. For round, nonpolarized cells (as the one shown in Fig. 31.3) it is best to set the scanning line through the middle of the cell. The positioning of the scanning line in polarized cells is more complex. The experimenter should be aware of the polarization of the cell type and place the line in such a way that calculations of the speed of a Ca^{2+} wave can be performed most accurately. It will be clear that, if a Ca^{2+} wave does not travel exactly along the x axis, then subsequent calculation of the Ca^{2+} wave will lead to an overestimation of the speed.

Adding test substances   By selecting the type of scanning, xy versus xt, and the size of the scanbox, the user can perform the actual experiment and add test substances as preferred. It should be kept in mind, however, that one is scanning through a fixed plane of the cells. This means that minor changes in the z position of the cells, due to application, can lead to major changes in measured fluorescent intensities. Especially for the single wavelength emission dyes, this can lead to a false positive Ca^{2+} signal. The user can check for this type of application artifacts in the same way as described in the section "Troubleshooting." For this purpose, I found the combined use of Fluo-3 and Fura-red extremely useful.

Troubleshooting

Although the experimental conditions for single wavelength dyes look straightforward, the experimenter should initially include a few tests to see whether the obtained fluorescence signals really correspond to $[Ca^{2+}]_i$, or whether they result from loading artifacts or dye movement or bleaching:

- Loading artifacts
- Loading the cells with different dyes. Let us take, for example, the cell shown in Fig. 31.4. This cell has been double loaded with indo-1/AM (5 µM) and Fluo-3/AM (5 µM). For indo-1 the ratio of the F_{405}/F_{485} is presented as gray values, whereas for Fluo-3 the fluorescence intensity is presented as gray values. When we look at Fluo-3, we see that the

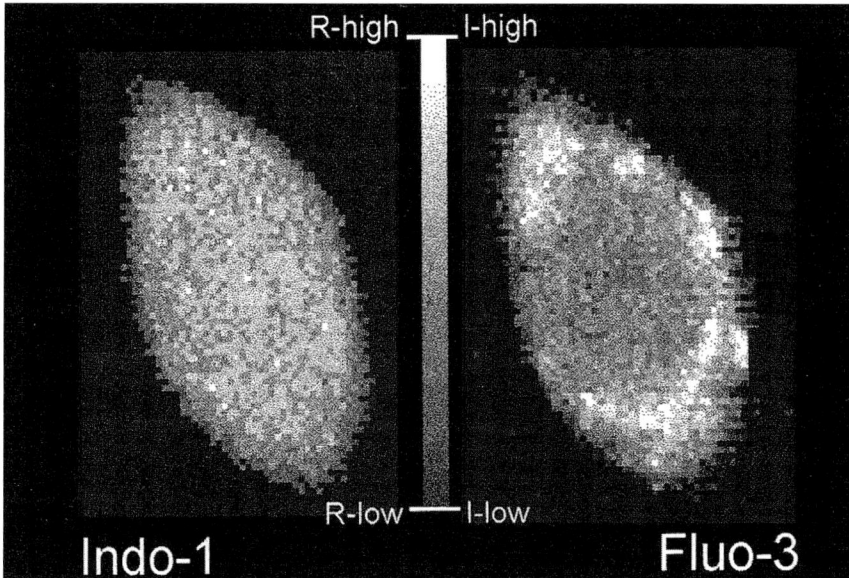

Fig. 31.4. Demonstration of loading artifacts. GH3 cells were loaded with both indo-1/AM (5 µM) and Fluo-3/AM (5 µM). The *left panel* shows the ratio of the indo-1 emission at 405 nm and 485 nm, the *right panel* shows the Fluo-3 intensity of the same cell. The experiment was performed on a Nikon RCM-8000

nucleus of the cell is less bright than the cytosol. Since Fluo-3 Ca^{2+} sensitivity is expressed as a difference in fluorescence intensity, these results could mean that the actual Ca^{2+} concentration in the nucleus is lower than that of the cytoplasm. However, since the indo-1 ratio values show a uniform picture, it can be concluded for this cell type that the loading of Fluo-3 is not uniform throughout the cell, and therefore the apparent difference between the $[Ca^{2+}]$ in the nucleus and cytoplasm represents a loading artifact. When no UV laser is available to perform this check, a combined loading of Fluo-3 (or calcium-green) with Fura-red can be performed. The emission spectra of these probes differ both in wavelength and in direction of the fluorescence signal upon Ca^{2+} binding.

- Movement artifacts
- Observing the change in fluorescence intensity in time. A combined loading of dyes not only can give the experimenter information about loading artifacts, but also provides information about whether observed temporal changes in fluorescence intensity actually are Ca^{2+} changes. For example, an increase in $[Ca^{2+}]_i$ should be accompanied

by an increase in the Fluo-3 intensity and a decrease in the Fura-red intensity. If both signals change in the same direction, then this change cannot be accounted for by changes in the $[Ca^{2+}]_i$, but rather, for example, by dye movement or cell movement. Double loading of these dyes can be very useful in studies of cell systems that undergo morphological changes during the experiment (Lipp and Niggli 1993).

- Calibration of $[Ca^{2+}]_i$ with single wavelength dyes
- Checking bleaching properties. In theory, an experimental condition can be found in which the results from these double-loaded probes can be ratioed, leading to semiquantitative results (Lipp and Niggli 1993). When the cells are loaded with the AM forms of the probes, extreme care has to be taken in quantifying the results since the bleaching rates of the two probes can differ considerably in cells, and also interexperimental loading variablity can form a problem. For example, in pituitary melanotrope cells it has been found that Fluo-3 bleaches about three times faster than Fura-red (Scheenen et al. 1996).

- Scanning artifacts
- When an xy scanning experiment is performed, it can happen that during one scan the fluorescence changes in the lower part of the cell, whereas the upper part remains unchanged. These type of changes must not be confused with intracellular differences in Ca^{2+} signaling since most likely they will be an artifact induced by a combination of the slow scanning speed of the system with a high speed of intracellular Ca^{2+} changes. Remember that the laser scanning during an xy scan goes from the top of the screen to the bottom. Suppose the laser scanning process is halfway in a cell when this cell undergoes a rapid change in $[Ca^{2+}]_i$, then only the lower part of the cell will have a changed fluorescence. This type of result indicates that the scanning speed is too slow to capture spatial differences in Ca^{2+} signaling. Therefore, if the experimental question addresses spatial differences in Ca^{2+} signaling, the scanning speed needs to be increased either by reducing the box size or by performing an xt scan.

Results

Analyzing the data   After an experiment is performed, the experimenter has a stack of frames containing either xy or xt information. The last step will be to analyze this stack of frames in order to obtain the desired xyt information.

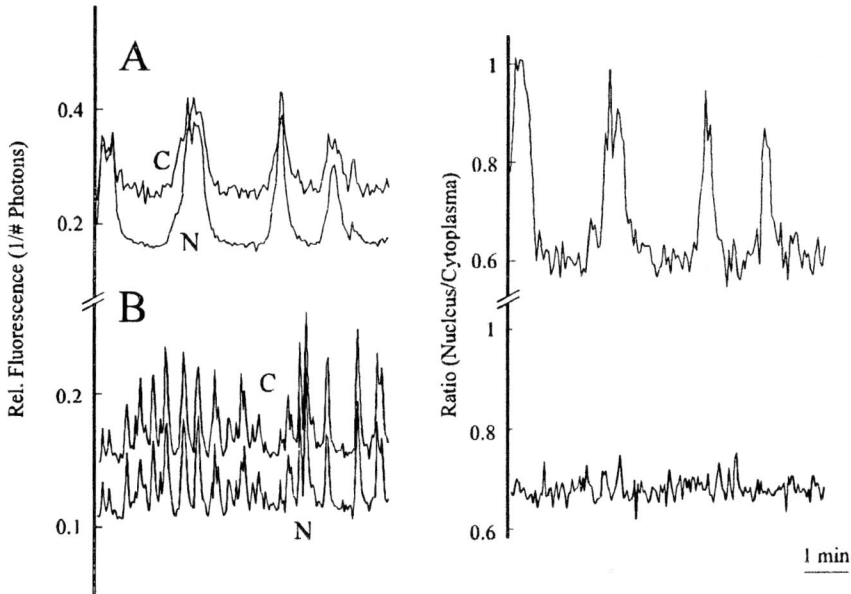

Fig. 31.5. Analysis of a stack of images acquired in the xy scanning mode on a Bio-Rad MRC-600. The cells were loaded with Fura-red/AM (6 μM). Analysis of the Fura-red signal is presented as reciprocal values. (Reprinted with permission from Scheenen et al. 1996)

For xy scanning experiments the analysis is a simple process, very similar to that performed for video imaging experiments. The user can place masks on the frame of regions of interest that need to be analyzed for each frame. The data derived in this way can be plotted in a graph. An example is shown in Fig. 31.5. In this experiment the question was whether there was a difference between the Ca^{2+} oscillatory pattern in the nucleus and the cytoplasm of the cell. The cells had been double loaded with Fura-red. Since for this experiment it was not necessary to obtain a high time resolution, it was performed in the xy scanning mode. After performing the experiment regions of interest were selected in the cytoplasm and nucleus of the cells and all captured frames were analyzed. The derived fluorescence intensity cannot be taken as a quantitative measure for the Ca^{2+} concentration since a single wavelength dye was used. However, what this experiment did show was that there can be a difference in the relative amplitude of the fluorescence changes between nucleus and cytoplasm.

The boxing of regions of interest for xt experiments differs slightly. Since a frame contains the time information in the y axis, analysis of

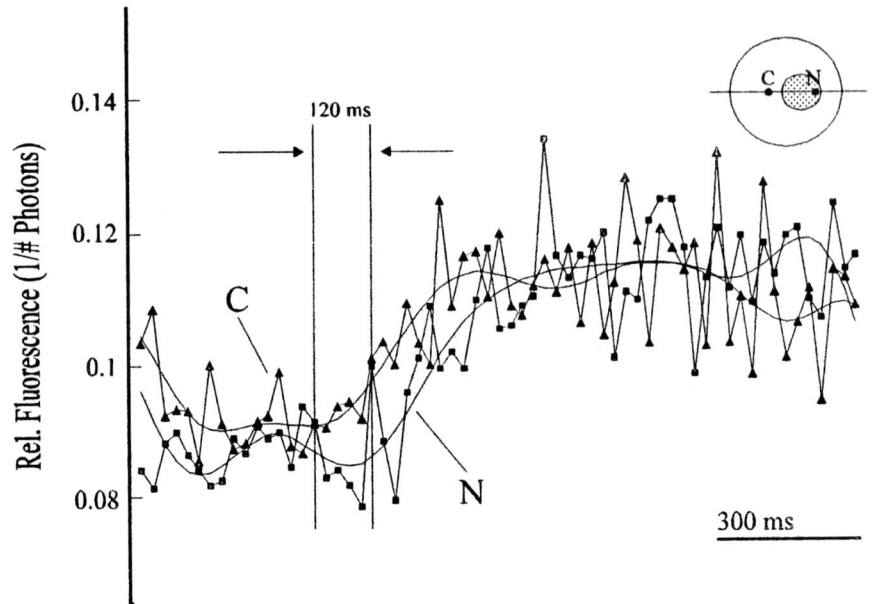

Fig. 31.6. Analysis of a frame acquired in the xt scanning mode on a BioRad MRC-600. Loading conditions were similar as those of Fig. 31.5. The *smooth line* was derived after subjecting the data to a fast Fourier transform. (Reprinted with permission from Scheenen et al. 1996)

each line needs to be performed. An example of a line scanning experiment is shown in Fig. 31.6. A frame of data has been analyzed for two different regions in the cell, one in the cytoplasm and one in the nucleus, both at the same distance from the cell membrane. The experiment was performed with a scan speed of 6 ms. In order to increase the signal-to-noise ratio, four x values and four time lines have been averaged, leading to a final time resolution of 24 ms (Scheenen et al. 1996).

As can be seen, the signal is still very noisy, and in order to obtain a better signal, the datapoints have been subsequently analyzed by performing a fast Fourier transform (FFT) (Lam et al. 1981). The reasoning for performing a FFT was that noise coming from the detection system will have a constant and high frequency, higher than the frequency of the Ca^{2+} signal. By transforming the data with a FFT, thereby visualizing the frequencies present, and subsequent cutting of the high frequencies a much smoother signal was obtained (dotted lines). The FFT filtering can be performed quite easily with the latest versions of popular spreadsheet programs like Borland Quattro Pro or Microsoft Excel.

References

Inoue S (1995) Foundations of confocal scanned imaging in light microscopy. In: JB Pawley (ed) Handbook of biological confocal microscopy. Plenum, New York, pp 1–17

Jaffe LF (1993) Classes and mechanisms of calcium waves. Cell Calcium 124: 736–745

Lam RB, Wiebolt RC, Isenhour TL (1981) Practical computation with Fourier transforms for data analysis. Anal Chem 53: 889A–900A

Lipp P, Niggli E (1993) Ratiometric Ca^{2+} measurements with visible wavelength indicators in isolated cardiac myocytes. Cell Calcium 14: 359–372

Millot J-M, Pingret L, Angiboust J-F, Bonhomme A, Pinon J-M, Manfait M (1995) Quantitative determination of free calcium in subcellular compartments, as probed by indo-1 and confocal microspectrofluorometry. Cell Calcium 17: 354–366

Pawley J (1995) Light paths of current commercial confocal light microscopes for biology. In: JB Pawley (ed) Handbook of biological confocal microscopy. Plenum, New York, pp 581–598

Poenie M, Alderton J, Steinhardt R, Tsien R (1986) Calcium rises abruptly and briefly throughout the cell at the onset of anaphase. Science 233: 886–889

Sandison DR, Williams RM, Wells KS, Strickler J, Webb WW (1995) Quantitative fluorescence confocal laser scanning microscopy (CLSM). In: JB Pawley (ed) Handbook of Biological confocal microscopy. Plenum, New York, pp 39–54

Scheenen WJJM, Jenks BG, van Dinter RJAM, Roubos EW (1996) Spatial and temporal aspects of Ca^{2+} oscillations in *Xenopus laevis* melanotrope cells. Cell Calcium 19: 219–227

Sheppard CJR, Cogswell CJ Gu M (1991) Signal strength and noise in confocal microscopy: Factors influencing selection of an optimum detector aperture. Scanning 13: 233–240

Webb RH Dorey CK (1995) The pixelated image. In: JB Pawley (ed) Handbook of biological confocal microscopy. Plenum, New York, pp 55–68.

Yao Y, Choi J, Parker I (1995) Quantal puffs of intracellular Ca^{2+} evoked by inositol trisphosphate in *Xenopus* oocytes. J Physiol (Lond) 482: 533–553

Subject Index